Standard Application of Mechanical Details

Standard Application of Mechanical Details

Jerome F. Mueller, P.E.

McGRAW-HILL BOOK COMPANY

New York   St. Louis   San Francisco   Auckland   Bogotá
Hamburg   Johannesburg   London   Madrid   Mexico
Montreal   New Delhi   Panama   Paris   São Paulo
Singapore   Sydney   Tokyo   Toronto

Library of Congress Cataloging in Publication Data

Mueller, Jerome F.
    Standard application of mechanical details.

    Includes index.
    1. Mechanical engineering.   2. Engineering design.
I. Title.
TJ145.M79   1985   621.8   84-21857
ISBN 0-07-043962-1

1234567890   HAL/HAL   898765

ISBN 0-07-043962-1

*The editors for this book were Betty Sun and Nancy Young,
the designer was Naomi Auerbach, and the production supervisor was
Sara L. Fliess. It was set in Times Roman by Byrd Press.*

Printed and bound by Halliday Lithograph.

Contents

Preface

In the preamble of one state's professional engineering licensing law there is a phrase that notes that engineering is a learned profession. Certainly part of the learning is the proper method of installing equipment that the engineer specifies. The principal purpose of this book is to aid in that effort.

As a practicing consulting engineer with over 1000 mechanical systems designed by me and members of my firm, I have many times met with installing contractors who, with various facial and verbal expressions, pointed out shortcomings in the installation details of plans bearing my professional seal. All of you who have been in this sort of embarrassing predicament know that when such a situation occurs there is no place to run and no place to hide. If your client happens to be at your side at such a moment, the situation is truly a frustrating, exasperating experience. The old cliche that an engineer builds his mistakes takes on a newer meaning. The engineer is caught trying, it sadly appears, to build a mistake.

Many of the details in this book may seem to be very similar to or the same as ones you have seen, have, or can acquire from another source. This is inevitable. While there are cases in which more than one method of installation is perfectly acceptable, there are also many cases in which the way depicted in this book is actually a field-tested way that you have seen before. Simply put, the methods of installation depicted in this book do work and have worked successfully.

Of all of the fields of engineering it is most often the mechanical engineering field in which the words "common sense" are frequently applied to a given solution. When these words are used by a skilled professional, I have no quarrel. Actually his or her usage implies that I am an engineer, you are an engineer, and our conclusion is common sense. There is nothing really common about this knowledge. The professional is knowledgeable and to him or her this knowledge is common sense, which brings us to the point of common sense in using the details in this book. In theory almost no sense at all is required if the designer, detailer, or engineer simply traces the detail. In fact both education and experience are required in selecting the proper detail. Further experience is required to modify the detail (and they can, on occasion, be modified) and to be certain that all of the parts of the detail will fit in the space allocated.

The real, ultimate purpose of this book is not only to provide usable information but also to aid in the education of the new employee. All of us are guilty at times of expecting our employees to perform tasks that are beyond their actual level of experience. Again using that pair of words common sense, it also is the essence of common sense to realize that the new employee may be long on theory but short on experience. This is the reason for the text that accompanies each detail. Experienced engineers need little, if any, explanation. For them this book is merely a readily usable and reproducible source of information. They do not need to be told how. Under pressure of a deadline, all they need and can get by using this book is a thought-out solution to their present detail problem. The book is purely and simply a profitable time saver.

The new employee solves the problem just as fast as his or her manager when he or she copies the detail. But copying is not learning. One of my purposes in preparing this book was to solve a problem for me as well as for you: how to further the education of my employees. Because of their engineering education background, I have not found the overall understanding of the design much of a problem to impart. However, if they are going to have detailers and drafters under their supervision, they simply must know what work to assign and why. The details in this book can be explained to them using, when appropriate, the words common sense and the written text or your own, expert version of same.

How to Use This Book

This book is arranged to create an orderly flow of details from the beginning to the end of a given system, and the chapter titles directly relate to this concept of the detail flow. Each category begins with what is called source equipment, covering boilers, chillers, water supplies, and the like. It is followed by source support equipment such as pumps, towers, and the like. Also included is "using" equipment such as pipes, radiation, etc. The final sections of the book, Chaps. 12 through 16, are the special categories.

As sources, water treatment plants, incinerators, and sewage treatment plants have been omitted as these are almost totally in the sanitary/environmental field and beyond the scope of this book. For example, a furnace commonly referred to as a forced warm air furnace has no particular detailing requirements. The connections to the furnace are supply and return ductwork with a canvas or fireproof flexible connection for vibration isolation. This is similar to the air handling unit connection. The coal or wood fired stove is simply a heating appliance with a smoke pipe or vent connector extending from the stove to the chimney. But, the boiler system delivering either steam or hot water requires a certain amount of detailing in order to be clearly represented on the drawings. In this broad sense every item of equipment related to the boiler or directly connected to the solar plant or heat recovery device comprise the heat source equipment. Detailing is required to separate and clarify the various connections and arrangements.

The question of whether to provide composite details or delineate on the plans a series of details illustrating the various parts is a subject of debate among many consultants. There is no really clear-cut answer.

The premise of this book is that the user would be best served if the parts of any proposed composite detail were clearly shown as separate details. Providing a master composite detail generally takes a very considerable area of the plans and may become an entire plan sheet or sheets of and by itself. When this sort of composite presentation is made, it usually is more of a flow diagram than a composite detail. Frequently it is used, and misused, in the presentation of the overall system.

Depending on the size of the system involved, the generally preferred method of presenting composite details or overall representations of all of the work in the equipment room is to provide a plan, sometimes a plan and elevation, at no scale. This is probably the best representation of a composite since it enables the designer to present the locations of the equipment in their proper relationship and position. It is sometimes difficult, and frequently confusing, to try to incorporate every single connecting piece to an individual item. The author of this book feels that the best way to do this is to provide partial plans at a larger scale plan and then provide accompanying details separately to illustrate precisely what the connections will be at each particular piece of equipment. In a no-scale detail presentation it is reasonable and possible to make every connection very clear to the installing contractor.

The book starts with details of a boiler and its surrounding piping only. This sort of detailing is followed by the fuel burning equipment, whether it is gas, oil, or other, to cover the combustion chambers. It then continues on to the related equipment which includes circulating pumps, boiler return pumps, expansion tanks, induced draft fans, breechings, economizers, central control panels, blowdown tanks, deaerators, pressure relief valves, and chemical treatment, and related items. The basis of this entire presentation is to logically progress from the boiler outward to the related peripheral equipment.

It is not required that the designer necessarily include each and every detail for any given job. Frequently in the less complex project, especially when the scheme has been presented in plain view at a larger scale, many of the connecting piping and the control arrangements become self-evident. There is no real need to detail the obvious. It should be a rule that nothing is duplicated on the drawings. By that we mean that there should be no situation in which what is already shown clearly on the plans is repeated in the same fashion, with no further amplification, on the accompanying detail. This is the sort of situation which creates problems. When you already have a clear-cut presentation at a large scale on the plans, your detail frequently, in repeating exactly what is on the plan, accidently omits or adds some extraneous item which becomes the source of future controversy. If there is any doubt, the safety and security of the plans would be enhanced by carefully noting that the work required includes the work described both on the plans and on the detail. However, the contractor finds this sort of statement extremely annoying.

For those of you who are already knowledgeable and experienced your method of using the book is probably to first consult the Index to ascertain whether or not the book has the detail you desire and to then simply proceed to the proper page, assuming the detail is listed, and make your own decision on usage.

The system of chapter arrangement should ideally serve the needs of project engineers, design engineers, designers, drafters, and detailers. They all have at their disposal, through the chapter and system arrangement, a built-in checklist. Certainly you are not going to need all the details on every job. But by comparing your design with the book's details, you can decide at each point whether or not to use a detail depicted in the book.

As was noted in the Preface, the object of the book is to educate, as well as to inform. When there is time, in the office or at home, the reader can take a given system and, by reading the related material chapter by chapter, create a mental walk-through of all the various points of system design and equipment installation. If this is related to a system design manual, the reader will have an excellent overview of the total, installed system.

List of Drawings

Standard Application of Mechanical Details

ONE

Heating Source Equipment

This is the first of the many chapters in this book in which we follow each system's details from the beginning to the end.

Boilers, for the purpose of this book, are divided into four categories. These are low pressure steam, high pressure steam, low temperature hot water, and high temperature hot water. To further define what is meant in this book by each of these categories, low pressure steam is defined as steam 15 psig and under, and high pressure steam is taken as 125 psig. The high temperature water system is considered to be water at 400°F, and the low temperature water is water at 200°F. Water is also used at other temperatures such as 260°F, 300°F, and the like. However, it was felt that the above four categories fairly clearly define the areas of detail delineation and to a large degree also cover all related variations in temperature or pressure.

The other basic sources of heating equipment are warm air furnaces, electric sources used for electrical baseboard heating, other electric panels, and direct electrical energy-using devices. These types of systems do not require any particular detailing at the source of energy, which is the furnace or the electrical panel. And none are shown in this book. Although the warm air furnace does have duct work to and from the furnace and possibly a fresh air intake, these connections are similar to duct connections in all types of air handling equipment. What applies in general to the air handling equipment details would apply to the warm air furnace and its supply, return, and fresh air connections.

For a large part of the United States solar energy is primarily used for domestic water heating and for backup to heating and hot water systems but not as a primary system for building heating. However, there are demonstration applications and actual installations of partial solar space and domestic water source heating in all parts of the country.

There are many situations in newer residential buildings, especially those with the proper orientation, in which the solar system can be the major heating system. On this premise we are presenting one solar detail in schematic outline form in this area of our book as heating source equipment. We will present other details under domestic water heating equipment.

When we describe automatic solid fuel burning systems, we are primarily discussing coal burning systems although obviously any form of wood, garbage, solid waste, or any other material other than a liquid or gas could be considered a fuel for a solid fuel burning system. Since World War II and especially since the late 1940s the idea of burning any fuel except oil or gas has seemed to be an almost prehistoric idea. Many engineers have little appreciation, understanding, or knowledge of solid fuel burning systems, commonly known as coal burning systems.

In this type of system the problem is in one sense the same as it is for gas or oil. You still have a boiler or furnace. You still have an automatic fuel device which in this case is a stoker. You still have a breeching and some form of draft inducing device. And you still have a chimney. These items are common and attached to the basic burner or furnace itself regardless of whether you burn solid, liquid, or gaseous fuel.

As far as this book is concerned and the details that may be involved, our problem is similar. We have to detail the fuel storage system which, in this case, is a coal bin; a method of transporting the coal from this bin to the boiler; and, one more problem not related to other fuels, removal of the ash residue. The designer's problem, which is not a detail problem, is how big should the container be that holds the coal.

Systems that burn solid or liquid waste were, until recently, primarily part of the civil-sanitary-environmental engineering design of an overall waste disposal program. But the standard batch-fed incinerator was a source of unacceptable air pollution, and its pollution solution was both difficult and expensive. Disposal of liquid waste by some sort of burning process is primarily of two types. Certain waste oils, under certain

conditions, can be used instead of regular fuel oil. Practically all other sources of liquid waste disposal require a special chemical treatment process and are not addressed in this book on mechanical details.

The burning of solid waste is a mechanical design project involving shredders, grinders, fuel transfer equipment, boilers especially designed to burn refuse, and ash handling and pollution control devices. All of these items are part of a very special system which involves customized detailing. Primarily, the detailing

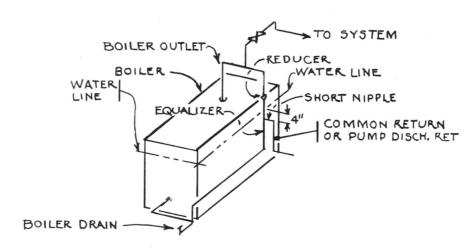

SINGLE BOILER INSTALLATION

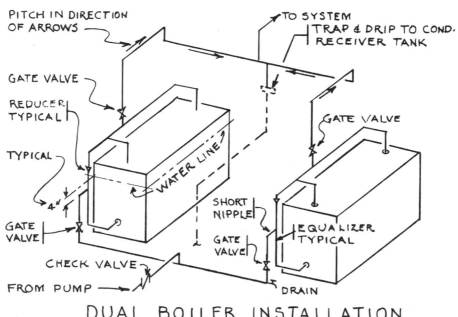

DUAL BOILER INSTALLATION
LOW PRESSURE STEAM SYSTEM
NO SCALE

FIGURE 1-1

in this case is related to plans and elevations of carefully dimensioned points of connection between the various system components.

Boilers and Furnaces

Low Pressure Steam Piping: The first of our details (Fig. 1-1) shows the basic piping of a single low pressure steam boiler in the upper part of the figure and a two boiler arrangement in the lower part of the figure. While the piping connections seem fairly obvious, there are certain items contained in the detail which should be made very clear. First, most boilers have more than one steam connection on the top of the boiler. This is generally true whether the boiler is used for hot water or steam heating purposes. When a multiple series of outlets is available on top of the boiler, at least two of them should be utilized, preferably one toward the front and one toward the rear of the boiler. This is because circulation in all boilers, especially cast iron boilers, is greatly improved if you use a number of outlets for the steam or hot water. The circulation through the boiler can flow freely, front to back.

Second, you will note that there is an indication in the upper part of Fig. 1-1 that the system is a gravity return system and requires the dripping of the steam header and the pitching of this header to a low point. The arrows that are shown on the dual boiler installation indicate the flow to a low point. This is because we are looking for a balance of pressure between the two boilers. You could have a situation in which a number of tappings were taken off of this main. In a pumped system condensate in the header is dripped through a drip trap, as shown. This is not the correct application if your system is a gravity return system. A gravity system requires dripping to the return main. Notice also that, off each boiler, the line that goes to the common header travels over and down into the header. This is to facilitate the flow of condensate and to preclude it from any building up back against the gate valve shutoff for the individual boiler.

On the back of the boiler there is the well-known Hartford Return equalizer connection. Normally this connection is a relief line that is sized at 50 percent of the supply line or a minimum of 2 in. The equalizer rarely functions satisfactorily when sized less than 2 in. In addition, while the connection of the equalizer to the return is self-evident, the really important part of this pipe is the connection of the returning condensate to the return line and not directly to the return header, or headers, of the boiler. The connection of the return condensate should be made 2 to 4 inches below the water line depending on whether this line is a pumped return or a gravity return. For smaller boilers the Hartford Steam and Boiler Insurance Inspection Company has specific rules for the location of this connector. The connecting point should be a tee fitting, not a wye fitting. A wye fitting in a pump connection would tend to allow condensate to travel up into the steam area, creating rapid condensation and possible water hammer.

If for any reason the pump discharge is run above the water line, a spring-loaded check valve in the return line at the return header will keep the discharge line flooded and prevent water hammer, which frequently follows a system shutdown. This is the sort of detail that is not easily shown on the usual boiler room plan drawing, regardless of scale. The piping situation is best covered by a detail such as Fig. 1-1.

Large Steam Boilers: Figure 1-2 is somewhat different from that of Fig. 1-1. The general impression created by Fig. 1-1 is that this obviously was a small system and not the size that is commonly part of a design-office project. However, it did enable us to illustrate both the supply and the return piping since details such as automatic feeders usually do not show the steam supply piping. Figure 1-2 follows the same rules as Fig. 1-1, but it is on a larger scale. It is still a low pressure boiler. What is not shown in Fig. 1-2 is the steam supply piping which would go to a steam header shown on the plan and not on the detail. What we really want to cover in Fig. 1-2 are the important parts of the typical larger plant which, in this case, is a two-pass oil fired scotch type steam boiler. This is usually what would be encountered in the typical design engineering detail. First, you will note that there is a low water cutoff shown and carefully piped. Second, there is a separate feeder and low water cutoff feeder. The low water cutoff and the feeder could have been combined into one instrument, a combination feeder and low water cutoff, which is sometimes done. The piping to the drain would have been one line instead of separate lines shown. The connection to the Hartford loop that is shown on the plans still would have been made. Note also that the pumped condensate return goes through check and gate valves before connecting to the Hartford loop. Finally, note that there are two relief valves, which is very common on larger boilers. As a matter of fact, there may be three if the boiler is very large.

The depicted items are referred to as standard trim. The following items should always be part of the minimum number of standard trim items: a high limit pressure stat, an operating pressure stat, a pressure gauge, a pump starter, a low water cutoff and alarm, an emergency water feeder, and a steam relief valve of the number and capacity required by the ASME code. The boiler stop valves shown on the drawing must be of the OS&Y rising stem type. The makeup water feed must be connected on the boiler side of all valves.

High Pressure Steam Boilers: In Figs. 1-3 and 1-4 we present a front and rear picture of a high pressure watertube steam boiler operating at 125 psig. We could have presented the detail of a firetube boiler in lieu of the watertube boiler. The scotch marine firetube type

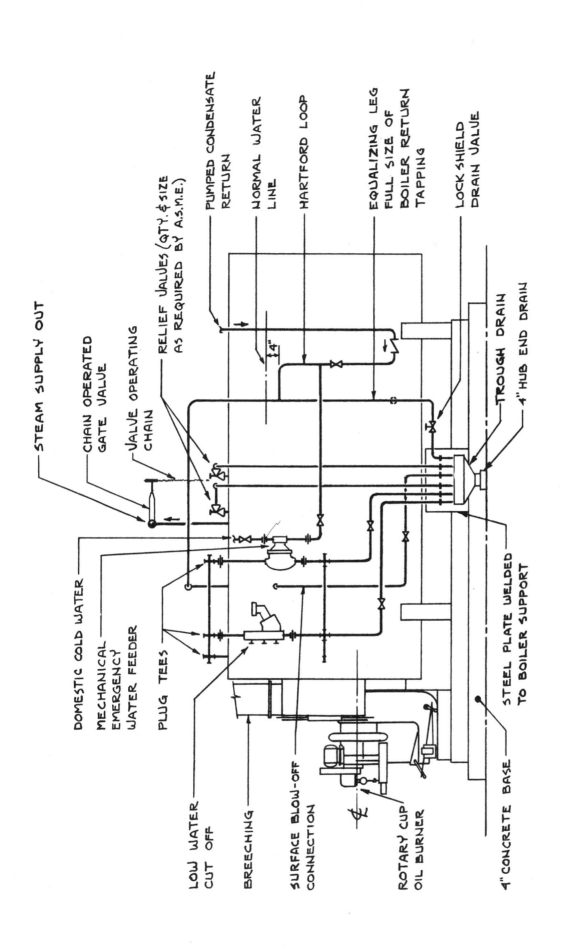

INSTALLATION of the TWO PASS, OIL FIRED, SCOTCH TYPE, STEAM BOILER

FIGURE 1-2

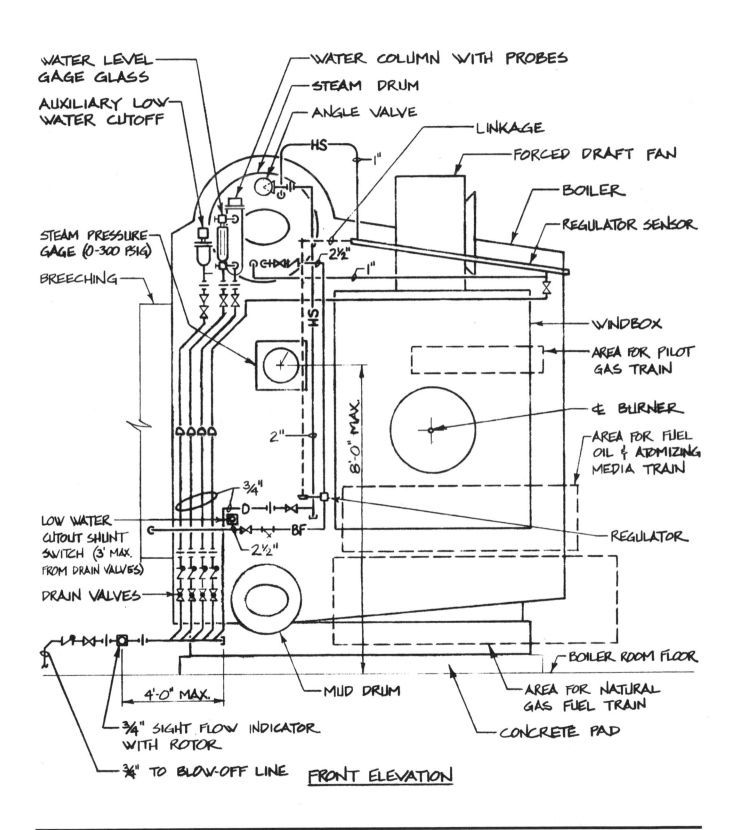

WATER LEVEL GAGE GLASS

AUXILIARY LOW WATER CUTOFF

WATER COLUMN WITH PROBES

STEAM DRUM

ANGLE VALVE

LINKAGE

FORCED DRAFT FAN

BOILER

REGULATOR SENSOR

HS

1"

STEAM PRESSURE GAGE (0-300 PSIG)

BREECHING

2½"

1"

HS

WINDBOX

AREA FOR PILOT GAS TRAIN

8'-0" MAX.

2"

¢ BURNER

AREA FOR FUEL OIL & ATOMIZING MEDIA TRAIN

¾"

D

LOW WATER CUTOUT SHUNT SWITCH (3' MAX. FROM DRAIN VALVES)

BF

2½"

REGULATOR

DRAIN VALVES

MUD DRUM

BOILER ROOM FLOOR

4'-0" MAX.

AREA FOR NATURAL GAS FUEL TRAIN

¾" SIGHT FLOW INDICATOR WITH ROTOR

CONCRETE PAD

¾" TO BLOW-OFF LINE

FRONT ELEVATION

FIGURE 1-3

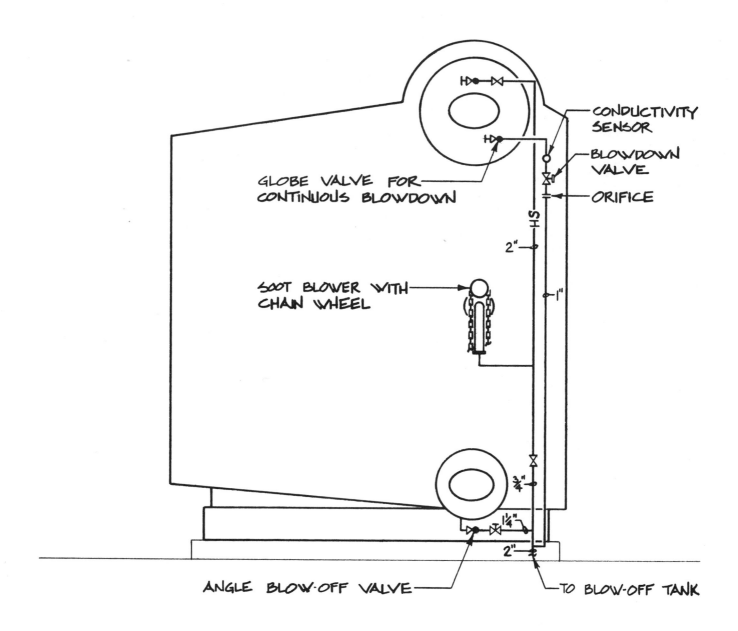

CONDUCTIVITY SENSOR

BLOWDOWN VALVE

ORIFICE

GLOBE VALVE FOR CONTINUOUS BLOWDOWN

HS

2"

1"

SOOT BLOWER WITH CHAIN WHEEL

¾"

1¼"

2"

ANGLE BLOW-OFF VALVE

TO BLOW-OFF TANK

REAR ELEVATION

FIGURE 1-4

can be readily be used at 125 psig. The piping that is shown is only a small part of the overall piping. In these two details no mention is made of the piping of steam out of the top of the top drum. This steam supply pipe passes through a stop valve (as it did in the low pressure boiler) which is normally chain operated from the floor and has a controlling gate valve after the stop valve. The line terminates in a connection to a main high pressure steam header which must be properly dripped.

We feel this steam piping is rather obvious and could be left off the drawing for the sake of clarity. What we do show, for example, is the water column with probes, the steam drum and the angle valve which tie to the Copes regulator sensor, and the forced draft fan on top of the burner that is depicted the front of the boiler. Associated with these items you will note arrows pointing to the wind box on the front of the boiler, the area for the gas pilot train, the center line of the burner, the area for the fuel oil and atomizing media train, and the area for the natural gas fuel train. These items are covered under separate details in Chap. 2.

On the boiler itself we have shown the various regulators that normally are supplied with it: the steam pressure gauge, the auxiliary low water pressure gauge, the drain lines from these items, and the drain lines from the Copes regulator. They all tie to a flow indicator through gate and check valves and terminate in a blowoff line. This is the piping which, along with the steam pressure gauge shown in the illustration, is normally on the front of a 125 psig watertube boiler. A continuation of the blowoff line is separately detailed in Chap. 2.

In Fig. 1-4 we show the piping that is commonly on the rear of the boiler, except for the obvious high pressure return piping, which is not shown simply because it is a line running along the side of the boiler and would not materially add to or improve our detail. Instead we have chosen to show on the rear of the boiler the continuous blowdown line that is common to all boilers of this type, as well as to the steam or air pressure operated soot blower. In Fig. 1-4 we show a steam operated soot blower. This device can blow the soot off the tubes in each of the sections of the high pressure boiler.

Thus what is shown in this detail is basically the piping that is on the boiler and not any of the piping that obviously goes into the boiler from the boiler return pump or from the outlet of the boiler to the steam header. These would vary from place to place depending on the situation and are better shown on the plans than on the detail itself since they are merely connecting lines. When drawn at a larger scale, these lines can easily and clearly be depicted without any further detailing.

Low Temperature Hot Water Boiler Piping: Figure 1-5 depicts the piping around any low temperature hot water boiler. If we were to show only the piping that directly related to the boiler, as we have done in steam piping details, we would have almost nothing to show. In the hot water application the boiler is hardly more than a heat exchanger. Consequently, about the only items that would be seen are a pipe going in, a pipe going out, an operating quastat, and an automatic water feeder. Instead, we chose to show something of a composite detail of the piping typical of a low temperature hot water installation.

There are a number of points to be made about the piping shown on this detail. The first refers to boiler shock, which is defined as a sudden thermal change inside the boiler causing rapid and uneven contractions which tend to loosen boiler tubes. Watertube and cast iron boilers have, in general, a high resistance to

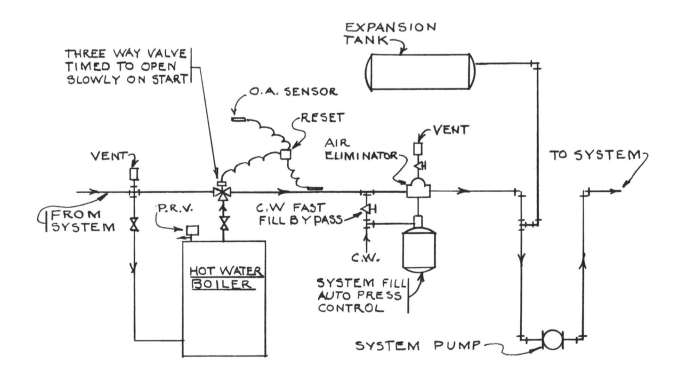

FIGURE 1-5

boiler shock. Larger firetube boilers are generally more susceptible.

Boiler shock frequently occurs in large systems in which weekend or night shutdown takes place; the sudden introduction of large amounts of cold water into the hot boiler during startup causes trouble. It could also occur in a three-pipe system that was switching from heating to cooling. Basically, to solve the problem the real trick is to change the incoming flow rate of the cooler water rather slowly so that boiler temperatures are changed slowly. There are a number of ways of accomplishing this. In our detail we show a timed three-way valve that is set to open slowly, taking some 20 to 30 min to go from closed to fully open.

The second point to be noticed in this detail is the location of the pump. Plans frequently show pumps discharging into the boiler and, in effect, into the compression tank. This pump location is improper for most higher resistance systems. If the pump pumps into the boiler, the pump therefore is pumping against the pressure in the pressure tank. Given a significantly high pump head, system pressure may drop to atmospheric, sucking air into the system through automatic vents and creating cavitation at the pump suction. Pumping into the boiler also creates a condition in which the pump operates at suction pressure equal to tank pressure minus the pump head. In the detail shown in Fig. 1-5 the pump is properly located, discharging away from the compression or expansion tank, establishing positive pressure and eliminating this potential trouble source.

Pumping Flow Arrangement: There is yet another point to be considered very seriously. Figure 1-5 shows that water goes into the bottom and out of the top of the boiler and, when the pump is base-mounted, back down again into the pump. Frequently there are situations in installations in which the engineer has allowed the flow to pump into the top of the boiler, out of the bottom of the boiler, and into the pump, thus supposedly saving piping and making for what appears to be a smooth flow of water from top to bottom. This is definitely not a desirable situation under any circumstance. The proper placement of the pump and compression tank in the boiler system is as shown on our details.

Sometimes the argument is made that reversing the flow in a cast iron boiler will increase the heat transfer from the flue gases to the boiler water. This might be true if heat was only transmitted by convection and conduction and if a true counter-flow from boiler water to flue gases existed. However, this is not the case. In a cast iron boiler, for example, for up to 60 percent of the primary heating surface, the heat is transmitted mainly by radiation. Water flow has no bearing on the heat absorption. In other words, reversing the water flow through the boiler will not apprecia-

bly change the heat transfer rate from the flue gas to the boiler water.

Several problems are created if the return is put into the top of the boiler and the supply is pulled out of the bottom. Since air must rise against flow direction, the elimination of air is a problem. The air released from the boiler water can be trapped and create air pockets. These pockets will often create hot spots and steam bubbles. Steam bubbles can also be caused by a poorly operating firing device. On their way to the top of the boiler steam bubbles rising from the bottom will come in contact with cool return water, causing them to collapse and producing water hammer which may be transmitted through out the piping systems. Air which can escape into the system can also cause noise and uneven heat distribution.

In addition, in low flow condition the water velocity in the boiler sections slows down considerably and by gravity may flow upward while being pulled downward. In other words it is possible that one part of the boiler flow can be pulled downward under low flow conditions while another part of the boiler can have a natural upward movement. This further aggravates air elimination control functions.

Finally, intermittent pump operation reverses the temperature pattern within the boiler each time the pump starts and stops, which puts an additional strain on the boiler. Controls for normal upward flow which have been predetermined by the manufacturer generally do not operate satisfactorily when used in reverse flow. It is also obvious that reverse flow definitely decreases the output of the tankless heater that is normally located in the upper part of the boiler.

The detail (Fig. 1-5) illustrates an expansion tank, related air eliminator air vent, and an automatically filling cold water bypass in an arrangement that is common to a pressurized tank installation. The other type of compression tank which is not pressurized is illustrated with its piping in a separate detail in Chap. 2.

High Temperature Hot Water Boiler Piping: Figures 1-6 and 1-7 illustrate the arrangements that would be used in a high temperature hot water heating system. Figure 1-6 illustrates a system containing a single pump for both boiler circulation and system circulation, and Fig. 1-7 illustrates two pumps, one for system circulation and one for boiler circulation.

Except for the fact that in the actual design all of the components are much bigger, the arrangement is very similar to the low pressure hot water system. However there are two basic differences. First, the compression tank is generally an inert-gas (nitrogen) pressurization arrangement, as noted on the details. Second and most important, a single pump should not be installed so that it pumps away from the boiler. Note that in a dual pump situation the boiler pump is in the same position as the single pump. In both cases the pump must pump

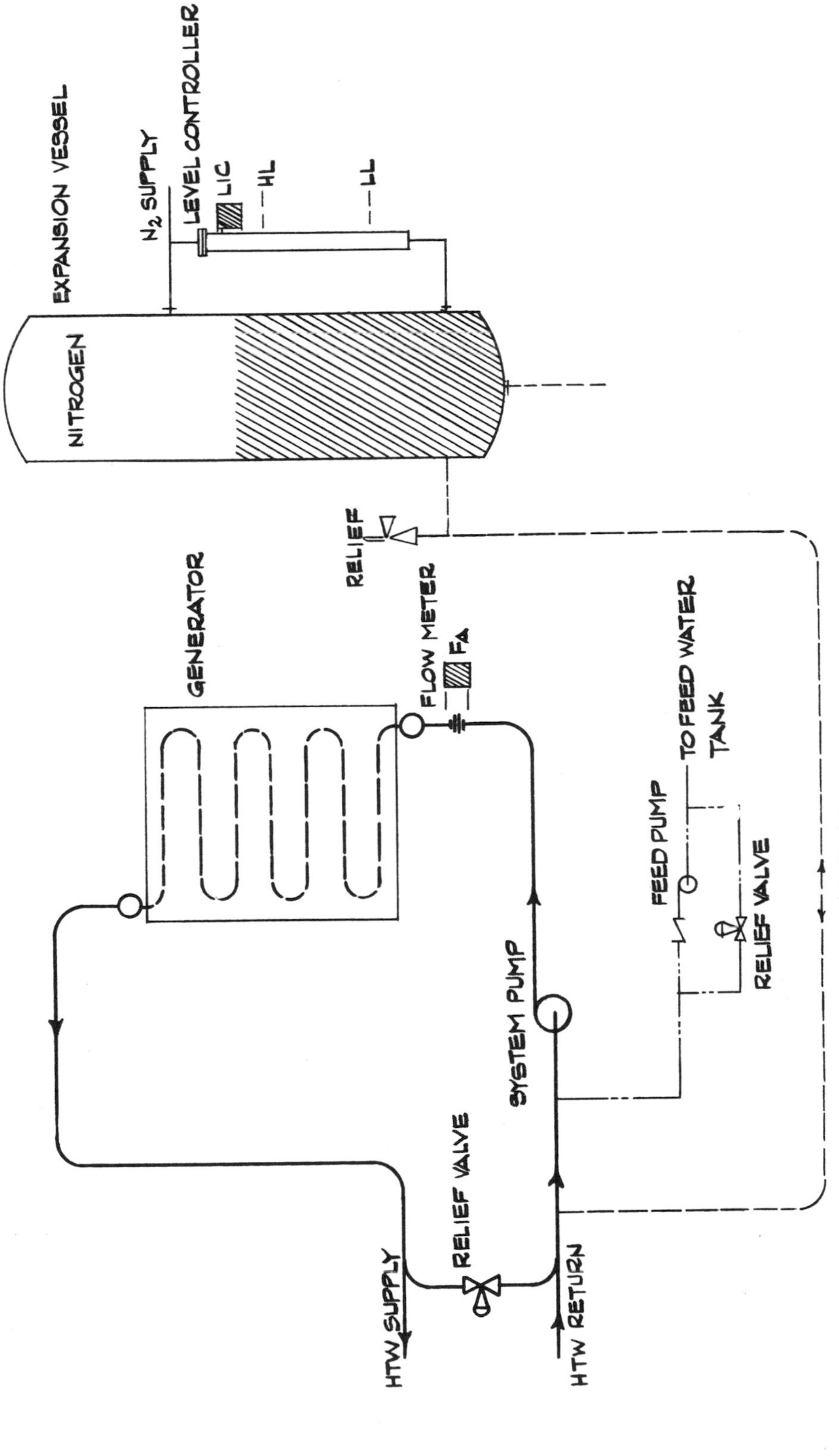

FIGURE 1-6

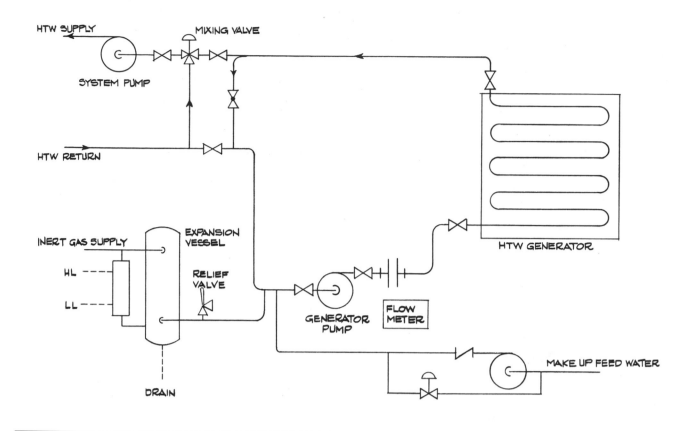

FIGURE 1-7

into the boiler and not out of it to maintain proper boiler circulation at all times.

Constant flow must be maintained in the boiler to help prevent burnout. As a consequence we cite a typical example in which pressurization is set to a minimum fill pressure equal to saturation pressure corresponding to 25°F above the leaving supply temperature. A 300°F supply temperature is, in effect, a 325°F saturation temperature, which requires about 85 psig minimum fill pressure. The boiler maximum operating pressure will be the minimum fill pressure plus the system pressure rise plus the pump head. The pump, as noted, must discharge into the boiler. In separate details in other parts of the book we cover the various elements of water treatment and system pumping. We also will illustrate in detail the nitrogen supplied pressurized expansion tank installation and all of its associated piping connections. Figures 1-6 and 1-7 do not pretend to show the complex piping of the nitrogen pressurized expansion tank.

Solar Heating

The solar collecting plate of the solar system can be used to heat air circulating over its plate or water circulating in coils in its plate. Basic to solar design is

the fact that there are approximately 8760 hr in a year. During half of these hours, on average, it is dark and no solar energy is present. To utilize solar energy to the maximum it is logical to have a storage system for the solar energy collected. The detail shown in Fig. 1-8 is a flow diagram rather than a specific detail. However, the piping and ducting would follow the basic piping and ducting rules for gravity flow or pumped flow of fluids similar to those in ordinary mechanical systems. Consequently, the real problem in solar heating is to calculate the amount of heating that you will get from the collector. This is not a subject for this particular book. It is covered in detail in the *ASHRAE Applications Handbook* and in the literature in a number of solar plate collector manufacturers. Generally the *average* plate output *per day* for 365 days is about 1000 to 1300 Btu/sq ft.

In general when the output of the collector is below 100 Btu/sq ft, the losses from the collector may exceed the heat that can be absorbed. This situation obviously varies with the temperature difference between the collector inlet temperature and the ambient air. Thus, provision must be made to disconnect the collector output from the system when, as is common in the winter, there is insufficient heat from the sun and the

SOLAR HEATING, AIR CONDITIONING & DOMESTIC WATER
— NOT TO SCALE —

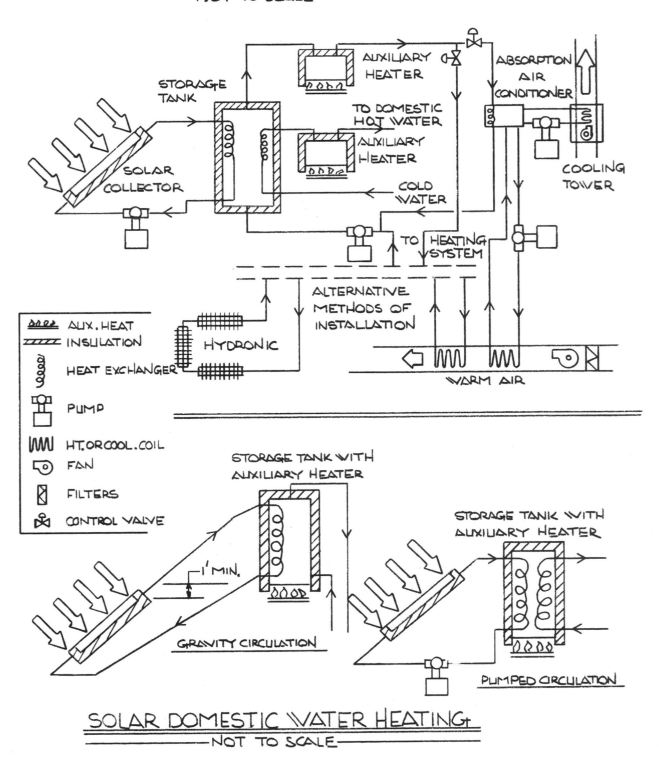

STORAGE TANK

AUXILIARY HEATER

TO DOMESTIC HOT WATER

AUXILIARY HEATER

SOLAR COLLECTOR

COLD WATER

ABSORPTION AIR CONDITIONER

COOLING TOWER

TO HEATING SYSTEM

Legend:
- AUX. HEAT
- INSULATION
- HEAT EXCHANGER
- PUMP
- HT. OR COOL. COIL
- FAN
- FILTERS
- CONTROL VALVE

ALTERNATIVE METHODS OF INSTALLATION

HYDRONIC

WARM AIR

STORAGE TANK WITH AUXILIARY HEATER

1' MIN.

GRAVITY CIRCULATION

STORAGE TANK WITH AUXILIARY HEATER

PUMPED CIRCULATION

SOLAR DOMESTIC WATER HEATING
— NOT TO SCALE —

FIGURE 1-8

collector actually becomes a heat-losing rather than heat-providing device. In Fig. 1-8 our flow diagram illustrates that there is an auxiliary heater on the system to provide for the necessary boost in system capacity. The storage tank for a hot water system would have limiting controls on the incoming and outgoing lines from the solar collector to compensate for lack of input when the temperature falls low enough to take heat out of the storage tank rather than provide heat for it.

Solar Collector Variations: You will note that in the lower half of Fig. 1-8 domestic hot water heating has been added to the system diagram. That is the only primary change in the detail. We are again presuming that cold water drawn into the system goes through the storage system of the solar collector and is warmed, fully or at least partially, before being discharged to the standard domestic water heating installation.

In all solar systems storage plays an important role. The most widely used storage systems use water storage tanks or rock systems to store heat. In terms of efficiency, water at the same volume can store approximately three times as much heat as rocks. Rocks, on the other hand, have the advantage that they are noncorrosive and can neither freeze nor boil. Leakage is only a minor problem. Rock storage is particularly suited for use in air systems and can provide a large amount of inexpensive heat transfer surface.

Specific heat storage systems in use today have demonstrated their ability to provide storage for peak-shaving and auxiliary uses of energy in many systems other than solar systems and are becoming a very logical part of any heating system. They should be seriously investigated by the designer. In general, calculations involving heat transfer through either water or air have to be based on very slow rates of transfer rather than the more rapid rates used in standard furnaces, coils, and boilers. As a consequence the air or water flow tends to approach the gravity velocities of either system.

Automatic Solid Fuel Burning Systems

Coal Storage: The amount of space for the coal storage is usually determined by the minimum required for timed truck delivery. Since storage usually is based on a 4-day period of delivery frequency based on a 12-hr daily operation at the maximum continuous rate of the plant, the requirement is to provide this volume of storage, that is, a 4-day supply at the maximum burning rate over a 12-hr period. How big should the ash pit be for the storage of ashes taken from the burning process? The basic rule of thumb is that ashes will be collected every 2 days and that the ash supply volume is approximately 10 percent of the coal supply volume. Therefore, if we know the 4-day coal supply volume, we can take 10 percent of the

volume burned and divide it by 2 to calculate the volume of the ash storage area pit or container.

There are situations in larger installations in which the delivery of coal and the collection of ashes may be at a more frequent rate. This would have to be resolved by the design group before the pit is designed. How much coal does it take to provide the necessary heat output? This also is a calculation required of the engineering group. In round figures a pound of coal is rated at approximately 12,000 Btu/hr. In rough calculations, 12 lbs of coal equal 1 gal of oil, and slightly more than 8 lbs of coal equal 100 cu ft of gas. These figures are all approximate and should be verified by the engineers in charge of the design of the solid fuel burning system.

Coal Fired Boiler Supply Systems: Figures 1-9 through 1-11 illustrate the design on the supply side of the solid fuel system. Figure 1-9 depicts a typical boiler room with the boilers on grade and the coal supply below grade. Figure 1-10 shows a typical boiler room with both the supply of coal and the boilers themselves below grade. The third detail, Fig. 1-11, shows a system in which both the fuel supply and the burner are on the same level, but the feed to the burner is a screw-type feed leading directly to the stoker. Customarily, as shown on Fig. 1-10 and 1-11, the fuel supply is taken upward by conveyor from the fuel storage bin to a holding container which then allows the coal to drop by gravity into the receptacle at the stoker from which it is fed through the stoker into the boiler.

Figure 1-9 depicts a boiler and storage arrangement that supplies coal at the rate of 25,000 lbs over a period of 50 hr, or approximately 560 lbs/hr. Generally 560 lbs of coal is considered equal to 7,000,000 Btu/hr input when coal is rated at 12,000 Btu/lb. In Fig. 1-10 the storage capacity is slightly larger and is capable of holding 57,600 lbs of coal. At a burning rate of 1144 lbs/hr you have an approximate 4-day storage. The capacity of the system in Fig. 1-11 is approximately equal to that of Fig. 1-9 in terms of storage and boiler burning capacity, which is approximately 560 lbs/hr or about 7,000,000 Btu.

Coal Fired Boilers: The boilers shown in Figs. 1-9 through 1-11 are commonly known as horizontal return tabular boilers. Inherent in their design is a large combustion space which lends itself to the economical burning of solid fuel. You will note that the details do not give the particular sizes or dimensions of many of the items which would be part of the design. The details shown in these three illustrations also do not describe the removal and disposal of the coal residue. There is much more to the design of a coal burning boiler than is shown on the details, but in this book we are not talking about design but about typical schematics or flow diagrams that may be put on the plans to accompany the basic plan design and serve as appro-

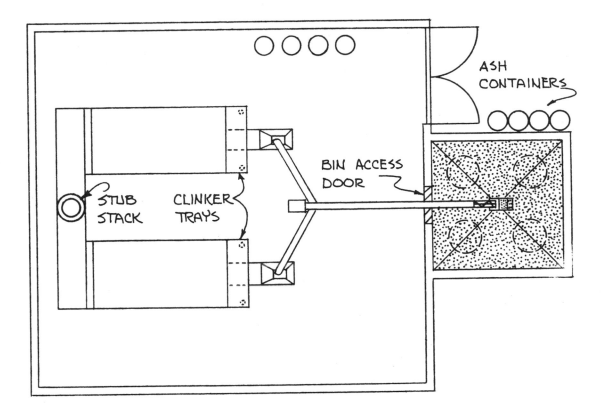

ASH
CONTAINERS

STUB
STACK

CLINKER
TRAYS

BIN ACCESS
DOOR

PLAN

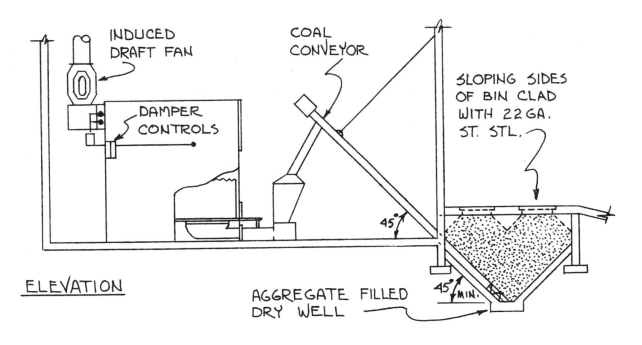

INDUCED
DRAFT FAN

COAL
CONVEYOR

SLOPING SIDES
OF BIN CLAD
WITH 22 GA.
ST. STL.

DAMPER
CONTROLS

45°

ELEVATION

45° MIN.

AGGREGATE FILLED
DRY WELL

TYPICAL COAL SUPPLY FOR
BOILERS ON GRADE WITH BELOW GRADE STORAGE
NO SCALE

FIGURE 1-9

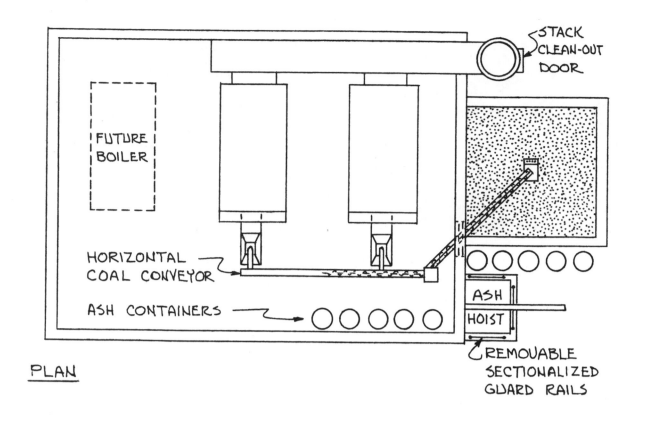

PLAN

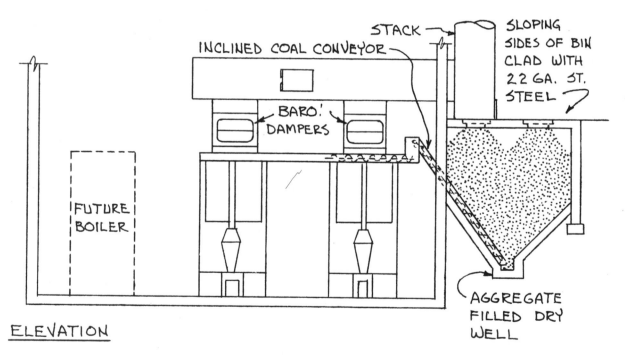

ELEVATION

TYPICAL COAL SUPPLY FOR
BOILERS AND COAL STORAGE BELOW GRADE
NO SCALE

FIGURE 1-10

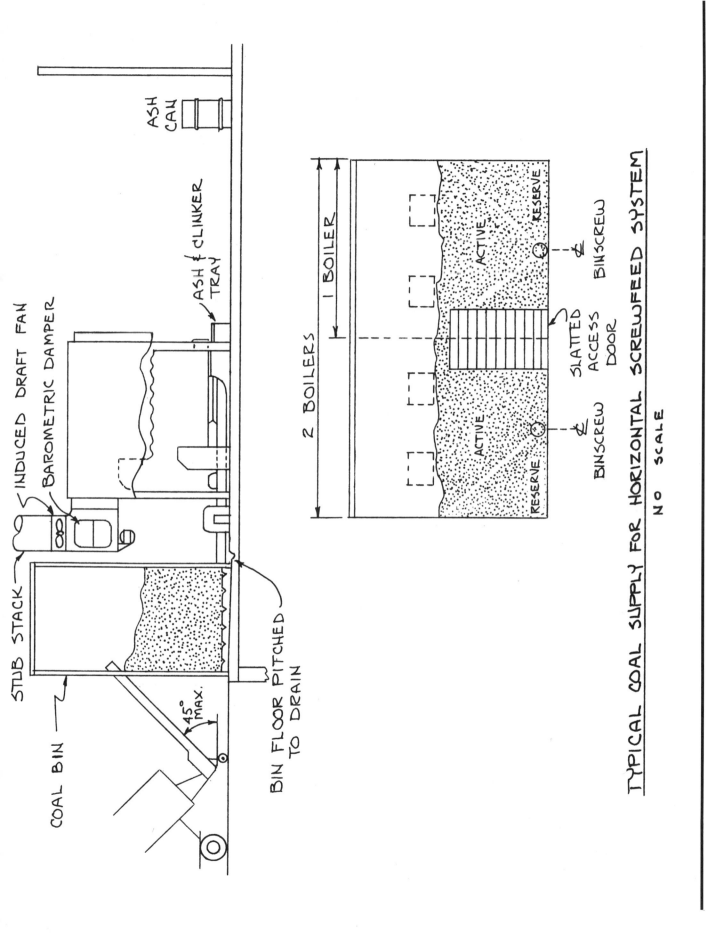

ASH CAN

INDUCED DRAFT FAN

BAROMETRIC DAMPER

STUB STACK

ASH & CLINKER TRAY

COAL BIN

45° MAX.

BIN FLOOR PITCHED TO DRAIN

2 BOILERS

1 BOILER

ACTIVE

RESERVE

BINSCREW

ACTIVE

RESERVE

BINSCREW

SLATTED ACCESS DOOR

TYPICAL COAL SUPPLY FOR HORIZONTAL SCREWFEED SYSTEM

NO SCALE

FIGURE 1-11

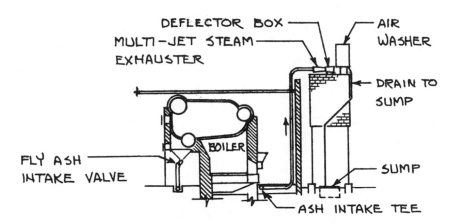

FOR UNDERFEED or TRAVELING GRATE STOKERS
(FLY ASH HAS LOW DUST CONTENT)

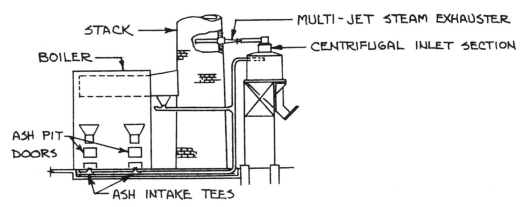

FOR SPREADER & PULVERIZED COAL FIRED BOILERS
(FLY ASH HAS HIGH DUST CONTENT)

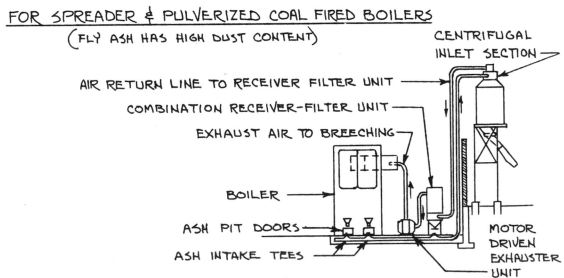

FOR LOW PRESSURE STEAM PLANTS

PNEUMATIC ASH HANDLING SYSTEMS
NO SCALE

FIGURE 1-12

priate details. These three details do serve that particular purpose.

Ash Disposal: Having burned the coal, the next problem is to get rid of the ash. In Figs. 1-12 and 1-13 we show typical pneumatic ash handling systems. In pneumatic ash handling systems an air stream induced by a mechanical steam or water exhauster transports ashes from the boiler ash hoppers to an elevated silo or a ground level pit. Such systems have great versatility in design. Pneumatic ash handling systems range in size from those suitable for small commercial heating plants to those for large utilities. For smaller systems such as we have been describing in which no fly ash is handled, a ground level storage bin of cinder block or other inexpensive construction can be used. This is illustrated in Fig. 1-13. Figure 1-12 shows three versions of a pneumatic ash handling system. The upper detail shows one with the capacity of 5 to 50 tons per hour and is usually used with underfeed or traveling grate stokers in which the fly ash has a low dust content. In the middle of Fig. 1-12 we show a simplified detail of a system suitable for spreaders and pulverized coal fired boilers in which the fly ash has a high dust content. And in the bottom of the same figure we show an ash system for low pressure steam plants in which vacuum is introduced by a motor driven exhaust unit.

Ash Disposal Rules: It is generally considered good engineering practice to install a system that will remove ash accumulated during an 8-hr period of full power operation in 1 hr or less. This is not a firm rule. Conveyor piping is usually supplied for medium or large pneumatic systems and normally has a diameter of 6 to 8 in. The use of 4-in diameter piping for fly ash, other than for branch lines, is normally not recommended except in smaller installations. If ash must be held to the intake, then quicker breaking is required. The ash handling capacity of the 6-in system is 5 tons per hour and the capacity of the 8-in system is 8 tons per hour. With free flowing ash, fly ash, and gravity discharge to the intake hopper, maximum capacities are normally 14 to 20 tons per hour. The maximum size clinker that can be handled by the 6-in system passes through a 3-in sq opening. The maximum size handled by an 8-in system normally passes through a grid of a 4-in sq openings. Pipe and ash handling can be one of three types: abrasive resistant sand cast pipe, centrifugally cast abrasive pipe, or heavy black steel pipe. Smaller systems generally have centrifugal cast pipe, and the exhauster-cyclone separator is usually a standard design.

The exhauster-separator may be factory or field assembled. The exhauster creates a vacuum of 5 to 12 in mercury in the conveyor lines. The induced air flow carries ash from the hoppers to the primary and secondary separators, which are designed to separate out as much as 95 percent of the untreated ash. When a steam exhauster is used, some of the remaining fly ash is caught by the condensed vapor and is dischargd into the stack. On a mechanically exhausted system for a large installation in which residual dust from the separators may damage the exhauster, an air washer separates out the dust at a point before the exhauster inlet.

In coal burning plants in which high pressure steam is not available to power a steam exhauster, the pneumatic ash handling system can be activated by a mechanical exhauster. This essentially is the same as a steam powered system except that an air washer is usually recommended to preclude the mechanical exhauster from excessive wear.

In the standard ash handling system ashes are normally stored in a cylindrical tile or steel elevated silo until final dispersal. Ashes may also be directly deposited in a low fill land if it is available. Cylindrical ash storage silos are generally favored because maximum utilization of structural material is attained, ash arching and dense storage are kept to minimum, and a stronger structure can be built. Vitrified glass tile and steel plate are the structural surface materials commonly used.

Ash Storage: In small coal burning plants in which a large investment in ash handling equipment cannot be justified a simplified system using a ground level cinder block bin instead of an elevated silo may be installed, as illustrated in Fig. 1-13. In this case a front-end loader is used to empty the bin periodically. There are many forms of fly ash treatment available to conform to environmental factors. At plants in which dust is a critical factor, such as those in residential areas, at service institutions, and in hospitals, dustless unloading may be achieved by specialized containers. The ash container normally consists of a motor driven rotary drum that mixes ash with water in proper proportions to moisten all the dust (ash) particles.

The subject of ash handling is a complex subject and further description is really beyond the scope of this book. There are many available pieces of literature; the best sources of information for the engineer can be obtained from the National Coal Association and the U.S. Department of the Interior. In addition, information can be obtained from manufacturers of stokers; they are listed in publicly available catalogs.

Refuse Burning Systems

Our single detail, Fig. 1-14, is a composite schematic, or flow diagram, overview of a basic refuse burning system and the use to which the resultant energy can be applied. If you can use heat that is available 24 hours a day, 365 days a year for a building, a process, or for electrical energy, the system may be applicable. The underlying premise is that the average heating value of solid waste refuse is approximately one-half that of coal on a pound for pound basis.

A AIR INTAKE

B ASH INTAKE

C FLY-ASH INTAKE

D STEAM EXHAUSTER UNIT

E TILE ASH-STORAGE UNIT

F PRIMARY SEPARATOR

G SECONDARY SEPARATOR

H AIR WASHER

I ASH CONDITIONER

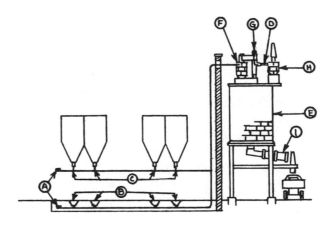

STANDARD PNEUMATIC SYSTEM SCHEMATIC

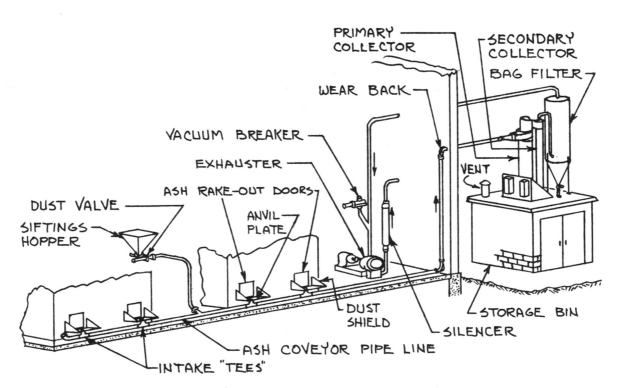

GROUND LEVEL CINDER BLOCK ASH STORAGE BIN

PNEUMATIC ASH HANDLING SYSTEM

NO SCALE

FIGURE 1-13

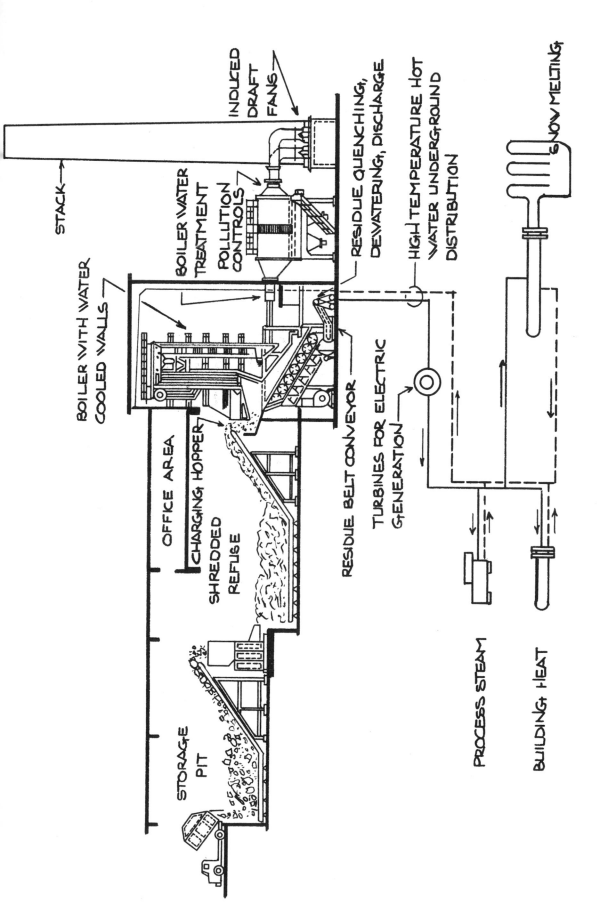

STACK →

INDUCED DRAFT FANS

BOILER WATER TREATMENT

POLLUTION CONTROLS

RESIDUE QUENCHING, DEWATERING, DISCHARGE

HIGH TEMPERATURE HOT WATER UNDERGROUND DISTRIBUTION

BOILER WITH WATER COOLED WALLS

OFFICE AREA

CHARGING HOPPER

SHREDDED REFUSE

RESIDUE BELT CONVEYOR

TURBINES FOR ELECTRIC GENERATION

STORAGE PIT

SNOW MELTING

PROCESS STEAM

BUILDING HEAT

SOLID WASTE SYSTEM STEAM GENERATION AND ELECTRICAL POWER PRODUCTION

NOT TO SCALE

FIGURE 1-14

Wood, bark, and wood chips have always been a source of energy, as have been paper and dry paper pulp. More than 80 years ago this was a common source of fuel. Today some boiler manufacturers are again listing in their equipment catalogs boilers with this fuel use capability. As far as the mechanical detailer is concerned, a boiler is a boiler. The boiler connection details are the same regardless of the fuel used. We do not show the type of fuel feeder as it is generally special and part of the standard boiler equipment. Conveyors, hoppers, and ash handling details would be similar to those previously shown for coal burning systems.

TWO

Heat Source Support Equipment

All the heat source equipment related to or supportive of the boiler plant, solar energy collection system, or electrical energy source is shown in this chapter.

Supportive equipment normally includes the fuel supply system, combustion chambers, circulating and condensate pumps, boiler return pumps, expansion tank systems, forced and induced draft fans, breechings, water treatment, the central control system, blowdown and flash tanks, deaerators, pressure relief valving, and the like.

Great strides are being made by a number of equipment manufacturers in the improvement of equipment covered in this chapter. For example, we depict field constructed combustion chamber details. This sort of installation has, for the most part, gone the way of gas lighting. Practically all boilers today come with pre-built manufacturer-supplied combustion chambers. However, as we note in the detail description of combustion chambers, the engineer works on both new and renovation projects.

It is very possible that this chapter may provide information that of and by itself makes the purchase of this book worthwhile. The automobile presents a similar situation. All of the usual discussion that occurs during a car-buying situation invariably focuses on the visable body and the engine under the hood. But it is the related, properly installed accessories that permit the engine and body to become a usable automobile.

For each of the more than 50 details in this book there is at least one variation that is currently installed and working on some project, somewhere. Those depicted are typical of their type in the situation involved.

Fuel oil Systems

Oil Transport Systems: Figures 2-1 through 2-3 present schematic composite details that show the installation of a complete system from the tank to the burner. The system could use an unheated oil, nos. 2, 4, or a hot no. 5; a heated oil, nos. 5 or 6; or a normal no. 6 oil. You will note that there are both similarities and differences in these figures. When the oil lines reach the burners themselves, the general sequence of items shown on the suction and discharge piping are similar for both heated and unheated fuel oil. There is, however, a very clear difference in the drawings. Note the separate suction lines for the unheated oil installation and the arrangement of piping from the fuel transport or transfer pumps in the heated oil installation to the suction of the given burner. As a general rule, a pump is required whether the oil is transported heated or unheated if the distance from the burner to the storage tank is greater than 100 ft.

The transport of any oil in a fuel oil system assumes that the bottom of the tank is not more than 15 ft in vertical elevation below the inlet either to the fuel pump or to the burner inlet. This is a pump suction limiting factor. Further, it is good practice to keep the suction pipes as short as possible and to design systems in which the tank is as close as possible either to the burner or to the primary pump. The piping should always be welded if black iron is used and soldered if copper is used. Oil lines to the suction side of a meter should be the same size as the fuel oil suction lines. The oil lines should have nonrising stem stuffing box gate valves on them and compound pressure gauges should be used. Swing check valves should be in-

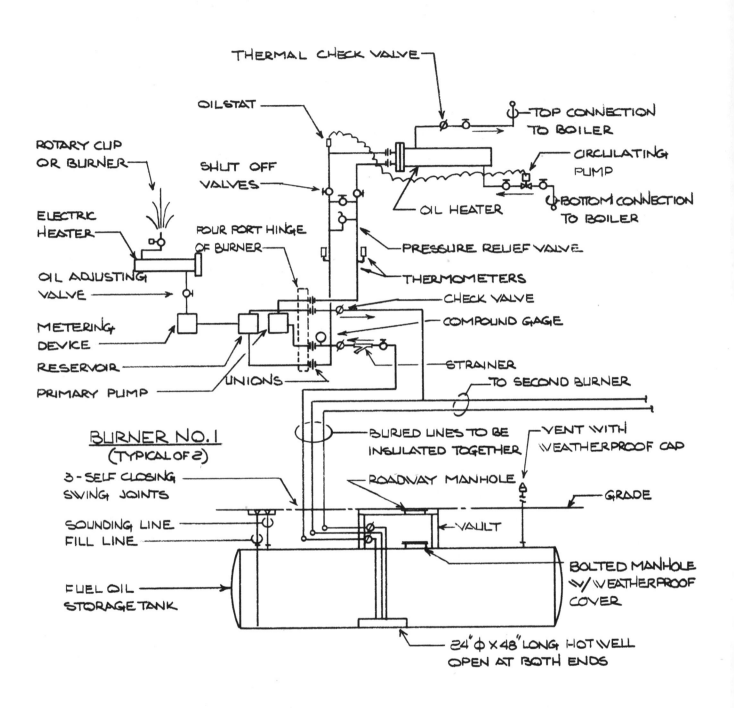

THERMAL CHECK VALVE

OILSTAT

ROTARY CUP OR BURNER

SHUT OFF VALVES

TOP CONNECTION TO BOILER

CIRCULATING PUMP

ELECTRIC HEATER

FOUR PORT HINGE OF BURNER

OIL HEATER

BOTTOM CONNECTION TO BOILER

OIL ADJUSTING VALVE

PRESSURE RELIEF VALVE

THERMOMETERS

METERING DEVICE

CHECK VALVE

COMPOUND GAGE

RESERVOIR

STRAINER

PRIMARY PUMP

UNIONS

TO SECOND BURNER

BURNER NO.1
(TYPICAL OF 2)

BURIED LINES TO BE INSULATED TOGETHER

VENT WITH WEATHERPROOF CAP

3 - SELF CLOSING SWING JOINTS

ROADWAY MANHOLE

GRADE

SOUNDING LINE
FILL LINE

VAULT

FUEL OIL STORAGE TANK

BOLTED MANHOLE W/ WEATHERPROOF COVER

24" Ø x 48" LONG HOT WELL OPEN AT BOTH ENDS

PIPING DIAGRAM FOR NO. 5 OIL BURNING SYSTEM
— NOT TO SCALE —

FIGURE 2-1

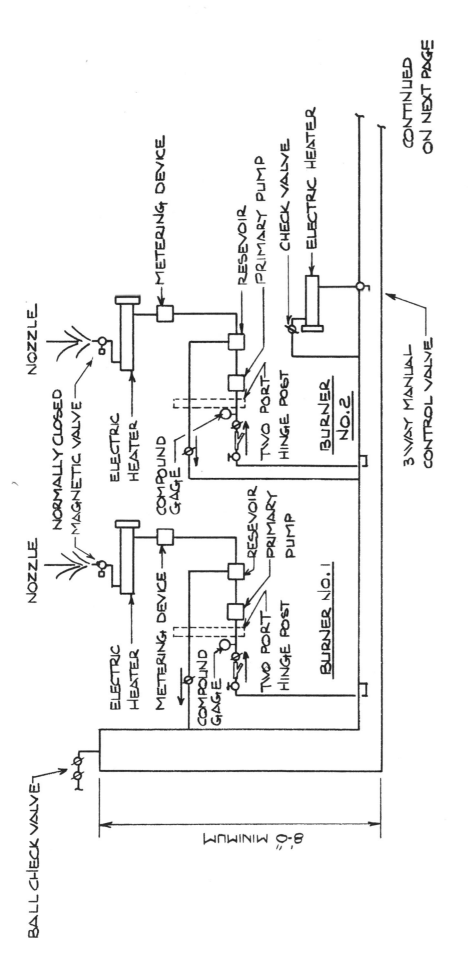

PIPING DIAGRAM FOR NO.6 OIL BURNING SYSTEM

SHEET 1 OF 2

NOT TO SCALE

FIGURE 2-2

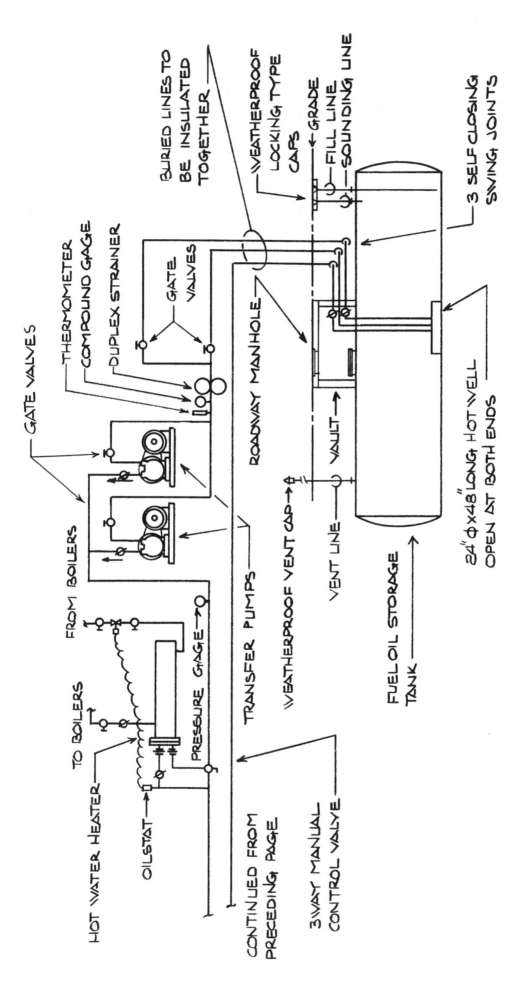

PIPING DIAGRAM FOR NO. 6 OIL BURNING SYSTEM

SHEET 2 OF 2

NOT TO SCALE

FIGURE 2-3

stalled in the horizonal line near the point at which the suction line(s) enter the tank to prevent oil from running back into the tank when the pump(s) is (are) stopped. A foot valve is not recommended. When a gas ignition is required, the line size should be as recommended by the burner manufacturer. The electrical heater shown at the burner is intended for use on cold startup only and is generally not needed for no. 2, no. 4, or properly selected no. 3 oil operation. The hot water or steam heat exchanger should be sized in accordance with carefully calculated requirements for the heating of the fuel. These requirements are determined by the flow of oil that is selected by the system engineer to provide the amount of energy needed for the necessary boiler output.

Oil Tank Installation: Once the basic overall flow diagram is established by a detail similar to Figs. 2-1 through 2-3, certain other basic parts of the oil system also become a standard part of the detailing. The first of these is the installation of the oil tank itself. Insofar as the oil tank is concerned, the rules of the National Board of Fire Underwriters and all local and state codes and authorities should be followed in locating and installing the oil storage tanks. Buried oil storage tanks should be located with ample clearance between the tank and any sewer, water, or gas lines. The fill line and the burner supply line should be as short as possible. Buried tanks should be covered with not less than 2 ft of earth or 1 ft of earth covered with a concrete slab that is at least 4 in thick.

The answer to the question of whether or not the tank should have the mat shown in the detail depends on the condition of the soil at the project site. Water in the ground can dislodge a buried tank. When this condition is possible or thought to be possible, the tank should be anchored to a concrete base weighing three-quarters of the weight of the oil that would be needed to fill the tank. All buried tanks should be coated with one heavy coat or two normal coats of asphalt tar or other suitable protective coating to reduce corrosion on the outside of the tank. The fill material around the tank should not contain ashes or cinders. The fill pipe opening should be at least 5 ft from any window or building opening, and the fill terminal, when not in use, should be capable of being tightly closed. The diameter of the fill pipe should not be smaller than 2 in on small tanks and should be larger on larger tank sizes as required by local codes and ordinances.

If the fill opening is not directly over the horizontal portion of the tank, the fill line must slope toward the tank. Every tank should be vented with a properly sized vent line. While certain codes may have special requirements, in general the vent line should be at least 2 ft away from any window or other building opening. The top of the vent must have a return bend or a proper vent cap and must extend above the ground to a distance sufficient to prevent obstruction by ice or snow.

Individual localities will usually specify in their codes the height of the vent riser pipe. Frequently, in larger tanks, such as the one depicted, it is 10 ft above grade. There are also various types of vents which permit the vent line to be run from within the building wall construction and to terminate in an approved vent brick rather than to be exposed on the exterior wall surface. Normally the 2-in × 8-in vent brick matches the exterior size of a standard brick.

Vent piping should be pitched slightly downward toward the tank and should be large enough to prevent air pressure from building up in the tank during the filling but should not be smaller than the size listed below.

Tank capacity (gal)	Pipe size (in)
500 or less	1¼
501 to 3000	1½
3001 to 10,000	2

Oil Tank Details: All of the comments previously made apply to Figs. 2-4 and 2-5, which are typical underground installation details of a no. 6 oil storage tank. They would equally apply to no. 2 or no. 4 oil storage tanks except that there would be no heating of no. 2 or no. 4 oil or, in some cases, no. 5 oil. Therefore the hot well would not be depicted. Note the manhole shown on the detail which is approximately 6 ft × 3 ft 6 in. This is not a firm dimension that applies to all situations. The manhole should always be sufficiently large to allow access to the piping that goes into the tank and to the 20-in access manhole that is commonly furnished with tanks.

In the detail shown in Fig. 2-5 note that both in plan and elevation there are two piping connections, the fill line and the sounding line. The fill line could, if necessary, also be used as the sounding line. Two lines are not always necessary. When the fill line is installed as is shown in the detail, a separate sounding line is needed since the fill line runs to the bottom of the tank as a solid pipe and terminates 6 inches from the bottom. Although it is not absolutely necessary for the fill line to go to the bottom of the tank, it is the better way to install this particular line.

Finally, we show in Fig. 2-5 the angle supports welded to the tank and tied to 1-in rods which are fastened to the reinforcing in the concrete tank mat. This should be very carefully detailed for a situation in which vibration from road traffic or movement caused by the underground water table may be a problem.

In many installations in lieu of six rods and six angle supports per side three may suffice; this should be

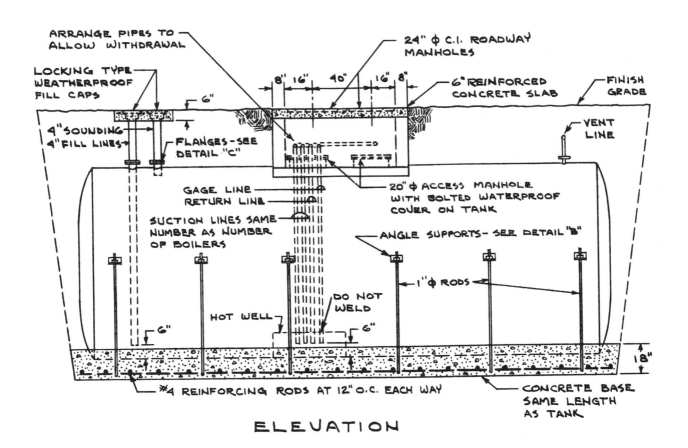

ELEVATION

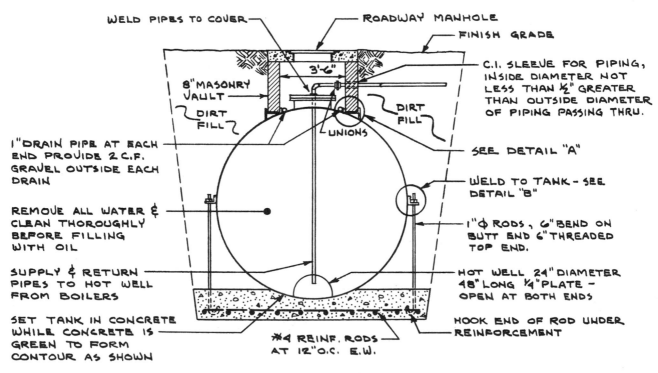

SECTION

TYPICAL UNDERGROUND INSTALLATION OF A No. 6 OIL STORAGE TANK
(SHEET 1 OF 2)

FIGURE 2-4

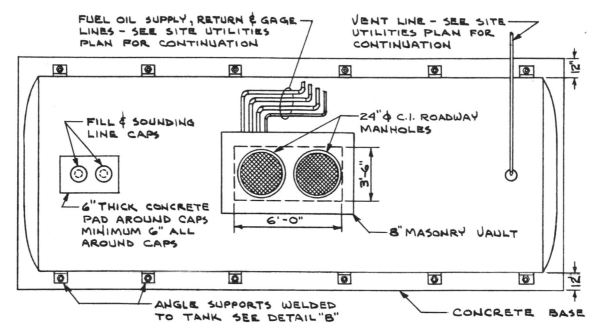

FUEL OIL SUPPLY, RETURN & GAGE LINES - SEE SITE UTILITIES PLAN FOR CONTINUATION

VENT LINE - SEE SITE UTILITIES PLAN FOR CONTINUATION

FILL & SOUNDING LINE CAPS

24"∅ C.I. ROADWAY MANHOLES

6" THICK CONCRETE PAD AROUND CAPS MINIMUM 6" ALL AROUND CAPS

8" MASONRY VAULT

ANGLE SUPPORTS WELDED TO TANK SEE DETAIL "B"

CONCRETE BASE

PLAN VIEW

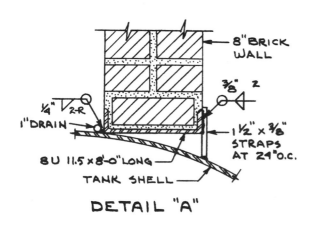

8" BRICK WALL

1" DRAIN

8U 11.5 × 8'-0" LONG

TANK SHELL

1½" × ⅜" STRAPS AT 24" O.C.

DETAIL "A"

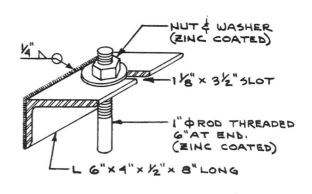

NUT & WASHER (ZINC COATED)

1⅛" × 3½" SLOT

1"∅ ROD THREADED 6" AT END. (ZINC COATED)

L 6" × 4" × ½" × 8" LONG

DETAIL "B"

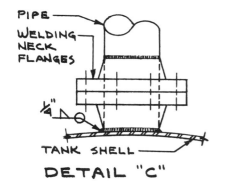

PIPE

WELDING NECK FLANGES

TANK SHELL

DETAIL "C"

NOTES:

1. OIL STORAGE TANK SHALL BE SUITABLE FOR UNDERGROUND USE & SHALL BE BUILT IN ACCORDANCE WITH UNDERWRITERS LABORATORIES INC. & SHALL BE SO STAMPED.

2. ALL SUCTION LINES SHALL BE SUCTION TESTED TO 100 MICRONS AFTER INSTALLATION, ALL RETURN LINES SHALL BE PRESSURE TESTED TO 50 P.S.I.G. AFTER INSTALLATION.

3. ALL UNDERGROUND PIPING SHALL BE WELDED.

TYPICAL UNDERGROUND INSTALLATION OF A No. 6 OIL STORAGE TANK
(SHEET 2 OF 2)

FIGURE 2-5

27

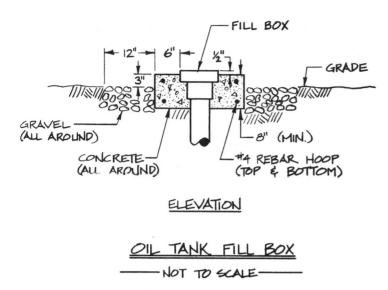

ELEVATION

OIL TANK FILL BOX

—— NOT TO SCALE ——

FIGURE 2-6

determined by a proper structural engineering calculation.

Fillbox: Figure 2-6 depicts a typical detail of an oil tank fill box. This is a detail that is commonly overlooked in many oil tank detailing schemes. The fill box should be surrounded by concrete whether it is in a roadway, in a driveway, or under a lawn. The premise that a lawn would not require concrete around the fill box is not logical since earth movement can frequently cause the fill box to become inaccessible or covered over by a growing lawn. The concrete, even in a lawn, is a safety precaution.

Oil Heating: Figures 2-7 and 2-8 show a method of fuel oil heating that uses high temperature hot water or steam. The electric heater on the burner in the no. 6 oil installation, and frequently in the no. 5 installation, is supplemented by a heater on the oil circulating line. This heater may be steam, hot water, or electric. In areas in which electricity is expensive, the steam or hot water heater should be sufficiently sized to provide

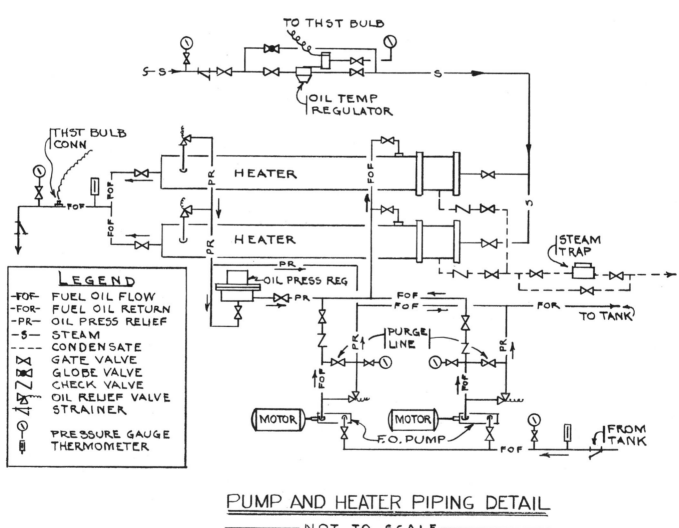

PUMP AND HEATER PIPING DETAIL

—— NOT TO SCALE ——

FIGURE 2-7

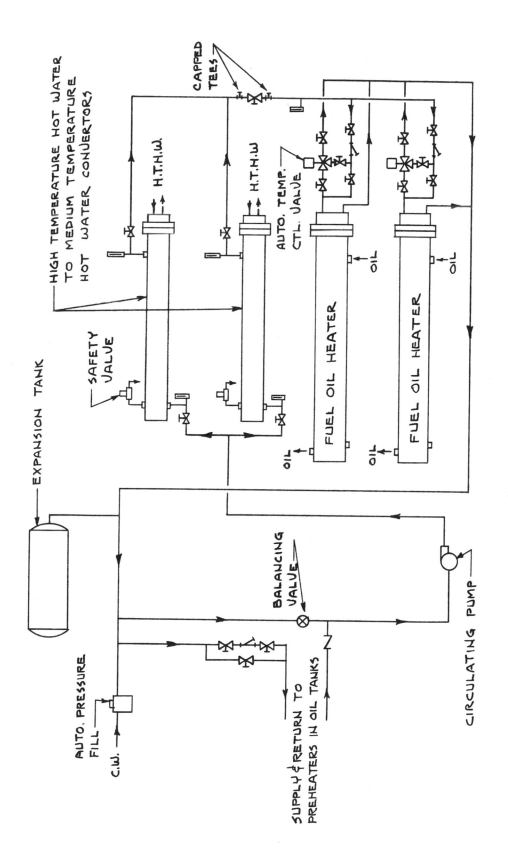

HIGH TEMPERATURE HOT WATER to MEDIUM TEMPERATURE
HOT WATER FUEL OIL HEATING PUMP & HEATER SET
PIPING DIAGRAM

FIGURE 2-8

all the heat after the boiler is operating so that electric energy use is limited. The heater may be connected in any of several positions in the piping system. Oil heaters should not be located in the suction line between the tank and the oil pump since the friction lost through the heater reduces the suction available for lifting oil and overcoming line friction losses. The hot oil also tends to vaporize on the suction side of the pump.

Figure 2-7 utilizes medium temperature hot water, which is water in the 280°F temperature range, to heat both the fuel oil line heaters in each line and the preheaters when they are required in the oil tanks themselves.

Return Line Heater: Figure 2-9 illustrates a steam or hot water heater in the return line. This is an alternative method of heating the fuel oil. In lieu of a separate hot water or steam source the steam source is combined with the return line electric immersion heater.

Tank Heaters: As is implied in our detail of the high temperature to medium temperature water conversion, Fig. 2-8, there is a supply and a return to preheaters in the oil tanks. There are basically three types of heaters that may be used to heat the oil in the tank itself. They are supplied by steam or hot water feed and, as shown in Fig. 2-10, can be horizontal coils, vertical coils, coil heaters in a hot well, or a suction bell type heater. All of these heaters perform with reasonable satisfaction. The horizontal coil heater is a more expensive and frequently more satisfactory heater when carefully designed and located in the bottom of the horizontal tank. It usually provides for smooth and even heating of all of the oil in the tank. The piping material of the heaters should be wrought iron. Vertical and suction bell heaters are standard devices that can be found in various manufacturers' catalogs. Steam heaters are used when continuous steam pressure is available. A steam regulating valve is required to prevent the oil from overheating. The condensate from the heater in smaller installations is

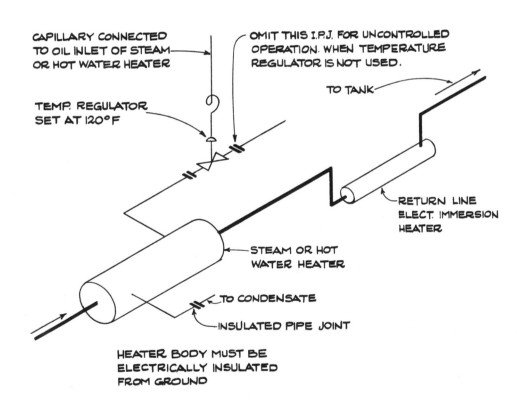

STEAM OR HOT WATER HEATERS IN RETURN LINE DETAIL

FIGURE 2-9

most frequently passed from the steam trap into a drain so that the oil leakage in the steam line will not be returned to the boiler. In larger installations the condensate from steam boilers may be run into an open receiver in which any oil leakage can be observed. If steam at higher pressure is used, a reducing valve in the line to the steam heater is desirable to facilitate the control of the oil temperature.

If the oil heating medium is hot water, it may be circulated either by gravity or with a pump. When gravity is involved, no oil temperature regulating device is needed, but with forced circulation a temperature sensor is required to operate the circulator.

All-Electric Oil Heating: In addition to the oil line heating shown in Figs. 2-7 through 2-9 there is available an entire electric pipe heating system for the oil burner installation, with or without remote heating pumps. This system is shown in a composite detail in

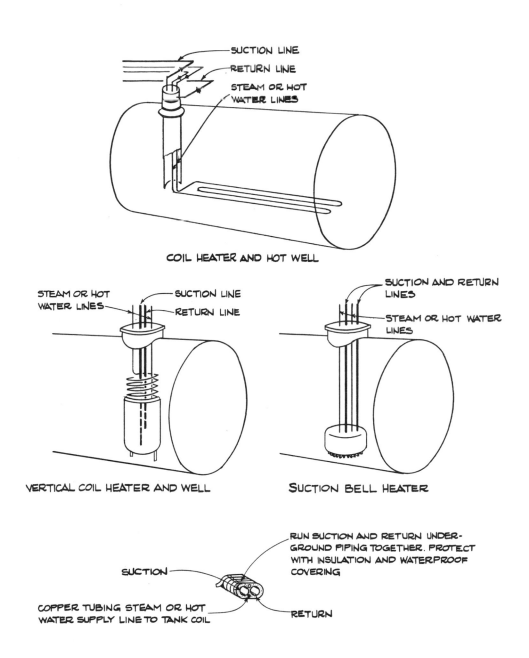

COIL HEATER AND HOT WELL

VERTICAL COIL HEATER AND WELL

SUCTION BELL HEATER

TANK HEATERS

FIGURE 2-10

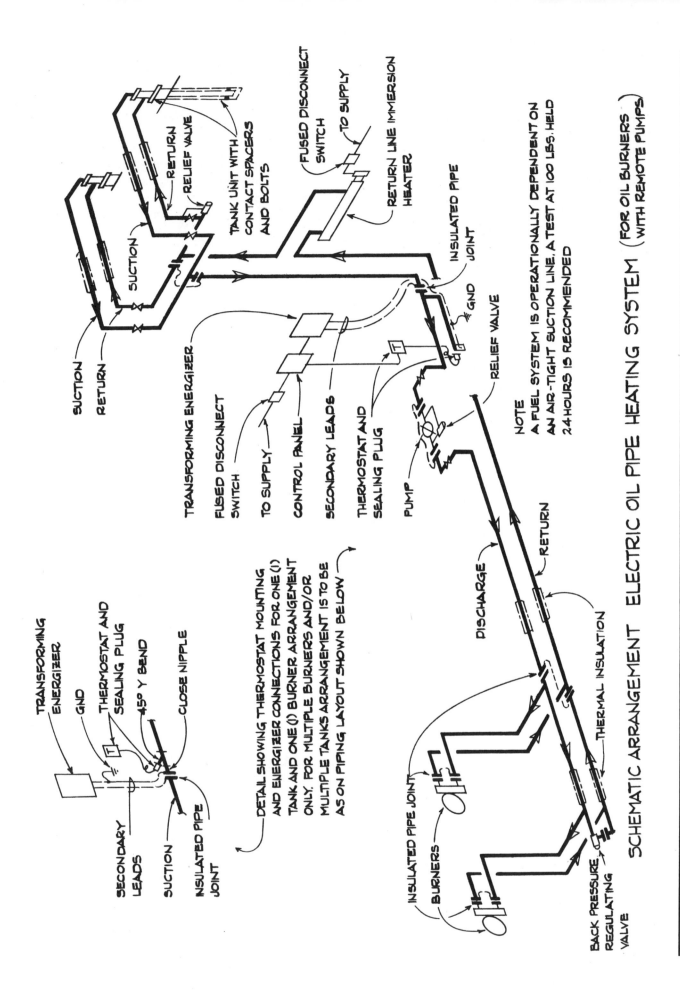

SCHEMATIC ARRANGEMENT ELECTRIC OIL PIPE HEATING SYSTEM (FOR OIL BURNERS WITH REMOTE PUMPS)

FIGURE 2-11

Fig. 2-11; Figs. 2-12 and 2-13 illustrate special items of the electric heat installation. In this particular design the entire oil line becomes an electrical resistance heater. The system uses electricity to heat the entire pipe line and, as a consequence, heats the oil passing through the lines. This system may be furnished with tank units which will also electrically heat the oil in the tank. In terms of operation this is an exceedingly accurate system although, because of the ever-rising cost of electricity, it is an expensive system. The energy conservation requirements of the 1980s have created a circumstance in which the use of fuels at lower viscosities, such as no. 4 and no. 2 oil, in certain situations are more than competitive to heated no. 5 or no. 6 oil.

In the electrically heated system, as in all other heated oil systems, one must insulate the oil lines that are heated whether they are heated with a steam or hot water heater or are heated by an electrical system such as the one shown in Fig. 2-11. The ability to heat lines is created by insulated pipe joints which provide the resistance-type heating arrangement to heat the pipe. These are noted on the figure as insulated joints and are connected to a transformer that is controlled by a line

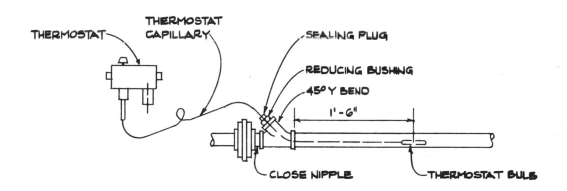

BULB MOUNTING IN PIPE RUN
FOR ONE (1) TANK & ONE (1) BURNER ARR'G'T.

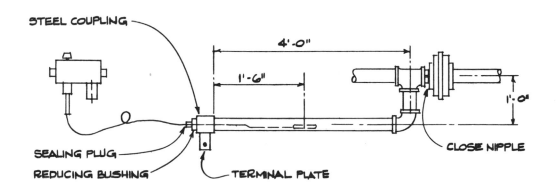

BULB MOUNTING IN PILOT STUB TAKE-OFF
FOR MULTIPLE BURNER AND/OR MULTIPLE TANK ARR'G'T.

FIGURE 2-12

thermostat. Note that there are insulated joints illustrated near the burners, at a midpoint in the lines, and near the lines going to the tank and in the tank itself. In addition, this electrical system employs a return line heater that is also electrically operated. When using this system, certain special connections, such as the thermostatic control, the sealing bulb mounting, and the sealing plug, are required. They are shown in Figs. 2-12 and 2-13. These connections are self-explanatory; however, careful note should be made of the distances required for temperature sensors.

The composite scheme and its details are intended to cover all parts of the most complex system. When the burners you select have integral pumps and the piping design conditions do not require booster or oil circulating pumps, the pump shown in Fig. 2-11 in the line from the tank may be omitted.

Gas Systems

Commercial and industrial gas systems are ones that commonly occur in the practice of most design engineers. The details are concerned primarily with two

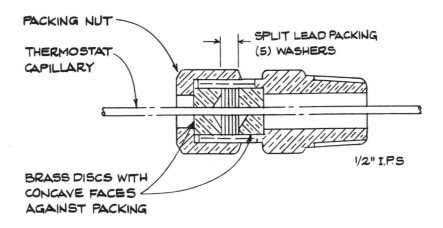

SEALING PLUG DETAIL

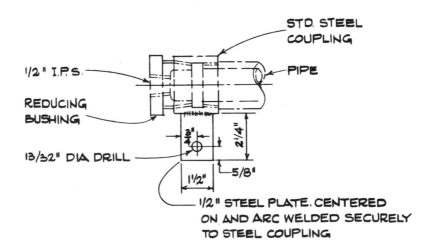

TERMINAL PLATE DETAIL

FIGURE 2-13

basic burners, the atmospheric and the power types, and with the piping of gas to the burner itself. Normally the piping of the gas supply to the burner is part of the plans. The connection to the burner is shown very easily and clearly on the plans, and the only detailing that is usually required is shown in Fig. 2-14. Gas piping is a matter of regulating the flow of gas to the system in a manner commensurate with safety and good operating practice. The two typical gas burner trains that are shown in Fig. 2-14 are not necessarily the only way to install the gas piping to the burner.

Note that there are lines going to the burner and to the pilot. This may, or may not, be the actual circumstance in every case; however, these details are in general conformity with what is expected on the plans. The details should be checked with the local code requirements and the supplier of the gas burner specified in your contract.

Combustion Chambers

In today's design departments combustion chamber design is a relatively unheard of task since almost all of

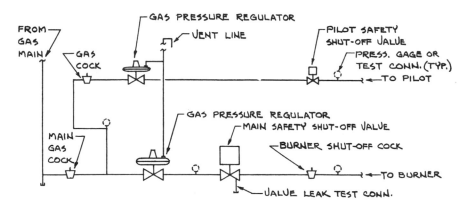

GENERAL PURPOSE

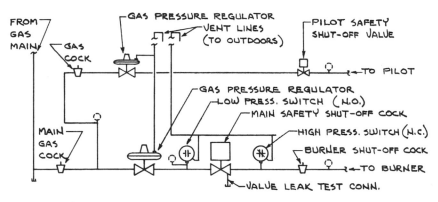

WITH GAS PRESSURE SWITCHES

TYPICAL GAS BURNER
GAS LINE TRAINS

FIGURE 2-14

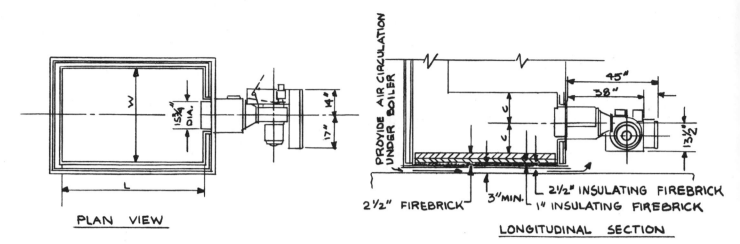

PLAN VIEW

LONGITUDINAL SECTION

2½" FIREBRICK   3"MIN.   2½" INSULATING FIREBRICK
1" INSULATING FIREBRICK

PROVIDE AIR CIRCULATION UNDER BOILER

BURNER INPUT M.B.T.U.H.	LENGTH L. INCHES	WIDTH W. INCHES	BURNER C. INCHES
4,500	52	26	13
6,750	60	30	15
10,500	72	36	18
15,000	84	36	18

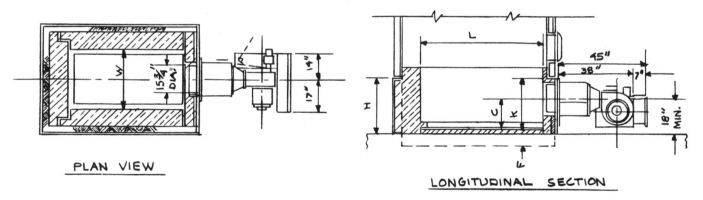

PLAN VIEW

LONGITUDINAL SECTION

BURNER INPUT M.B.T.U.H.	LENGTH L. INCHES	WIDTH W. INCHES	BURNER C. INCHES	SETTING OUTSIDE K. INCHES	HTG. MIN. INSIDE K. INCHES	FLOOR THICKNESS F. INCHES
4,500	52	26	13	26**	21△	*
6,750	60	30	15	26**	23△	*
10,500	72	36	18	26**	26△	*
15,000	84	36	18	32	26△	*

NOTE: * CHAMBER FLOOR THICKNESS (F.) REQUIRED WILL BE DETERMINED BY FIRING RATE FIRING CYCLE, MATERIAL USED, & PROTECTECTION REQUIREMENTS OF BOILER ROOM FLOOR AND BOILER FOOTINGS.

** MINIMUM FOR BURNER- FLOOR CLEARANCE OUTSIDE SETTING
△ MINIMUM WATERLEG TO CHAMBER FLOOR.

BURNER REFRACTORY FURNACE DIMENSIONS

FIGURE 2-15

the boilers and furnaces that are picked by the engineers responsible for the overall design are complete manufactured items with factory-supplied combustion chambers that require no design input. Detailing by the consultant is not required. However, since combustion chambers still will be required on occasion, especially in older boiler installations which the engineer may be required to renovate and modify, the data contained in details that follow is given as a sort of an overall estimate of the possible designing that would be required if the renovation required new burners and combustion chambers.

Figure 2-15 gives tables of probable combustion chamber sizes taken from actual manufacturers' data and illustrations of what to expect in terms of overall sizes. For example, a burner with a 10.5-million-Btu input requires a combustion chamber some 6 ft long, 3

ft wide, and 18 in to the center line of the burner, creating an overall height of 36 in. The plan views of the longitudinal section show a couple of choices, with and without a back wall and with and without air circulation provided by ventilation bricks. These details and the appropriate dimensions could be used directly on your plans.

Combustion Chamber Floors: In many older installations there is a problem of heat being transmitted from the floor of the combustion chamber to the building floor. As we show in Figs. 2-16 and 2-17, there are a number of ways in which ventilating air can be supplied underneath the floor by natural means to solve this problem. Again, this situation is not a problem in new installations, but these details are an excellent source of information when and if the problem arises.

Combustion Chamber Application: Finally, it may be

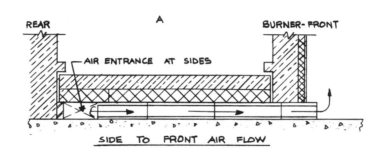

SIDE TO FRONT AIR FLOW

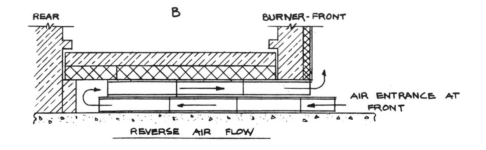

REVERSE AIR FLOW

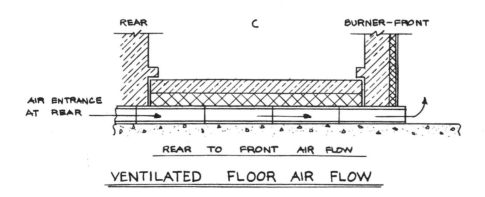

REAR TO FRONT AIR FLOW

VENTILATED FLOOR AIR FLOW

FIGURE 2-16

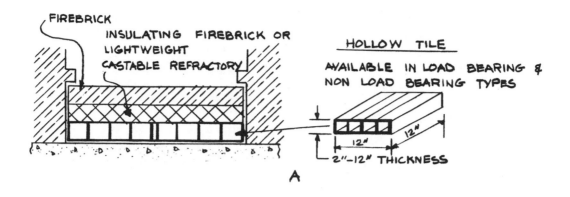

FIREBRICK
INSULATING FIREBRICK OR LIGHTWEIGHT CASTABLE REFRACTORY

HOLLOW TILE
AVAILABLE IN LOAD BEARING & NON LOAD BEARING TYPES

12"
12"
2"-12" THICKNESS

A

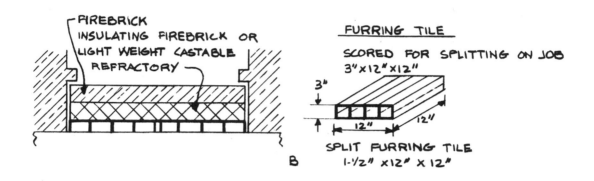

FIREBRICK
INSULATING FIREBRICK OR LIGHT WEIGHT CASTABLE REFRACTORY

FURRING TILE
SCORED FOR SPLITTING ON JOB
3" x 12" x 12"

3"
12"
12"

SPLIT FURRING TILE
1-1/2" x 12" x 12"

B

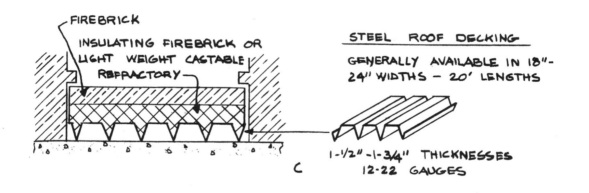

FIREBRICK
INSULATING FIREBRICK OR LIGHT WEIGHT CASTABLE REFRACTORY

STEEL ROOF DECKING
GENERALLY AVAILABLE IN 18"-24" WIDTHS - 20' LENGTHS

1-1/2" - 1-3/4" THICKNESSES
12-22 GAUGES

C

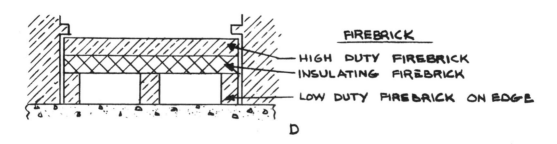

FIREBRICK
HIGH DUTY FIREBRICK
INSULATING FIREBRICK
LOW DUTY FIREBRICK ON EDGE

D

VENTILATED FLOOR SUPPORTS

FIGURE 2-17

asked how these details would be applied. In Figs. 2-18 through 2-20 we illustrate an application from an actual job, a typical installation. These three figures depict the installation of a combustion chamber to convert an old coal fired boiler into an oil burning boiler. There simply was not room enough in the old short section combustion area to provide a proper length combustion chamber for an oil fired installation. As a consequence in Fig. 2-20, in elevation, you can see what is described in Figs. 2-18 and 2-19. They show the combustion chamber as it is to be built but imply that there is a certain peculiarity in the system. The old coal combustion chamber somehow has to be lengthened. After calculating the proper size of combustion chamber, the problem became how to fit this combustion chamber into the old boiler. This was done by constructing a "dutch oven," which is an extension below the floor line and out from the face of the boiler.

To provide both the length and height of the combustion chamber needed to fit the new oil fired installation, the width of the combustion chamber had to remain within the width limits of the existing boiler. This is a practical illustration of how these details and the information contained on them, though seldom used, can be invaluable when needed.

In general, to avoid heat and expansion damage to a boiler room floor, the use of an isolating means such as a ventilated floor under a combustion chamber is always recommended. The general ratio of combustion chamber length to width varies from 1.5 to 3.0 over the range of most firing rates that would be encountered in normal practice. As far as base installation is concerned, the boiler should not be supported on the combustion chamber refractory. For conventional steam boiler bases separate structural steel supports are common. Brick or concrete supports should be

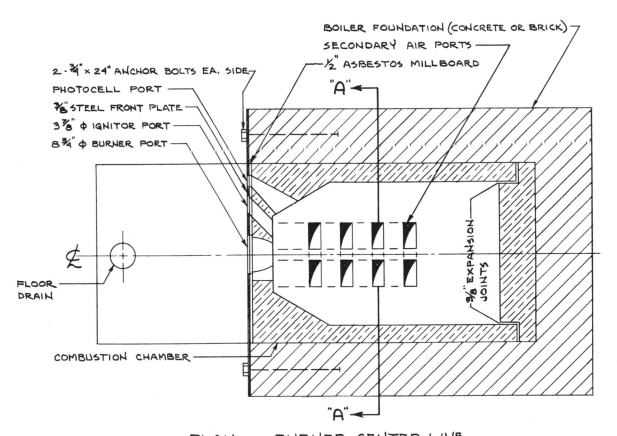

PLAN at BURNER CENTER LINE

TYPICAL OIL FIRED COMBUSTION CHAMBER on EXISTING BOILER
SHEET 1 of 3

FIGURE 2-18

used to carry the boiler weight. The vertical height of ventilation floors for most normal installations is commonly in the range of 2½ to 4 in.

Finally, on the details the type of brick that should be used for the construction of the combustion chamber is noted and should be followed in all cases.

Pumping Systems

As detailers we are not primarily concerned with the type of pump required by a particular application. The horsepower, the head against which the pump discharges, and other things that are related to the application are decided by the designer. We are concerned with showing this pump in our system. What sort of details do we require for this particular pump application?

Here we are concerned with pumps that form an integral part of the heat source system for either steam or hot water applications. As such we are concerned with low and high temperature hot water pumps and condensate pumps of all types. In many low tempera-

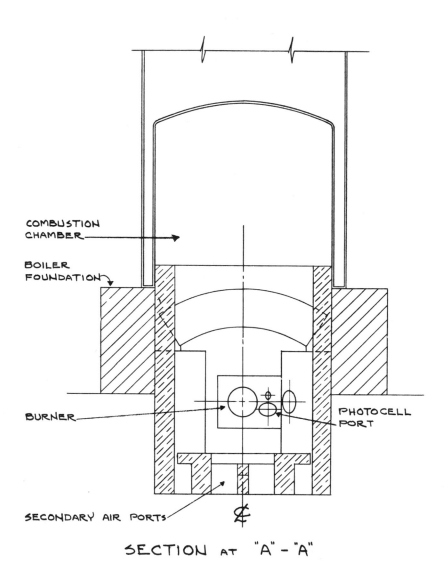

SECTION AT "A" - "A"

TYPICAL OIL FIRED COMBUSTION CHAMBER
ON EXISTING BOILER

SHEET 2 OF 3

FIGURE 2-19

ture and high temperature hot water systems it is not necessary to detail the pump or the surrounding appurtenances because they can be readily illustrated on the plans. However, we would like to point out that in typical floor-mounted high and low temperature hot water pumping applications any pump, especially a pump for a circulating hot or chilled water heating system, should always be installed with gate valves on both sides of the pump so that the pump can be removed without disturbing the rest of the system.

Circulating Pump: In Fig. 2-21 we show the typical arrangement of a floor-mounted pump. Except in situations in which transmission through the floor requires special treatment of the pump and motor mounts, the normal pump installation consists of a pump mounted on a base, connected to its supply and

return connections by properly selected flexible connections. This connection and its size and material are part of the system engineering design.

Generally, the standard installation has gate valves on the supply and return of the pump, as well as pressure gauges which enable pump performance to be read directly. The gauges, their sizes, psig range, and connection can all be spelled out in the specifications; therefore our separate gauge detail could be omitted. Our detail also depicts a typical pump foundation bolt.

Condensate Pump: In steam systems the pump is part of a condensate system that is generally detailed. The system itself is part of the boiler return system. In Fig. 2-22 we show a schematic piping diagram for a typical low pressure condensate system. This is a common standard for most installations of this type

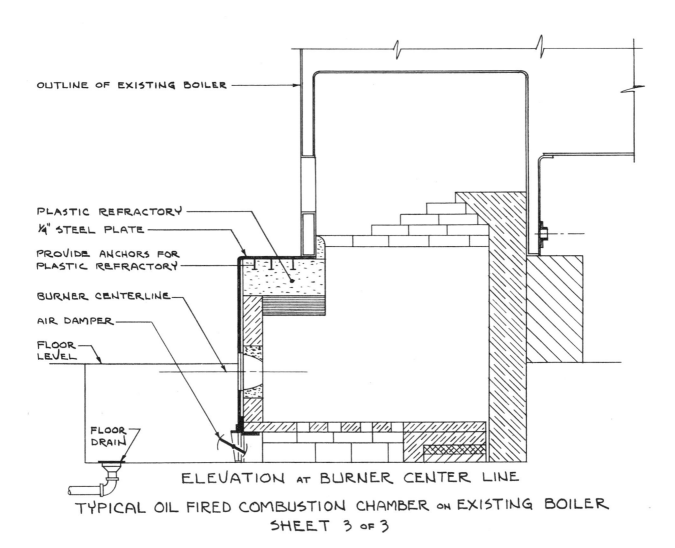

OUTLINE OF EXISTING BOILER

PLASTIC REFRACTORY
¼" STEEL PLATE
PROVIDE ANCHORS FOR PLASTIC REFRACTORY
BURNER CENTERLINE
AIR DAMPER
FLOOR LEVEL
FLOOR DRAIN

ELEVATION AT BURNER CENTER LINE
TYPICAL OIL FIRED COMBUSTION CHAMBER ON EXISTING BOILER
SHEET 3 OF 3

FIGURE 2-20

regardless of size. The pertinent factor to note in the situation is that this is for a low pressure, 15 psig or less, installation.

Again, as detailers we are not concerned with the materials of construction, the ratings of the condensate receiver, or the pump specification. The system is referred to as a condensate pump or as a condensate pumping system. We have to clarify things noted separately such as the condensate receiver tank and the boiler feeder pump which are normally part of a single condensate unit. The air that is present in most condensate returns would be brought back to the condensate receiver tank. When a tank vent is carried up to the ceiling of the boiler room and open ended

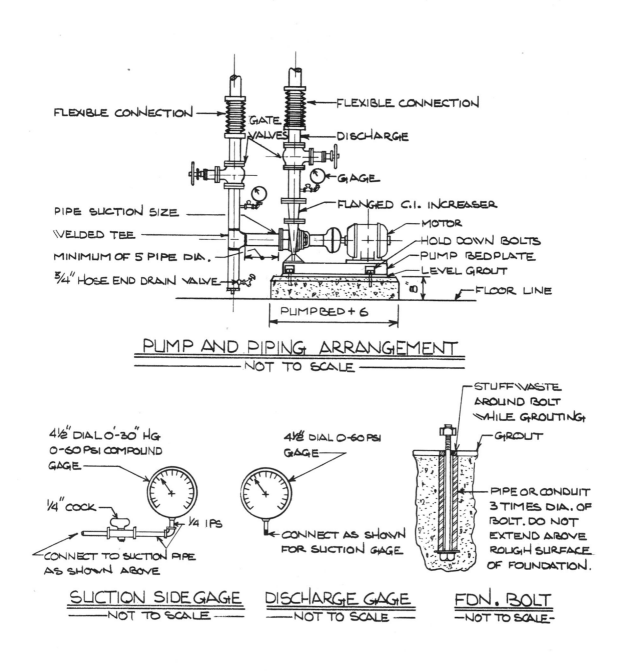

FIGURE 2-21

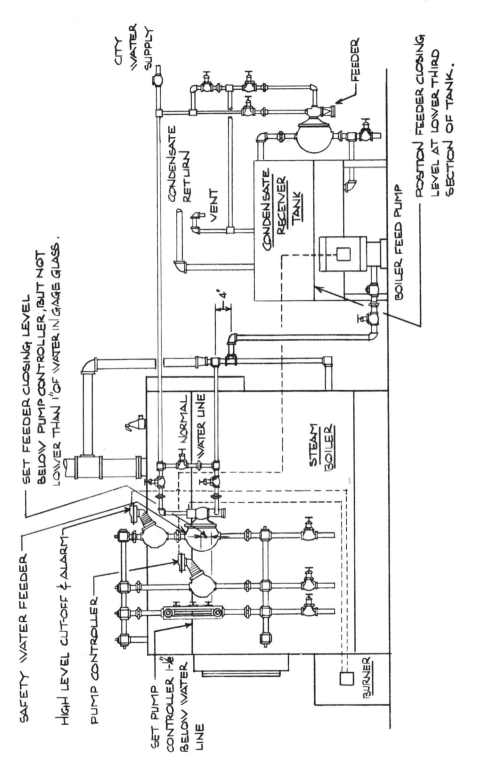

SAFETY WATER FEEDER

HIGH LEVEL CUT-OFF & ALARM

PUMP CONTROLLER

SET PUMP CONTROLLER 1-½" BELOW WATER LINE

SET FEEDER CLOSING LEVEL BELOW PUMP CONTROLLER, BUT NOT LOWER THAN 1" OF WATER IN GAGE GLASS.

NORMAL WATER LINE

STEAM BOILER

BURNER

CITY WATER SUPPLY

CONDENSATE RETURN

VENT

CONDENSATE RECEIVER TANK

FEEDER

BOILER FEED PUMP

POSITION FEEDER CLOSING LEVEL AT LOWER THIRD SECTION OF TANK.

4'

LOW PRESSURE CONDENSATE SYSTEM

NOT TO SCALE

FIGURE 2-22

with elbows turned down, as we show in our detail, makeup water to the condensate system, when required, is normally arranged with feeder level controls on the condensate tank somewhat as detailed in Fig. 2-22.

The discharge of the condensate pump is connected to the upper part of the Hartford Loop and is normally approximately 3 inches below the boiler line. The discharge connection is made with a tee, not a wye, connection. Further, in our detail we show the devices on the boiler that control the operation of the condensate receiver. We do not want condensate pumped indiscriminately into the boiler. The purpose of the condensate pump and its properly-sized receiver tank is to provide a source of liquid to the boiler which will enable the boiler to generate steam required by the system and maintain its normal water level operating line. Consequently, there are three basic items normally on the steam boiler that are related to the condensate pump. The first of these is a boiler water level gauge, which is to the left of the three items in Fig. 2-22. Next to this is the pump controller. The pump controller causes the boiler feed pump in the condensate system to deliver condensate whenever the water level line drops. Finally, there is a combination safety and low water feeder for the boiler. These two devices are located to the right of the pump controller and are set one over the other. They frequently are combined into one device. These devices provide the ultimate safety in the supply of water to the boiler. If the makeup feeder in the condensate or the condensate return itself does not work as designed, there is a combination automatic feeder that brings water into the boiler. In addition there is a high level cutoff and alarm so that if the condensate feeder fails to operate properly and too much condensate is accidentally pumped to the boiler or if the pump control on the boiler fails, the supply of water to the boiler is stopped; the pumping is also stopped by this high level cutoff and alarm.

Boiler Feed Pumps: In Fig. 2-23 we depict a group of three pumps that are described as boiler feed pump connections. This detail shows all of the piping connections to a set of three pumps, two electric driven and one steam driven, that are used in 125 psig steam system return applications. Not shown in this detail but shown in Chap. 1 in Figs. 1-3 and 1-4 are the connections of the two lines that pass from the boiler feed in the upper right-hand corner of the detail to the 125 psig boilers which are also not shown. In the bottom right of the detail it is indicated that the supply to the pumps comes from a feedwater heating source. This is normally the deaerator, which we have detailed in Figs. 2-44 and 2-45. Condensate from all sources is collected in a separate condensate receiver and pumping system that is shown separately; it is pumped to the deaerator. Thus, there are a number of systems related to this system that are depicted elsewhere in

this book but are part of the total system.

The first item to note in Fig. 2-23 is that the supply and return systems are parallel, which is customary in this type of application and which creates a flow path to any or all of the pumps in any order selected by the plant operator. This applies to both the return and supply systems. Secondly, the return and supply systems have pressure gauges that carefully delineate the pressure that is delivered to the system. Normally, these pumps are floor mounted and pump up to the ceiling, across the ceiling, and down to the boilers themselves. In certain situations these lines between the boiler and the pumps may need to be drained. This is done through globe valves shown on the discharge on each of the pumps. Immediately after the check valve there is a globe valve drain line. Careful control of the discharge pressure is part of the system of operation.

If things do not work out as planned, a way must be found to relieve the excess pressure. This is accomplished by a pressure relief valve connection from the pump discharge to a gate valve drain. The pump may be drained through a gate valve. The pressure to the boilers is carefully maintained by monitoring the pressure through pressure gauges on the discharge lines. Pressure gauges are also shown, placed immediately after the pump discharge. When the pumps are steam controlled, there is a series of control devices.

This elaborate pumping connection detail serves a number of purposes above and beyond the detail application depicted. We previously mentioned the detailing required in a low temperature or high temperature hot water application. Viewing this particular detail, one may easily relate the various features of the detail to normal low temperature or high temperature hot water applications. The same sort of arrangement could be used in a high temperature hot water application if required. In addition this detail could be used as the basis for detailing any low temperature pumping application. The mechanical design department could decide whether there should be a single or dual feed.

The pressure gauge could be supplemented with a temperature gauge for low temperature or high temperature hot water application. The gate valve arrangement on the supply and return side would be very similar. The drains could be utilized or not for the pressure relief bypass. In brief the two electrically driven pump piping connections could be used, with modifications as required, for any low or high temperature hot water system. If the low temperature hot water system consisted of in-line pumps mounted on the pipe itself, we would then be returning to the premise of the gate valve on both sides of the pump and a temperature gauge on the line in place of the pressure gauge. Thus, this detail not only provides a solution to the boiler return system in a 125 psig system, but it also is the basis of design for the pumping of low temperature or high temperature hot water applications.

Steam Driven Pumps: In Fig. 2-24 the detailing of the steam piping to the boiler feed pump is shown as a duplicate set of steam driven boiler feed pump systems for an application of uncontrolled steam supply. In the detail the high pressure steam from the boiler is reduced to low pressure steam and passed through a purifier in order to drive the turbine. In this particular detail the operation of the turbine is semiuncontrolled. Steam is always flowing. When the load imposed on the turbine by the boiler feet pump is light, the turbine does not use much in the way of steam energy. As a consequence most of the steam passes through the turbine directly to the deaerator. This sort of detail may be used in some applications in which there is not any large variation in the load on the steam driven turbine or there is a standard load in which steam, for the most part, is used to drive the turbine.

In highly varying loads the system is inefficient in terms of steam usage and excess steam is constantly discharged into the deaerator. If the deaerator becomes overloaded, the steam must be vented to the atmosphere. Figure 2-24 is not an application for a varying load. In Fig. 2-52 we show another type of steam turbine control which would be used under varying load conditions.

Expansion Tanks

In any hydraulic system there is a basic requirement to keep system pressure at a desired normal operational level. The primary objectives of the expansion tank

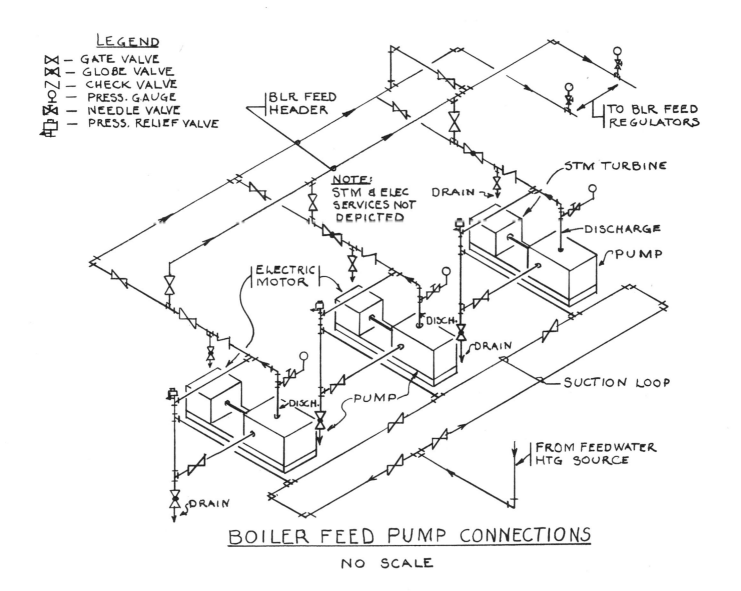

BOILER FEED PUMP CONNECTIONS

NO SCALE

FIGURE 2-23

system are to limit the pressure of all the equipment in the system to the allowable working pressure, to maintain minimum pressure for all normal operating temperatures, to vent air, to prevent cavitation at the pump suction and the boiling of system water, and to accomplish of all this with a minimum addition of water to the system.

In this book the most common basic types of

expansion tank systems are depicted. The first type is a system sized to accommodate the volume created by the water expansion, with sufficient gas space to keep the pressure range within the design limits of the system. Air is normally used and the system must remain as tight as possible because every recharging cycle introduces additional oxygen with the air and thus promotes corrosion. In the initial start-up, the

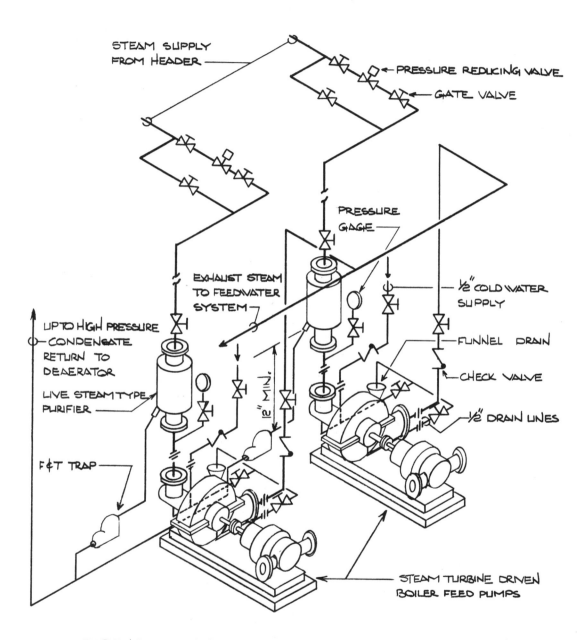

PIPING CONNECTIONS FOR STEAM DRIVEN
BOILER FEED PUMPS
——— NOT TO SCALE ———

FIGURE 2-24

oxygen reacts with the system components, leaving basically a nitrogen atmosphere in the expansion tank. The initial fill pressure sets a minimum level of water in the tank. The compression of the gas space because of water expansion determines the maximum system pressure.

An alternative to this arrangement is the diaphragm type expansion tank, which is precharged to the system fill pressure and sized to accept the normally expected expansion of water. The tank's air charge and the system water are permanently separated by a flexible elastomer diaphragm. For a high temperature hot water system the pressurization may be either by steam or inert gas, commonly nitrogen. Other details will depict nitrogen pressurization systems for high temperature hot water applications.

For all expansion tank systems the location of the system pump in relation to the expansion tank connection determines whether the pump pressure is added to or subtracted from the system static pressure. This is due to the fact that the junction of the tank with the system is the point of no pressure change whether the pump is in operation or not.

In our details, which depict common expansion tank situations, we show the expansion tank on the suction side of the pump. The reason for this location is that when the pump is discharging away from the boiler and the expansion tank, the full pump pressure will appear as an increase at the pump discharge. All points downstream will show a pressure equal to the pump pressure minus the friction loss from the pump at that point. The fill pressure needs to be only slightly higher than the system static pressure. If, on the other hand, the pump discharges into the boiler and the expansion tank, which is common in small residential and small commercial systems, full pump pressure is reduced and system fill pressure is increased. Therefore, until the fill pressure is higher than the pump pressure, a vacuum can be created in the system. Normally, a pump into the boiler and a pressure tank system is used only in low-rise buildings, small systems, or single-family residences in which the pumps need to have only a low total head capability.

Sizing of expansion tanks is generally based on a system determined from the standard ASME formula that is shown in the *ASHRAE Systems Handbook* and other publications. Without going into the details of this engineering calculation, we note that the size of the expansion tank is determined by the volume of water in the system; the range of water temperatures normal to the operation of the system; the pressure of air in the expansion tank when the fill water first enters the tank; the relation of the height of the boiler to the high point of the system, which is usually but not always the item in the system with the lowest working pressure; the characteristics of the expansion tank; the high point of the system; the pressure developed by

the circulating pump; and the location of the circulating pump with respect to the expansion tank and the boiler. Finally, it should be noted that anytime there is a change in water temperature in a hot or chilled water system, an expansion tank should be used. Expansion tanks are therefore needed not only in hot water systems and run-around coil heat recovery systems used for energy conservation designs but also in chilled water systems.

Factory Pressurized Tanks: Figures 2-25 through 2-27 illustrate the application of a factory pressurized tank with an expandable diaphragm. Figure 2-25 shows a typical expansion tank installation in a system that supplies either hot or chilled water to the system. The boiler pump shown is seemingly inconsistent with the arrangement described in our previous discussion of expansion tank location, but in this application the pump is performing a specific task. It is used to provide circulation in the boiler under certain special conditions.

A system pump by definition supplies the system. Note that on the chiller side we again have a special circulating pump for the chiller. The object of these two separate pumps is not merely to serve the system but also to balance and maintain the flow from within the chiller and the boiler under varying overall flow and temperature conditions. Note that in this detail, as in many others, the items related to the expansion tank are described. For example, in the supply of makeup water there is a filling control unit tied into the return side of the system. The system water passes through an automatic air purger with a manual or automatic vent. Finally, note that the pressurized tank is tied into the suction side of the system pump.

Chilled Water Expansion Tank: An expansion tank is needed in a chilled water system, as well as in a hot water system. Figure 2-26 shows the application of an expansion tank to a chilled water system only. This application is common when the chilled water and hot water are in separate piping systems or are otherwise separated. Again, the return fluid is passed through an air purger and vent to remove air. There is an automatic control of makeup water in the filling control unit. The pressurized tank, shown in previous details, is illustrated again in this detail.

In Fig. 2-27 we show the application of an expansion tank in the usual steam or hot water converter installation. The converter is a steam converter. We have shown the basic steam piping connections to the converter. At this point, whether the device is a converter or a boiler, the basic premise of the system is still as shown. System water is pumped into the heat exchanger. Frequently the way to make the heat exchanger perform most efficiently is to create flow under pressure. The vent on system is at the high point on top of the heat exchanger. It might also be noted that the pumping system for the converter is often on

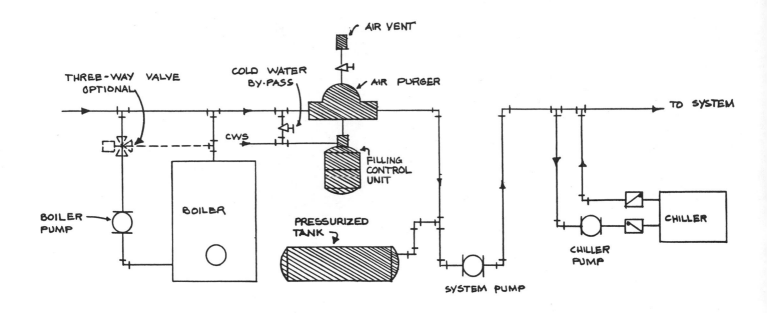

FACTORY PRESSURIZED TANK INSTALLATION
WITH DUAL WATER TEMPERATURES & SEPARATE BOILER
& CHILLER PUMPS

FIGURE 2-25

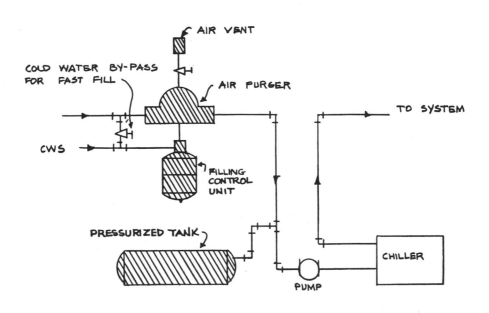

FACTORY PRESSURIZED TANK SYSTEM
FOR CHILLED WATER SUPPLY

FIGURE 2-26

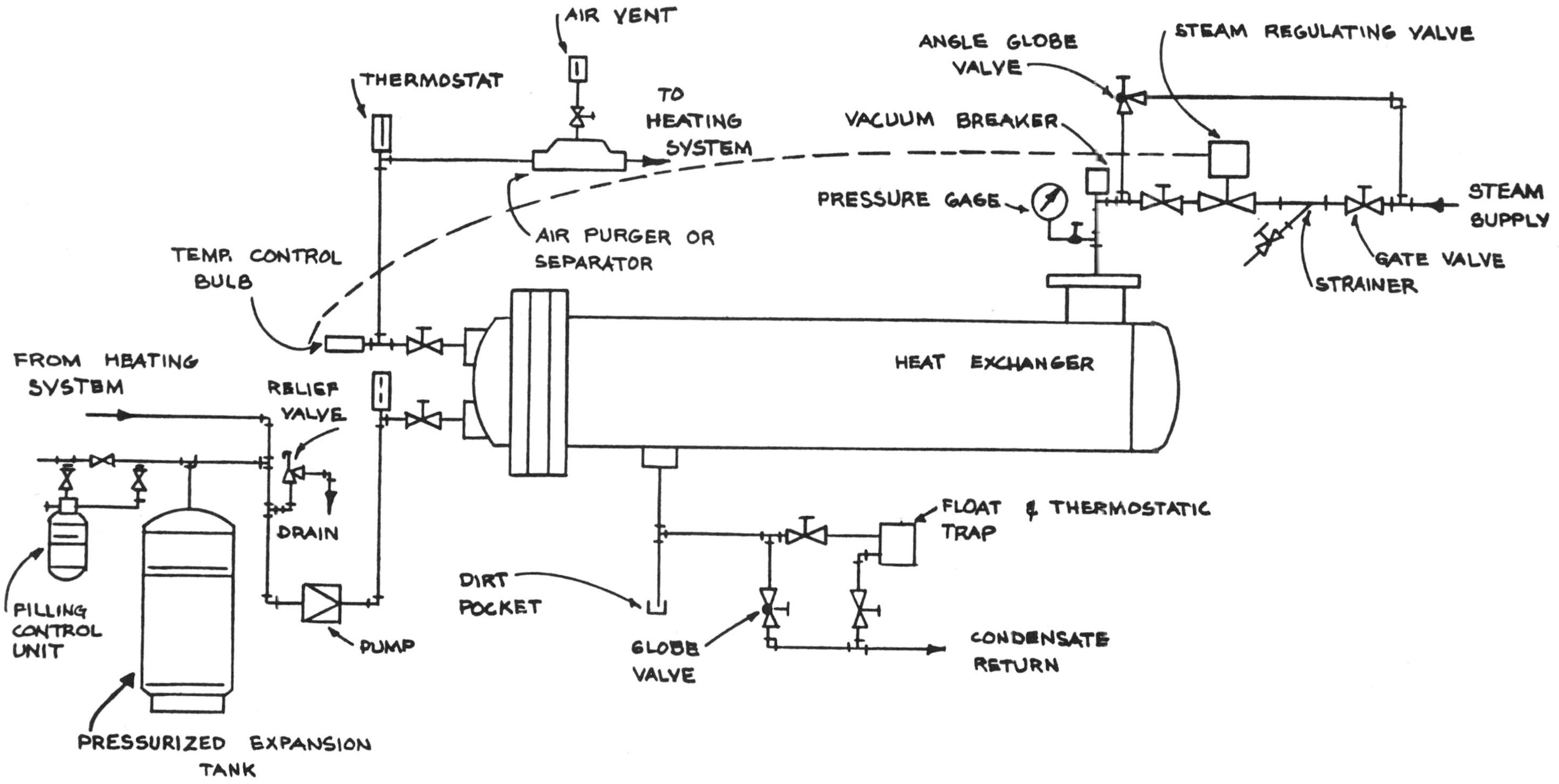

CONVERTOR/FACTORY PRESSURIZED TANK INSTALLATION

FIGURE 2-27

STEAM SUPPLY

GATE VALVE

STRAINER

STEAM REGULATING VALVE

ANGLE GLOBE VALVE

VACUUM BREAKER

PRESSURE GAGE

HEAT EXCHANGER

FLOAT & THERMOSTATIC TRAP

CONDENSATE RETURN

GLOBE VALVE

DIRT POCKET

AIR VENT

TO HEATING SYSTEM

AIR PURGER OR SEPARATOR

THERMOSTAT

TEMP. CONTROL BULB

RELIEF VALVE

DRAIN

PUMP

FROM HEATING SYSTEM

FILLING CONTROL UNIT

PRESSURIZED EXPANSION TANK

the floor of the equipment room, and the heat exchanger, sometimes separated from the converter by a considerable distance, is at the ceiling.

Air Source Expansion Tank: In Figs. 2-28 and 2-29 we illustrate the air source expansion tank. Here there is no diaphragm separating air from water. This arrangement has been common for expansion tank installation in many applications for a very long time. The boiler pumps and other devices are not clearly shown but are implied with the notation to pump suction and with the obvious notation that the air strainer is on the suction line of the pump. As can be seen, all of the connections rise vertically and pitch up to the expansion tank and to the cold water fill line. The makeup water and expansion tank lines are tied

into the air separator, which is designed for this particular type of system. Both the air separator and the expansion tank have drain valves to drain excess water from the expansion tank. The cold water fill line shown has a pressure reducing valve to supply makeup water to the system at an acceptable pressure.

Figure 2-29 is a nearly identical installation that is used for a special system. When snow melting or run-around energy conservation coils are used, the system fluid is a combination of gycol and water. The air type of system has the advantage of not having a diaphragm that can be affected by the corrosive nature of this commonly used nonfreeze solution. The detail is similar to the one in Fig. 2-28 except that there is no cold water makeup. Instead there is a gycol makeup which

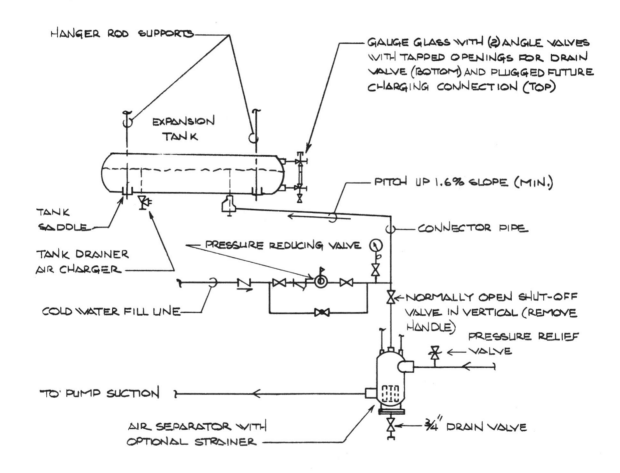

AIR CONTROL & PIPING CONNECTIONS FOR

WATER SYSTEMS
— NOT TO SCALE —

FIGURE 2-28

comes from a special pumped gycol-water solution tank that is not shown in this detail but depicted in a snow melting detail in Chap. 9.

Nitrogen Pressurization: In Fig. 2-30 we show the complete piping detail around a nitrogen-fed high temperature hot water expansion tank system. This type of system is special to the high temperature hot water installation. The piping shown on this system has some unusual points. First, as can be seen, the expansion tank is still on the suction side of the high temperature hot water circulating pump. Second, in a high temperature hot water system there usually is a soft water service. We note in our detail where the connection of this treated water goes into our expansion tank system. Controlling the amount of water in

the high temperature hot water expansion system is extremely important. In the lower right side of the detail the makeup pump with its low level switches and controls carefully adds water to the expansion tank. As a further safety precaution there are high and low pressure cutoffs and alarm switches. Knowing the system's limits is also very important.

Finally, high pressure water will, if the pressure is released, turn instantly into high pressure steam. Thus, the pressure relief valve used is similar to that used in a high pressure boiler. The connection of the pressure relief valve drip pan and its exhaust pipe through the roof is nearly identical to that of a 125 psig steam system. If there is a loss of pressure in the expansion tank, the result will cause the water to flash

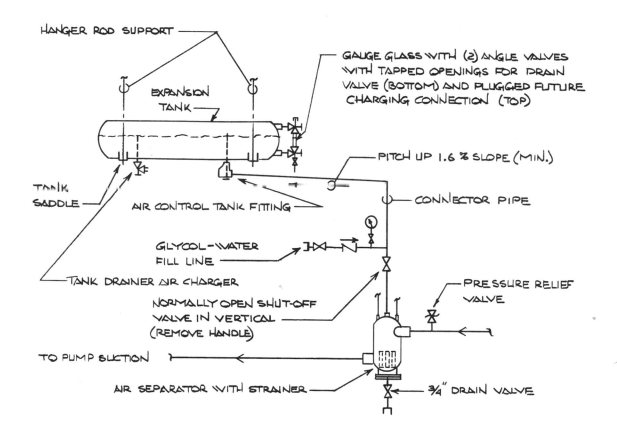

AIR CONTROL AND PIPING CONNECTIONS FOR GLYCOL-WATER SYSTEMS
— NOT TO SCALE —

FIGURE 2-29

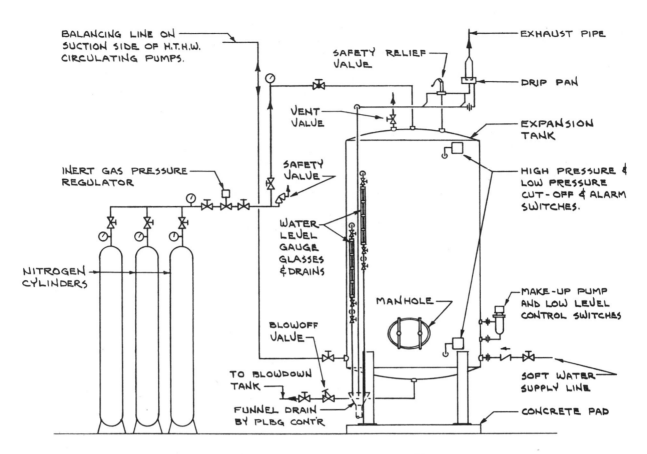

BALANCING LINE ON SUCTION SIDE OF H.T.H.W. CIRCULATING PUMPS.

EXHAUST PIPE

SAFETY RELIEF VALVE

DRIP PAN

VENT VALVE

EXPANSION TANK

INERT GAS PRESSURE REGULATOR

SAFETY VALVE

HIGH PRESSURE & LOW PRESSURE CUT-OFF & ALARM SWITCHES.

WATER LEVEL GAUGE GLASSES & DRAINS

NITROGEN CYLINDERS

MANHOLE

MAKE-UP PUMP AND LOW LEVEL CONTROL SWITCHES

BLOWOFF VALVE

TO BLOWDOWN TANK

SOFT WATER SUPPLY LINE

FUNNEL DRAIN BY PLBG CONTR

CONCRETE PAD

TYPICAL HIGH TEMPERATURE HOT WATER EXPANSION TANK PIPING

FIGURE 2-30

into steam. Note the line that goes to a blowdown tank; it provides protection against this occurrence. The pressurization for this system is normally provided by specially piped dry nitrogen cylinders. These cylinders and their output are controlled by a pressure regulator that has a safety valve to insure that the pressurization to the system is maintained at the proper value at all times. Finally, there are times when excess nitrogen does have to be vented. This is accomplished through a small vent valve in the top of the tank.

The system designer wants to reduce the area of contact between a gas and water, thereby reducing the absorption of gas in the water. This is why the tank is shown installed vertically, which is the generally preferred arrangement.

The ratings of fittings, valves, piping, and equipment generally are based on a minimum pressure, which is about 25 to 50 psig above the maximum saturation pressure. An imposed additional pressure head above the vapor pressure must be sufficient to prevent steaming in the high temperature hot water generator at all times, even under unusual flow conditions, such as firing rates at which the created flow of two or more generators is not evenly matched. This is a critical condition since a gas pressurized system does not have separate safety valves. The pressure varies with changes in water level in the expansion vessel. When the system water volume increases because of a temperature rise, the expansion of the system water into the vessel compresses the inert gas, raising the system pressure. The reverse condition takes place on a drop in system water temperature. The pressure is permitted to vary from a minimum point above saturation to a maximum that is determined by the materials used in the system. The expansion tank itself can be sized for the sum of the volumes required for pressurization plus the volume required for expansion and the volume required for sludge and reserve.

Breechings

The subject of breechings is referred to in the engineering and contracting field by a number of titles. These include smoke pipe, flue gas vent, and vent connector, all of which refer to the pipe that extends from the boiler smoke outlet to the chimney.

This book does not cover the design of the chimney itself since there are many excellent design texts, such as *Architectural Graphic Standards* and others, that detail construction of a chimney and since the chimney is usually part of the building. For prefabricated chimney detailing there are many manufacturers of prefabricated chimneys who supply complete details with their catalog data, which is readily available.

The breeching is part of the chimney system. The basic chimney system design involves the balancing of the forces which tend to preclude the passage of the flue gases against those forces which tend to exhaust them. The difference between these two forces is called theoretical draft. Available draft is the draft required by the appliance to provide for proper burning of oil, gas, coal, or whatever fuel is used in the combustion process. This is a negative draft requirement although there are appliances, particularly gas appliances, in which the available draft could be negative, positive, or neutral. The theoretical draft is the draft available at the appliance that is created by the combination of the breeching and the chimney. In our details the basic premise of the breeching is to provide a passage from the appliance to the chimney. The chimney is the vertical portion of this passage which could, in certain instances and with certain requirements, be merely a vertical extension of the breeching itself.

There are general considerations that should be part of every breeching design. Obviously, the pipe must be reasonably airtight so that none of the products of combustion leak into the equipment room space. Second, the pipe must be of the proper type and thickness of material to preclude pipe collapse. Third, there must not be any restriction of the flow of flue gases. The breeching is an integral part of the chimney system and frequently a very considerable part of the overall resistance of the system. When the smoke pipe is made of galvanized iron, connections are made with a slip joint-type construction, and screws or rivets are used to hold the two pieces of metal that are overlapping each other firmly in place. In heavier gauges joint connections are commonly welded. Because there is very little data available about what the thickness of the material should be, the following items are repeated from the Building Owners and Contractors Association (BOCA) Basic Mechanical Code.

Diameter of connector (in)	Galvanized sheet gauge (no.)
Less than 6	26
6 to less than 10	24
10 to 12	22
14 to 16	16
Greater than 16	16

It should be noted that these dimensions are based on round pipe diameter. Appropriate square or rectangular equivalents can be obtained from ductwork conversion tables in the *ASHRAE Fundamentals Handbook*.

There are also certain basic sizing requirements that should be followed. For example, when you are connecting two or more connectors into a common connector, the common connector should be sized to take the flue gas from both. That should be fairly obvious. Secondly, connectors servicing gravity vented appliances should not be connected to a chimney served by a power exhauster unless the connection is made on the pressure side of the power exhauster. The connector (smoke pipe or breeching) should be as short and as straight as possible. The horizontal run of the connector that is not insulated should not be more than 75 percent of the height of the vertical portion of the chimney or vent above the connector unless it is part of an engineered venting system. The horizontal run of an insulated connector to a natural draft chimney serving a single gas fired appliance should not be more than 100 percent of the height of the vertical portion of the chimney unless it is part of an engineered system. The breeching should pitch upward at a rise equal to ¼ in/ft of run. Finally, you should carefully check all codes to be sure that the clearances between your breeching and any combustible material are in accordance with those codes.

Complete combustion of any gas or oil used as fuel generally requires approximately 1 cu ft of air for each 100 Btu, or approximately 1400 cu ft of air for 1 gal of oil, and if there is an allowance of approximately 50 percent excess air, this is equal to approximately 2100 cu ft of air per gallon of oil. All too often the designer takes, for example, three 12-in round openings from three boilers and ties them in to a common breeching, enlarging the breeching by the sum of the areas. These shortcuts should not become habit. The sizing of the breeching all too often is not given the attention it deserves, and sooner or later such lack of attention will become a costly error. The sizing information in Chap. 26 of the *ASHRAE Equipment Handbook* is an excellent place to find information about the proper engineering required.

Multiple Boiler Breeching: Our first detail, Fig. 2-31, illustrates a one boiler installation and a three boiler gas fired boiler-burner installation. The draft hood, as shown, could have been a barometric damper if the installation were oil. And it could have been a combination of a barometric damper and induced draft fan if it were a larger system. The dimensions shown on Fig. 2-31 are intended to bring to your attention a sometimes overlooked fact. Insofar as draft, of which the breeching system is a part, is concerned, the point at which the flue gases exhaust is the effective draft position, not the overall height of the chimney, which may be some 5 ft or more higher. As can be seen on this detail, we noted vent height, effective height draft, and chimney height as three separate entities. These drawings show breechings that would undoubtedly be galvanized pipe of the thickness shown in our table of sizes and related gauges. The hangers should be on approximate 3-ft centers and should normally be simple strap hangers.

Breeching Details: Figure 2-32 shows a somewhat larger commercial type boiler installation in which the available chimney draft is simply not great enough. Here the boilers are equipped with induced draft fans. This is a very common situation. The induced draft fan is the solution to many flue gas problems. Ideally, the induced draft fan should be located directly on the smoke outlet of the boiler, furnace, or whatever appliance is involved. The installation of this device should be made in accordance with the manufacturer's recommendations.

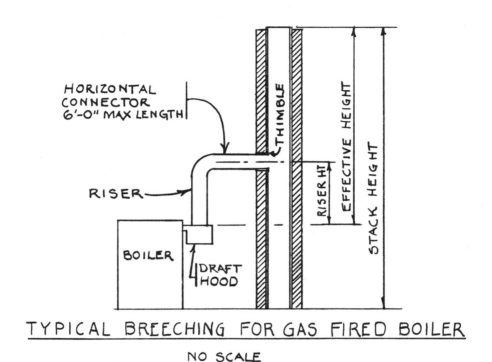

TYPICAL BREECHING FOR GAS FIRED BOILER

NO SCALE

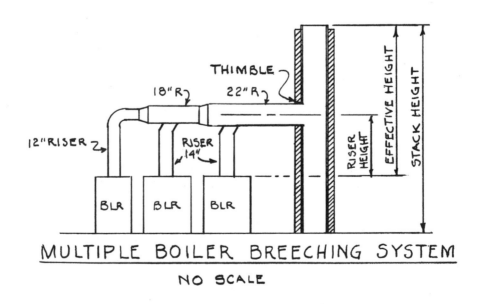

MULTIPLE BOILER BREECHING SYSTEM

NO SCALE

FIGURE 2-31

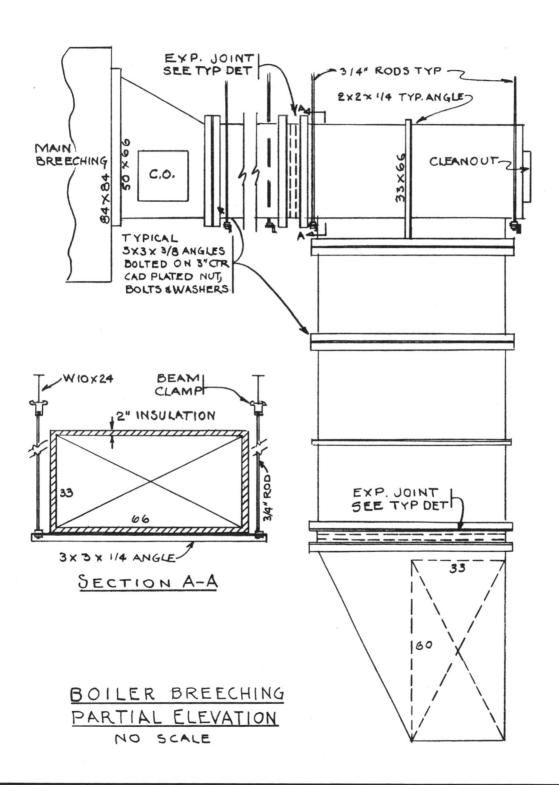

EXP. JOINT SEE TYP DET

3/4" RODS TYP

2x2x1/4 TYP. ANGLE

MAIN BREECHING

84X84

50 X 66

C.O.

A

33X66

CLEANOUT

TYPICAL 3x3x 3/8 ANGLES BOLTED ON 3" CTR CAD PLATED NUT, BOLTS & WASHERS

W10x24

BEAM CLAMP

2" INSULATION

33

66

3/4" ROD

3 x 3 x 1/4 ANGLE

SECTION A-A

EXP. JOINT SEE TYP DET

33

60

BOILER BREECHING
PARTIAL ELEVATION
NO SCALE

FIGURE 2-32

In most situations there is a general requirement of a barometric damper. While there may be times when this damper is not actually required, these times are very difficult to determine. Normally the induced draft fan suppliers include, as part of their package, a connection for a barometric damper to properly adjust to all draft situations.

The detail depicts the basic components of a large breeching installation. Here you have a situation in which the construction of the breeching is 16-gauge black iron. The joints are welded. Since there usually is an expansion problem in this type of system, it must be very carefully detailed. There is a requirement to support the breeching itself. This is shown resolved with the angle irons and rods that support the breeching from the overhead construction.

These items become extremely important when you are resolving the design of a large breeching of this type since the weight of the breeching itself is considerable. The weight of the breeching at 16 gauge is approximately 2.7 lbs/sq ft, and the 2-in insulation applied to the breeching weighs approximately 0.8 lbs/sq ft. This means that the breeching, whose dimensions are approximately 5 × 3 ft, weighs some 54 lbs/lin ft.

The plan presented in Fig. 2-32 is an elevation plan of the breeching, showing its vertical run with its relatively long horizontal run cut by a breakline so that it could be incorporated in this book. It is very important to properly calculate the amount of expansion that will occur in a breeching of this type. Be certain you have made the proper calculations for expansion and that you have allowed for the proper number of expansion joints. Generally, except in very long runs, the expansion can be taken up in one or two expansion joints. Normally, there are one or possibly two expansion joints in the horizontal run and usually at least one in the vertical run.

You must support the entire breeching and contain its movement between the expansion joints. Equally important, between every fixed set of supports there must be provision for expansion.

Expansion Joints: Figure 2-33 presents specific details for two different types of expansion joints, although the material used in each is essentially the same. Both joints are made from an elastomer material, which is essentially a combination of rubber and plastic that has been treated to withstand the temperature involved.

The joints are described as to their specific temperature limitation, which should always be part of your detail notations. There are also situations in which the scrubbing action of the flue gas may adversely affect the joint. To resolve this problem the detail in Fig. 2-33 shows joint material beyond the inside of the flue duct by some 4 in and protected by a special baffle of ³⁄₁₆-in

material mounted at a 10° angle to divert the flue gas away from any direct contact with the joint material. Both the angles holding the expansion material and the arrangement and mounting of the baffle should always be carefully described.

Fresh Air Intake: We have not presented any details on fresh air louvers since this is a fixed device. If a detail is required, it is readily available from a number of manufacturers' catalogs. Most codes describe that any screening applied to the louver should not be smaller than ¼ in; that is, the free opening should not be smaller than ¼ in. The reasoning here is that the dust buildup on smaller openings can easily clog the louver and preclude the free passage of air.

There are many ways to approach the sizing of the *free* air opening. One of the common rules is that the free air opening should be twice the size of the chimney flue. If, for example, the flue is 12 × 12 in, the louver's free air area would be 24 × 12. Since the normal louver is constructed so that its free air area is about 50 percent of the gross area, the *gross* area of the louver would become 24 × 24 which is twice the 24 × 12 free area. Thus the *gross* area of the louver is about *four* times the area of the chimney flue.

For gas or oil burning appliances another general sizing rule is that it takes 1 cu ft of air to produce 100 Btu. For 140,000 Btu/gal of oil this translates to 1400 cu ft of air for *perfect* combustion. Allowing for the normal 50 percent excess air, the value becomes 1400 plus 50 percent of 1400, or 2100 cu ft of air per gallon of oil. If 1 gal of oil is burned per hour, this is 2100 cfh or 35 cfm. Again, if we assume an air velocity of 2.5 mph, which approaches still air, the 2.5 mph equals 130 fpm. Dividing 35 cfm by 130 fpm, we arrive at approximately 0.25 sq ft, or some 36 sq in. Some codes require 50 sq in of louver per gallon, which equates to a 100 percent excess air condition.

Bringing air into the boiler room in large quantities can create very uncomfortable conditions during the winter months. It is not unusual to find that either heating the incoming air or specifically heating the building's occupants is a requirement. Water pipes in the path of the boiler room fresh air intakes can also freeze.

There are a number of code restrictions on breeching sizes and on fresh air intakes in most communities in the United States. These code restrictions should be carefully investigated before proceeding with any design. This is especially true in the matter of fresh air intakes. Rules such as 50 sq in/gal, high and low intakes, and the like all affect the location of louvers. In addition there are rules concerning the clearances between the breeching and combustible construction which may be at variance with what the firm is accustomed to, especially if the project is in a different part of the country. Finally, there are fire codes to be investigated. The best suggestion that can be offered in this situation is a review of the applicable sections of the NFPA.

Boiler Water Treatment

The basic purpose of boiler water treatment is to preclude corrosion. Corrosion is defined as the de-struction of a metal or alloy by chemical or electro-chemical means in reaction with the liquid contacting the metal. No corrosion occurs in dry air. In most all normal environments there is sufficient moisture in the air to demonstrate that dry air is seldom actually encountered. What is encountered is a mixture of various corrosive chemicals which in an air-water

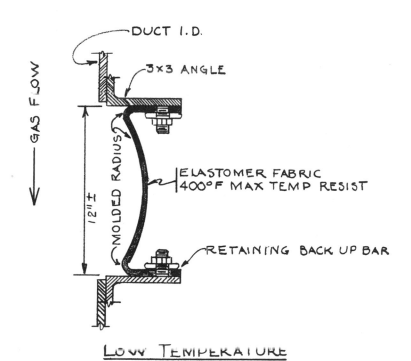

LOW TEMPERATURE

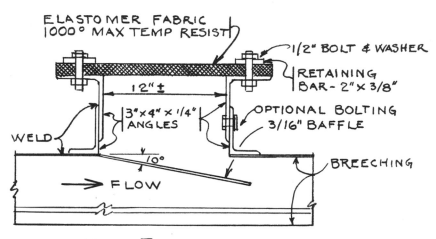

HIGH TEMPERATURE
EXPANSION JOINT CROSS SECTIONS
NO SCALE

FIGURE 2-33

atmosphere or in an all-water atmosphere tend to cause disintegration of the surface. Oxygen in its pure form will combine with most metals to form an oxide. As a result oxygen in any system will cause corrosion.

Chemical Treatment: While the subject of corrosion is a large and complicated one, the detailer is not concerned with finding a solution for corrosion but rather with detailing the application methods the treatment may require. Figure 2-34 illustrates a typical chemical feeder and the piping that would commonly be associated with this device. The chemical feeder usually has some sort of tank, some method of supplying the tank solution to the system, and some method of receiving the treated material that is to be delivered to the system. In Figs. 2-34 and 2-35 note that the tank is described as a soft water tank which receives a specially treated water from another source at which

the chemicals are mixed. From there the chemical treatment solution is pumped to the existing hot water system. This is a common form of treatment in all types of steam and hot water systems and may be used, with proper pumping pressures, for any system.

Water Softening: In Fig. 2-35 the supply of the soft water mentioned in Fig. 2-34 is covered in more detail. Water softening is a special treatment applied to existing water supply systems to insure that all calcium, iron, and other chemicals which make water hard are removed. The water delivered to the chemical treatment system is thus free of these chemicals and therefore is called "soft." The system operates most efficiently if it is allowed to run in continuously. Consequently, at times it may supply more soft water than may be needed in the boiler system. Commonly a soft water storage tank is supplied with the soft water

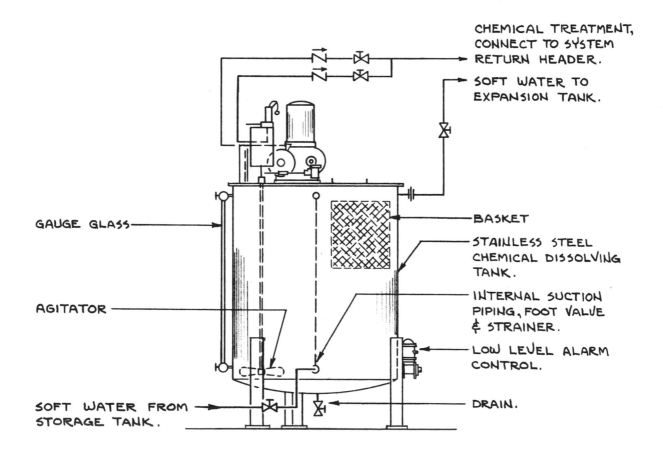

TYPICAL HIGH TEMPERATURE HOT WATER
CHEMICAL TREATMENT SYSTEM

FIGURE 2-34

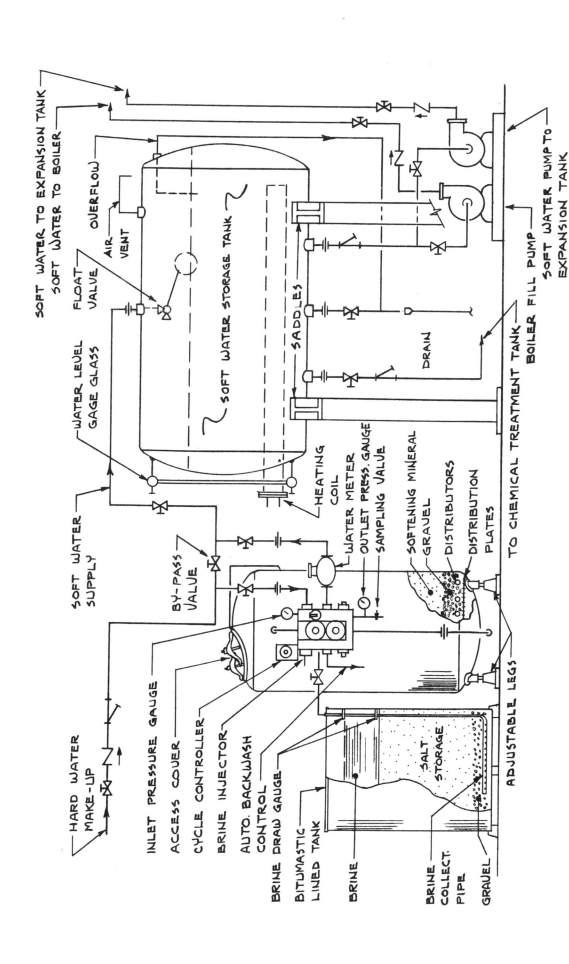

WATER SOFTENING SYSTEM FOR
HIGH TEMPERATURE HOT WATER SYSTEM

FIGURE 2-35

treatment system. The soft water passes into the tank by pressure created from the hard water intake to the water treatment tank and its associated salt storage tank. This softened water is then delivered to the storage tank in a continuous fashion until the storage tank is filled to a fixed level that is controlled by the float valve in the tank. When the tank is sufficiently (approximately nine-tenths) full, the float valve closes and stops the supply of soft water to the tank. The float valve shuts off the cycle controller on the water treatment system.

Chemical Feeder: Some boiler chemical treatments can be relatively simple devices. One such simple device is shown on Fig. 2-36. In this particular device the chemical is fed into the system by a simple funnel arrangement, and the chemicals, once in the funnel, can be sent directly to the boiler or to the boiler condensate feed line. This piping arrangement is typical of a simplified chemical injection feeder. In this particular situation, chemical analysis must be regularly made for control of chemical concentration. Instructions must be given to the system operators so that chemicals are not indiscriminately delivered into the boiler system, creating possible problems in the boiler steaming process.

Deaerator: As mentioned in the water treatment discussion, dissolved or free oxygen can attach to any metal and cause corrosion. Figure 2-37 shows a deaer-

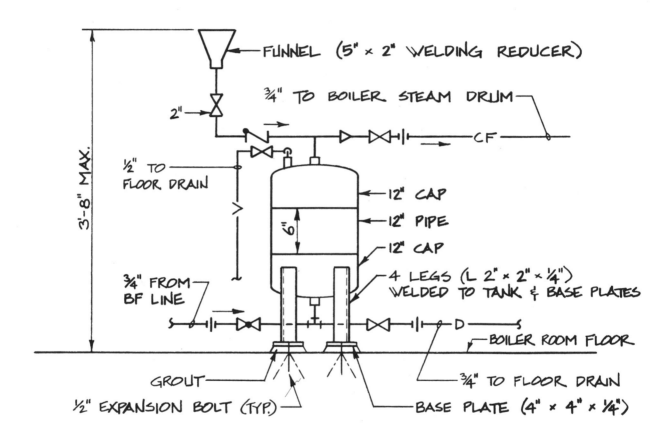

ELEVATION

BOILER CHEMICAL FEEDER
— NOT TO SCALE —

FIGURE 2-36

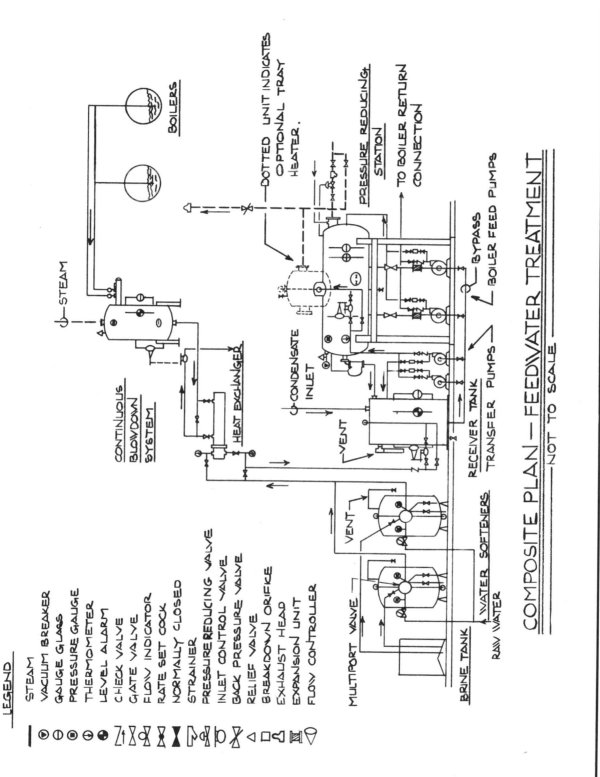

LEGEND

STEAM	
VACUUM BREAKER	
GAUGE GLASS	
PRESSURE GAUGE	
THERMOMETER	
LEVEL ALARM	
CHECK VALVE	
GATE VALVE	
FLOW INDICATOR	
RATE SET COCK	
NORMALLY CLOSED	
STRAINER	
PRESSURE REDUCING VALVE	
INLET CONTROL VALVE	
BACK PRESSURE VALVE	
RELIEF VALVE	
BREAKDOWN ORIFICE	
EXHAUST HEAD	
EXPANSION UNIT	
FLOW CONTROLLER	

COMPOSITE PLAN—FEEDWATER TREATMENT

NOT TO SCALE

BOILERS

STEAM

CONTINUOUS BLOWDOWN SYSTEM

HEAT EXCHANGER

DOTTED UNIT INDICATES OPTIONAL TRAY HEATER.

PRESSURE REDUCING STATION

TO BOILER RETURN CONNECTION

CONDENSATE INLET

VENT

BYPASS

BOILER FEED PUMPS

RECEIVER TANK

TRANSFER PUMPS

VENT

MULTIPORT VALVE

WATER SOFTENERS

BRINE TANK

RAW WATER

FIGURE 2-37

ating heater and its related connections. The deaerating heater is designed to remove oxygen and carbon dioxide from all boiler feedwater. It is very well adapted for deaerating low pH water which is high in oxygen content. The system heats the returning water by mixing it with low pressure steam. This raises the water to the boiling point and boils off gases from the returning water (condensate) solution. The deaerator is effectively a steam kettle that boils off oxygen and other gases. The deaerator is also an ideal location for recovering heat from medium and high pressure condensate traps. These traps commonly are dripped into the deaerator. In addition, the deaerator prevents corrosion in feed lines and economizers and it also provides storage for condensate from other sources.

The piping shown for the deaerator heating is fairly self-explanatory. Usually there are connections for steam and arrangements for venting excess steam to the atmosphere, as well as for the introduction of makeup water to the deaerator as required.

The relationship of the various treatment details to each other and to the system can be confusing. Figure 2-37, a composite detail of the items mentioned in this area of the book, is a large-scale composite detail that has been considerably compressed.

Since more than feedwater treatment is involved in a truly composite plan, we have also shown the continuous blowdown system, both the condensate receiver tank and transfer pumps, and the boiler feedwater pump.

In Fig. 2-36 we illustrated a small manual feeder. This is not shown on Fig. 2-37. Normally this sort of manual feeder would be installed, one per boiler, on the boiler return line just before this line connects into the boiler.

If you use this detail, we have a number of suggestions. First, note that not all of the connections to each item are shown on the composite detail. Second, since the detail is diagrammatic, it does not indicate or imply actual distances between items shown on your floor plans. Third, connections shown on the composite are not necessarily exactly where they would be shown on your individual details. The details take precedence over the composite, which is shown only for explanation and installation guidance.

Finally, we strongly recommend that you enlarge the items shown and separate them so that your final composition is about 11 × 17 in, which is what Fig. 2-37 was before reduction.

Boiler Plant Monitoring and Control

Before entering into a description of the various details presented in this area of our book, we would like to note certain precautions. Generally, in today's manufacturing processes the burner, whether it is part of a boiler or furnace package or purchased separately, comes complete with a factory-supplied control package. Detailing relates only to remote actuators of the

burner such as start-stop switches, aquastats, pressure controls, and its limit controls. As the system gets larger, the controls generally are arranged in a separate control panel mounted on, or adjacent to, the burner. A proper specification covers function and material requirements, and no detailing is required.

Burner Control Panels: In Fig. 2-38 we present a burner control panel for a large watertube 125 psig boiler. You will note that there is no control schematic. That is covered by the specification and the schematic wiring which is prepackaged and not part of any detailing requirements. However, the specification requirements become elaborate in this situation. The following is a list of possible specification items that may require some indication on the control panel.

1. Low water
2. High water
3. Low water cutoff
4. Low atomizing pressure
5. Flame failure
6. Limits satisfied
7. Purge in process
8. Purge completed
9. Ignition power
10. Main flame proven
11. Main fuel valves closed
12. Post purge
13. Alarm silencer (switch)
14. Blower manual purge
15. Plant air (or steam) to burner
16. Selector switch (gas-off-oil)
17. Low water reset (switch)
18. Burner stop (switch)
19. Burner start (switch)
20. Flame signal strength meter
21. Wind box pressure gauges
22. Furnace pressure gauge
23. Boiler outlet pressure gauges

The above list contains a few decision items but, in general, provides skilled plant operators with visual information about the operation or point of failure of the burner.

It could be specified that the collection of information should be put in a reasonably sized panel and no detail provided. However, to clarify the bid documents the generally accepted practice is to detail some sort of panel such as the one depicted with the size and arrangement as noted. Normally, a panel such as this has a rear-side access door the full height and width of the panel.

Furnace and wind box pressure gauges will have scales as required by the boiler and burner manufacturer. The boiler outlet draft gauge, if provided, must be coordinated with the economizer draft loss if an economizer is part of the boiler system. Normally, any flue gas oxygen analysis equipment is located on the

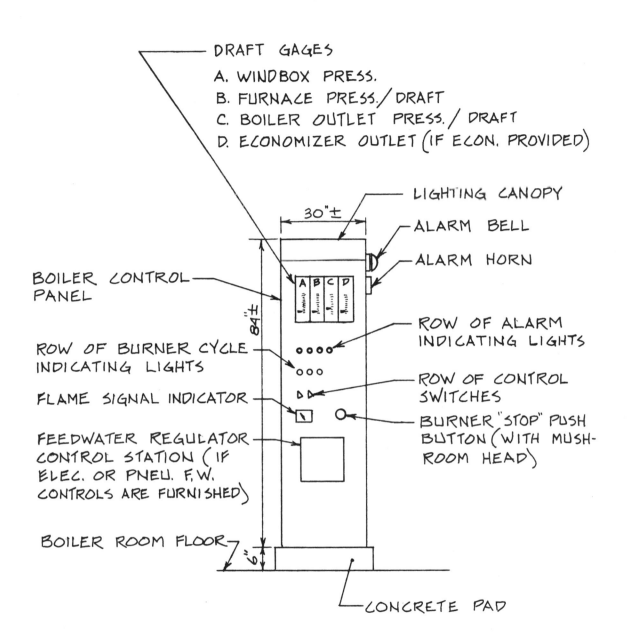

DRAFT GAGES

A. WINDBOX PRESS.
B. FURNACE PRESS./DRAFT
C. BOILER OUTLET PRESS./DRAFT
D. ECONOMIZER OUTLET (IF ECON. PROVIDED)

LIGHTING CANOPY

ALARM BELL

ALARM HORN

30"±

BOILER CONTROL PANEL

84"±

ROW OF ALARM INDICATING LIGHTS

ROW OF BURNER CYCLE INDICATING LIGHTS

FLAME SIGNAL INDICATOR

ROW OF CONTROL SWITCHES

BURNER "STOP" PUSH BUTTON (WITH MUSH-ROOM HEAD)

FEEDWATER REGULATOR CONTROL STATION (IF ELEC. OR PNEU. F.W. CONTROLS ARE FURNISHED)

BOILER ROOM FLOOR

6"

CONCRETE PAD

ELEVATION

BURNER CONTROL PANEL

NO SCALE

FIGURE 2-38

side of the panel. These panels are supplied, one per boiler, prewired by the burner control manufacturer. The plans should show the conduits required for connection of panel to boiler and burner.

Instrumentation Panel: Figures 2-39 and 2-40 present a boiler plant instrumentation panel. This is a large,

complex panel for a large and complex system, and it is not necessarily the panel that you will use on your project. This detail was used for a large hospital complex that had its own laundry facility. The arrangement and type of controls depicted are the result of many conferences and discussions with the client.

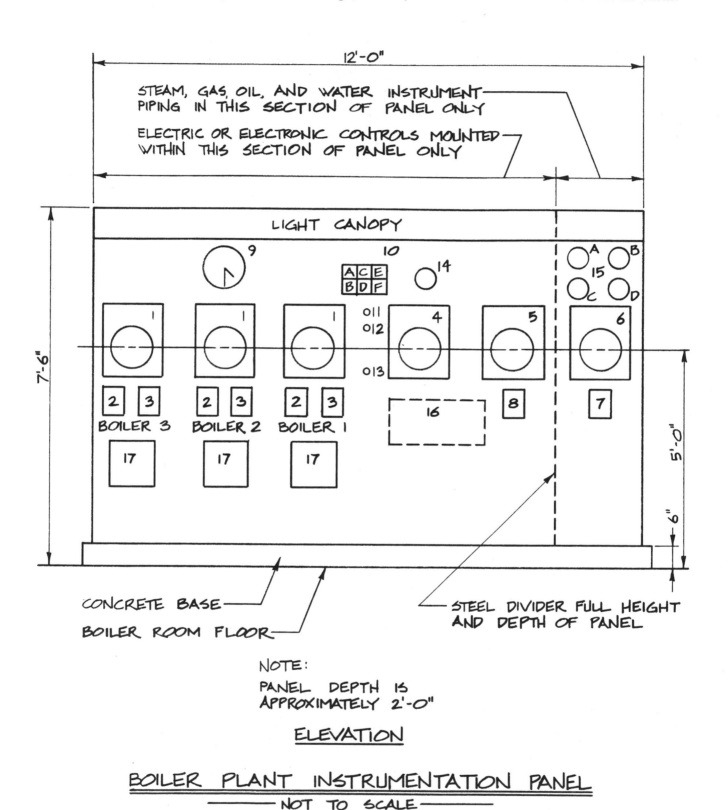

ELEVATION

BOILER PLANT INSTRUMENTATION PANEL
——— NOT TO SCALE ———

FIGURE 2-39

For each boiler there is a steam flow recorder, as well as a flue gas temperature and oxygen content recorder. The flow of steam for the entire building and, separately, for the laundry is recorded. There is a recorder to cover steam header pressure, boiler feed-water temperature, and outside air temperature. There is a master steam pressure controller. The panel comes with a 12-in clock and a canopy with fluorescent lighting for all instruments.

The system is large and complex. As may be noted in item 10 of the equipment list in Fig. 2-40, there is an alarm annunciator panel which provides both visual

---

BOILER PLANT INSTRUMENT PANEL
EQUIPMENT LIST

ITEM   DESCRIPTION
1. Boiler operation meter-recorder
   a. Steam flow: indicate, record, integrate (0 to 50,000 lb/hr, orifice design pressure 125 psig)
   b. Flue gas temperature: record (0 to 1000°F)
   c. Flue gas oxygen content: record (0 to 20 percent O_2)

2. Boiler submaster (manual/automatic station, boiler bias control)

3. Boiler air/fuel ratio control (manual station)

4. Steam flow recorder
   a. High pressure steam distribution: record, integrate (0 to 50,000 lb/hr, orifice design pressure 125 psig)
   b. Medium pressure steam distribution: record, integrate (0 to 25,000 lb/hr, orifice design pressure 100 psig)

5. Steam flow recorder
   a. Laundry steam distribution: record, integrate (0 to 25,000 lb/hr, orifice design pressure 125 psig)
   b. Boiler plant steam: record, integrate (0 to 10,000 lb/hr, orifice design pressure 125 psig)

6. Boiler plant operation recorder
   a. Steam header pressure: record (0 to 300 psig)
   b. Boiler feedwater temperature: record (0 to 300°F)
   c. Outside air temperature: record (−30°F to +120°F)

7. Master steam pressure controller
   Use this location if controller components *are* in a sealed case, protected from fluid penetration.

8. Master steam pressure controller
   Use this location if controller components *are not* in a sealed case. Provide pressure transmitter in sealed case located in section of panel where steam instrument piping is located. Pressure transmitter shall provide an electric or pneumatic signal to master pressure controller.

9. Clock (12-in dial)

10. Annunciator with six displays
    a. Condensate storage tank high level
    b. Condensate storage tank low level
    c. Feedwater heater high level
    d. Feedwater heater low level
    e. Water softener regeneration required
    f. Emergency gas valve(s) closed

11. Annunciator silencing button

12. Annunciator test button

13. Emergency gas safety shutoff valve switch

14. Annunciator buzzer

15. Pressure gauges (6-in diameter)
    a. Steam header (range 0 to 200 psig)
    b. Boiler feed header (range 0 to 400 psig)
    c. Fuel oil header (range 0 to 200 psig)
    d. Natural gas header (range 0 to 15 psig)

16. Start-stop buttons and indicating lights for pumps

17. Smoke density monitors

FIGURE 2-40

and audible alarms for certain specially selected items. This alarm system requires test and silence provisions.

The gas system and the boiler return pumps can be controlled at the panel. In item 15 of Fig. 2-40 there is pressure measurement at certain select locations. Finally, to comply with Environmental Protection Agency requirements, there is a smoke density meter which indicates the condition of the flue gas leaving the chimney.

We cannot overemphasize the fact that you absolutely cannot blindly copy this detail as is on your detail sheet. There is, however, a high degree of probability that you will require most of these items in your particular project. These details in Figs. 2-38 through 2-40 came from a specific project. The total written specification for the burner and instrumentation panels of this detail covered some 15 single-spaced pages.

Generally, the remote devices are a combination of electronic and pneumatic instruments. Thus, the panel items have sources of air and electricity. Pressure-electric switches are required to relate signals received to panel instruments. Most of the devices are of the direct reading, on-off type.

Orifice Metering: In Fig. 2-41 we illustrate a type of transmitter used to generate a flow signal. As can be seen in this detail, an orifice is used to provide a flow signal. The orifice must be carefully sized to relate to the range of flow signals that will be transmitted and the pressure of the steam. An orifice is calibrated at a given pressure and will not work at some other pressure. For example, an orifice calibrated to work at 125 psig will not provide accurate readings at 105 psig. The manufacturer can give you a modifying value which you can use to manually multiply your recorded results to secure the correct answer. But this is a tedious and sometimes unsatisfactory process.

An orifice may be placed in any vertical or horizontal steam line provided you allow for the length of straight pipe before and after the orifice as noted in the elevation section of the steam line. Sometimes the boiler feedwater regulator has an element that requires

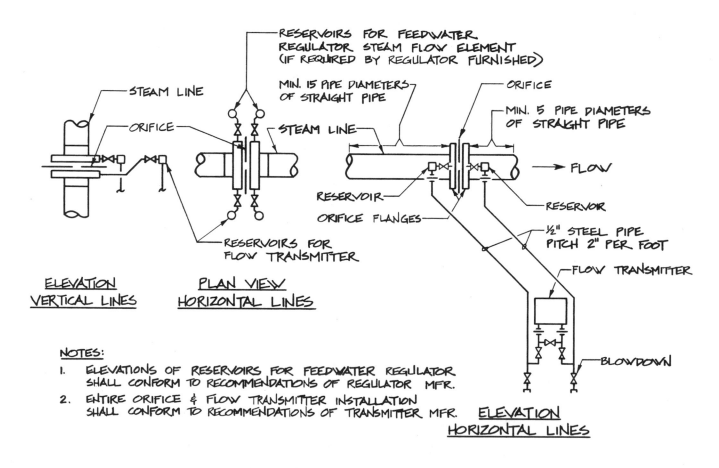

ORIFICE METERING INSTALLATION
——— NOT TO SCALE ———

FIGURE 2-41

the input of a related steam flow readout. In the center of the three details depicted in Fig. 2-41 we show a separate set of connections for this purpose.

The transmitter is normally located on a building column or separately supported on a rigid frame below the orifice connections. The pressure lines to the transmitter are sending pressure signals that record a drop across the orifice. There lines have valved connections to occasionally blow out the lines to clear them of collected impurities. The blowdown lines are piped to a blowdown tank that is not shown in the detail.

Storage Tank Control: In a condensate storage tank there are low water cutoffs to control water level and related pumping required to maintain a given water level. A gauge glass is also supplied to enable the operator to visually note the water level. Figure 2-42 depicts a typical installation for a remote water level signal. The transmitter is not a control but merely a fixed position indicator. Going back to Fig. 2-40, you will note under item 10 that there are high and low

level lights and horns for both the condensate storage tank and the feedwater heater. Figure 2-42 is a typical detail for the installation of such a transmitter on the storage tank. It is a standard detail.

In closing it should be noted that the operation of a large plant could be computerized. Even though it does take time to warm up a large boiler and the decision to have one, two, or three boilers is judgmental, computerization is still a reasonable option. However, when things go wrong and fast start, stop, or rotate boiler decisions are required, many users still prefer a human operator.

Pressure Relief Valve Piping

In Chap. 1 we provided details of equipment that is normally on a boiler and which requires some detailing to clarify additional work to be performed by the installing contractor. Among the items noted was the piping to a floor drain of the discharge of the pressure relief valve. This is standard accepted practice for a low pressure installation. But for a high pressure

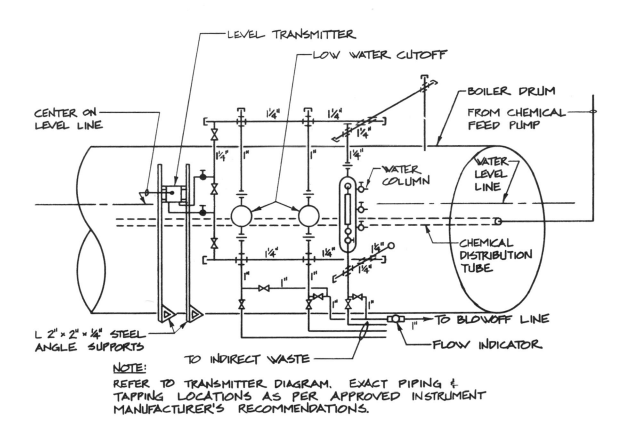

NOTE:
REFER TO TRANSMITTER DIAGRAM. EXACT PIPING &
TAPPING LOCATIONS AS PER APPROVED INSTRUMENT
MANUFACTURER'S RECOMMENDATIONS.

WATER COLUMN, LEVEL TRANSMITTER, AND
LOW WATER CUTOFF DIAGRAM
—— NOT TO SCALE ——

FIGURE 2-42

steam boiler or a high temperature hot water boiler, this could produce a very dangerous situation. Any pressure relief valve will emit steam (even one from a hot water boiler when the temperature is above 212°F) when the internal pressure exceeds the valve rating, and for higher pressures a way must be found to dispose of this blast of steam in a safe manner.

Figures 2-43 and 2-44 are directly related to each other, and in actuality Fig. 2-44 is merely an enlargement of the overall system detail shown in Fig. 2-43. An allowance must be made for the very rapid expansion of the relief valve discharge line.

Figure 2-43 shows an enlargement of the piping and system arrangement that is directly connected to the relief valve discharge opening. As shown, a short threaded piece is connected into the relief valve discharge opening. A special drip elbow is connected to this short pipe. The steam passes through this pipe into a steam riser which slides over the short pipe section but is not connected to the short discharge pipe. This allows for ease of rapid expansion and contraction.

Because the connection is not tight, some steam will drip back. This is caught in the drip pan which has two ¾-in drain connections, as detailed. This small amount of condensate is piped to a drain. As can be seen in Fig. 2-44, this oversized pipe carrying the discharge steam is piped through the roof and to a height of 6 ft or a height as required by local code. The same code may require a sound muffler on the pipe.

The pipe is held in place by pipe clamps at the roof, and being free at both ends, it can easily expand and contract. This is a detail that can be used for most high pressure safety relief valve exhausts. The piping cap material through the roof is usually Schedule 40 black iron pipe with welded joints.

Deaerator Piping

The theory of deaeration of boiler feedwater is a relatively simple straightforward application of basic principles. All water contains a certain amount of dissolved gases and chemicals which can corrode metals. In addition steam condenses, and by the time

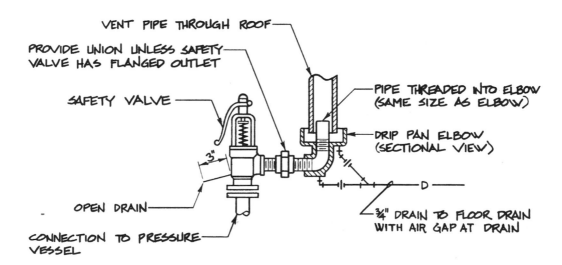

NOTE:
VENT PIPE SHALL TERMINATE
MIN. 6'-0" ABOVE FINISHED ROOF

VENT PIPE THROUGH ROOF

PROVIDE UNION UNLESS SAFETY
VALVE HAS FLANGED OUTLET

PIPE THREADED INTO ELBOW
(SAME SIZE AS ELBOW)

SAFETY VALVE

DRIP PAN ELBOW
(SECTIONAL VIEW)

OPEN DRAIN

CONNECTION TO PRESSURE
VESSEL

¾" DRAIN TO FLOOR DRAIN
WITH AIR GAP AT DRAIN

ELEVATION

STEAM SAFETY VALVE PIPING CONNECTIONS
— NOT TO SCALE —

FIGURE 2-43

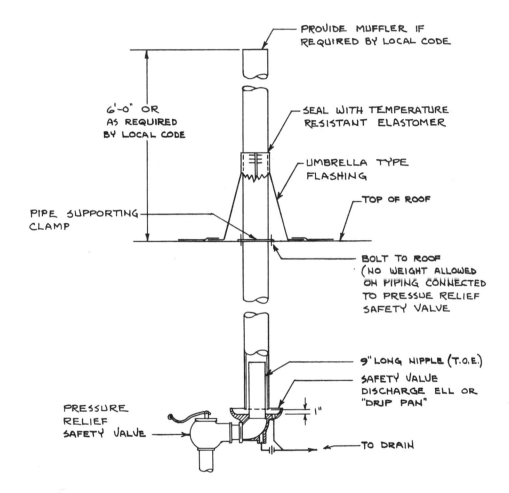

PROVIDE MUFFLER IF REQUIRED BY LOCAL CODE

6'-0" OR AS REQUIRED BY LOCAL CODE

SEAL WITH TEMPERATURE RESISTANT ELASTOMER

UMBRELLA TYPE FLASHING

TOP OF ROOF

PIPE SUPPORTING CLAMP

BOLT TO ROOF (NO WEIGHT ALLOWED ON PIPING CONNECTED TO PRESSUE RELIEF SAFETY VALVE

9" LONG NIPPLE (T.O.E.)

SAFETY VALVE DISCHARGE ELL OR "DRIP PAN"

1"

PRESSURE RELIEF SAFETY VALVE

TO DRAIN

TYPICAL HIGH PRESSURE BOILER SAFETY VALVE DISCHARGE PIPING

(HIGH PRESSURE BOILERS I.E. STEAM BOILERS OVER 15 P.S.I. AND HOT WATER BOILERS OVER 160 P.S.I. OR OPERATING OVER 250°F.)

FIGURE 2-44

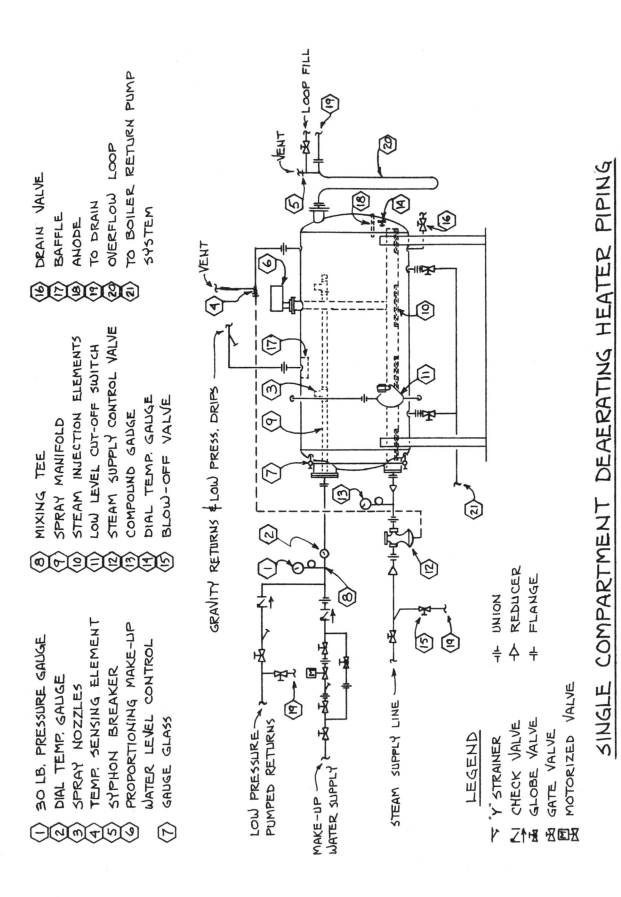

LEGEND

⅃ "Y" STRAINER
⅃ CHECK VALVE
⅃ GLOBE VALVE
⅃ GATE VALVE
⅃ MOTORIZED VALVE

—‖— UNION
—▷— REDUCER
—⊹— FLANGE

① 30 LB. PRESSURE GAUGE
② DIAL TEMP. GAUGE
③ SPRAY NOZZLES
④ TEMP. SENSING ELEMENT
⑤ SYPHON BREAKER
⑥ PROPORTIONING MAKE-UP WATER LEVEL CONTROL
⑦ GAUGE GLASS

⑧ MIXING TEE
⑨ SPRAY MANIFOLD
⑩ STEAM INJECTION ELEMENTS
⑪ LOW LEVEL CUT-OFF SWITCH
⑫ STEAM SUPPLY CONTROL VALVE
⑬ COMPOUND GAUGE
⑭ DIAL TEMP. GAUGE
⑮ BLOW-OFF VALVE

⑯ DRAIN VALVE
⑰ BAFFLE
⑱ ANODE
⑲ TO DRAIN
⑳ OVERFLOW LOOP
㉑ TO BOILER RETURN PUMP SYSTEM

SINGLE COMPARTMENT DEAERATING HEATER PIPING

NO SCALE

FIGURE 2-45

the condensate returns or can be returned to the boiler, it has become too cool for the efficient operation of a high pressure (125 psig) boiler.

The deaerator cannot resolve the problem of chemical impurities in the water, but it can, and does, very effectively resolve the problems of gases (oxygen) in the water. It is the primary, efficient way to reheat the condensate to a temperature (usually about 217°F) that makes for efficient operation of the 125 psig boiler.

The accompanying detail, Fig. 2-45, illustrates a common type of deaerator and the piping to it. It is an enlargement of the type noted in our discussion of Fig. 2-37. There are a number of variations on this basic deaerator design which vary somewhat in their connection details, but the basic system, as we show it, is representative. From a condensate storage tank, not shown in the detail, the gravity and pumped return condensate is again pumped to the deaerator, as noted in the top left area of Fig. 2-45. If there is insufficient condensate available, a cold water makeup line provides additional liquid. In the lower left of the detail, based on a temperature sensing element, a controlled amount of steam is introduced into the tank, raising the temperature some 5°F above the boiling point. The boiling water now gives off its dissolved air, and as directed by the boiler return pump controls, flows by gravity to the boiler return pumps from which the properly heated, air-free liquid is pumped back into the boiler to be turned into high pressure steam and start the cycle over again.

There are other connections not shown that may be part of the overall system. Frequently, in a high pressure steam system there are drip traps on the high pressure header. The lines from the traps contain liquid at high pressure which cannot be passed directly into the low pressure return system. They can be, and commonly are, discharged into a flash tank which, by atmospheric venting, wastes a large part of the high pressure liquid's potential and available energy. It is not unusual to have these lines tapped into spare tappings at the top of the deaerator. The condensate will flash into steam in the deaerator and perform a useful function of providing part of the heat the deaerating process requires.

In addition, a high pressure steam plant may have one or more pumps driven by a steam turbine. The exhaust steam leaving the turbine may be maintained at 5 psig or may be allowed to vary, rising up to as much as 30 psig under light load operating conditions at the turbine driven pump. This exhaust steam can be put to a number of uses. One is to allow this exhaust steam to be part of the source of steam required by the deaeration process.

Because we wanted our deaeration detail to be clear, we did not show alternative connections, such as condensate lines from high pressure traps and steam lines from the exhaust of steam turbine driven pumps. There are other items to be noted. If the total flow becomes sufficiently unbalanced to provide insufficient condensate to the boiler return pumps, a way must be found to stop the pumps. This is done by a low water cutoff. If the boiler return pumps stop, the low water cutoff in the boilers will shut down the overall system.

Again if for any reason the deaerator overfills, a way must be found to drain off the excess. Our detail depicts a vented and trapped drain loop. This *may* permit direct connection to a drain line. However, the final connection may still need to be arranged as an indirect waste system with an appropriate air gap. If this is a requirement, do not remove the loop from your detail. When the tank is vented without restriction to atmosphere, the seal created by the overflow insures that hot air or hot water is not inadvertently vented into a building drain line.

Deaerators, such as the one shown or similar types, are a common standard part of a 125 psig system. In the engineering and specifying of the deaerator care should be taken to modify this detail where necessary to meet the exact needs of the specified unit. Usually the modifications will be relatively minor but, even if minor, should be precisely resolved so that the final specified unit is not contradicted by the detail on the plan.

Feedwater Regulator: In our discussion of boiler plant monitoring and controls we tried to keep the subject as narrow as possible. This admittedly created a gray area of items that could have been included in that chapter. But we feel the series of items we will now cover could really have been put in a number of other locations as well.

The first of these items is the feedwater regulator installation depicted in Fig. 2-46; it is a typical single-element regulator. This particular regulator is designed for reasonably steady steam loads. In installations in which rapid or wide load changes are encountered, a two-element regulator is usually specified. The two-element regulator is commonly seen when a superheat is involved and the system is a large, special purpose power and/or utility plant.

The actuating element is a thermostatic expansion tube in which the water level moves up and down as the boiler water level rises or falls. When the water level drops, the steam temperature causes the tube to expand. Since it is fixed on one end only, the head piece moves. A bell crank lever at the head multiples the movement to operate the control valve. Through an adjustment at the fixed end, the desired water level in the boiler drum can be achieved while the boiler remains in operation.

This type of feedwater regulator is very common on 125 psig systems. The diagram in Fig. 2-46 is an enlarged schematic. The boiler usually comes with properly located tappings to permit this sort of control application. Too often the specification may not clearly define the feedwater regulator and its application.

Even if there is a poor definition, the inclusion of the detail will go a long way toward solving this system control requirement.

Centrifugal Separator: We have referred a number of times to blowoff or blowdown piping in the details we have previously presented. The statement has been made that these blowoff or blowdown lines vent to a piece of apparatus depicted elsewhere; Fig. 2-48 shows this apparatus and a possible way to conserve the energy being blown away.

There are a number of reasons for blowing down a 125 psig steam boiler plant. Detailers are not primarily concerned with the engineering analysis so let us say the primary purpose of the controlled blowing off of steam is to remove impurities in the boiler water. Our

concern is how to pipe the flashing mixtures of water and steam that occur in boiler blowoff lines, in lines from the evaporator, and in feedwater-heater drain lines. In accordance with basic laws of pressure and temperature, water will flash into steam anytime it is exposed to a temperature less than the saturated pressure corresponding to the temperature of water entering the line. For example, hot water at 260°F requires 35 psig to keep it liquid since it is above the 212°F boiling point. Open a valve on such a line and the pressure will drop immediately to atmospheric and the 260°F water will become steam instantly.

Since the steam-liquid mixture of the blowoff from the blowdown must be cooled to a readily disposable liquid, a common device called a blowdown separator,

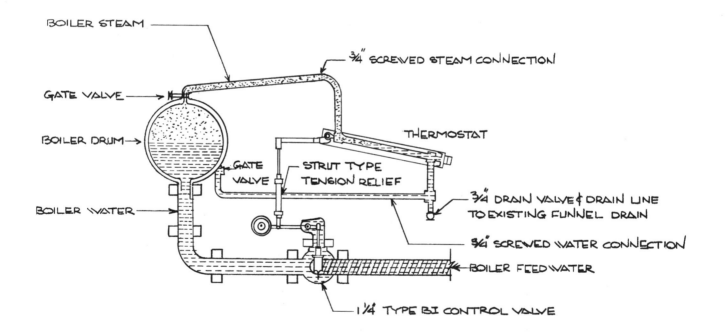

SCHEMATIC FLOW DIAGRAM FOR THERMOSTATIC & MECHANICAL MODULATING FEEDWATER REGULATORS
— NOT TO SCALE —

FIGURE 2-46

as shown in Fig. 2-47, is employed. In essence the steam can be vented, usually through a steam muffler, to atmosphere, and the hot liquid can be piped to a sewer. Frequently, the hot liquid is still at 212°F and can pose problems for drainage systems. As can be seen, a thermostatically controlled city water line provides additional final cooling.

Obviously, we have in the above descriptions and as depicted in Fig. 2-47 rid ourselves of the blowoff products in a safe manner. We have also wasted energy. Studies should be made to determine how this wasted energy might be saved. Figure 2-48 is a generalized schematic detail of a possible solution. The hot steam-liquid mixture can be intercepted on its way to the blowoff tanks and piped instead through a surge

tank and heat exchanger. The system could also provide a sample cooler arrangement, which will be discussed in a separate detail. In our detail the suggestion of use for boiler blowdown steam is to provide heat for cold water makeup to the system. This is not the only use for this blowdown steam, but it is an obvious one-for-one exchange of hot and cold water. In the operation of this system an electric valve is placed in each blowdown line to open and close with the on-off operation of the boiler feed pump. The surge tank provides a reservoir for the probable imbalance of makeup and feedwater flow. The surge tank may possibly be eliminated if a large feed tank is used, its level does not change appreciably when the feed pump operates at full load, and a float valve is used on the

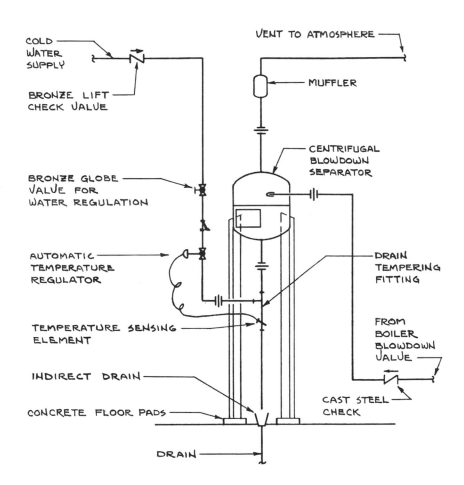

TYPICAL HIGH TEMPERATURE HOT WATER
BOILER BLOWDOWN PIPING FOR
CENTRIFUGAL SEPARATOR

FIGURE 2-47

feed tank makeup water inlet. The float valve so used must operate over a large range of modulating flow control. The best system is a small feed tank or a short-range makeup float valve and a surge tank.

The above system is to be used when blowdown is continuous. If blowdown is performed only on fixed, limited occasions, the system depicted in Fig. 2-48 should be used. Note that in our continuous system the blowdown liquid has been cooled to a sufficient degree to be wasted (through an indirect waste connection) to the sewer.

As we mentioned in our discussion of Fig. 2-48, the

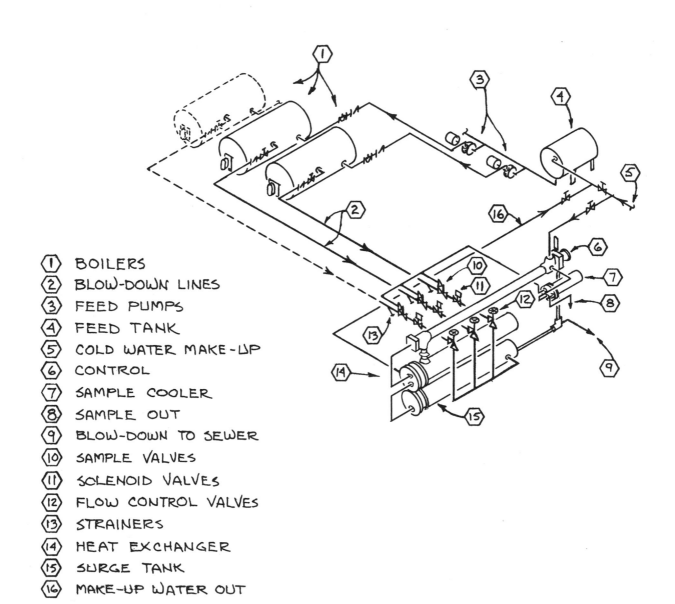

1. BOILERS
2. BLOW-DOWN LINES
3. FEED PUMPS
4. FEED TANK
5. COLD WATER MAKE-UP
6. CONTROL
7. SAMPLE COOLER
8. SAMPLE OUT
9. BLOW-DOWN TO SEWER
10. SAMPLE VALVES
11. SOLENOID VALVES
12. FLOW CONTROL VALVES
13. STRAINERS
14. HEAT EXCHANGER
15. SURGE TANK
16. MAKE-UP WATER OUT

AUTOMATIC APPORTIONING OF BLOWDOWN TO MAKE-UP WATER

NO SCALE

FIGURE 2-48

sample cooler is a separate item. In Fig. 2-49 we depict the piping to a sample cooler. Samples of water are normally taken from the drums of each boiler and from the feedwater supply line to the boilers (called FWS in our detail). All of these lines are carried over to the sample cooler, which is basically a shell and tube fluid cooler. Even though cooling and system liquids are separated, note the antisiphon vacuum breaker on the water line that precludes cross-contamination. Generally, the system as depicted is a manual operation with sufficient cooled water introduced to reasonably cool the small sample taken for chemical test purposes.

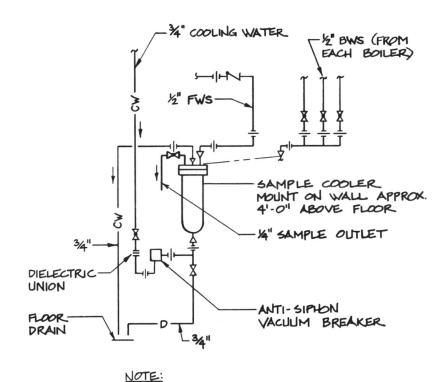

ELEVATION

BOILER WATER, OR FEEDWATER SAMPLE COOLERS
————NOT TO SCALE————

FIGURE 2-49

Flash Tank: In Figs. 2-50 and 2-51 we depict a typical detail in diagrammatic form of each piping to a flash tank and then the piping in and around the flash tank itself. High pressure steam systems generally have three types of condensate. These are condensate at 125 psig, condensate at an intermediate pressure of some 50 psig, and finally low pressure atmospheric condensate that comes back from many sources. The piping diagram shows two, not three, lines of condensate returns. These should be labelled with the high and intermediate pressure valves, i.e., 125 psig condensate and 50 psig condensate. Low pressure condensate comes out of the flash tank and ties to other low pressure lines.

You cannot permit medium or high pressure condensate to come into direct contact with low pressure condensate. The effect is the same as described in blowdown piping. The higher pressure condensate will immediately flash into steam if this occurs, and the result will be disastrous. The ensuing water hammer

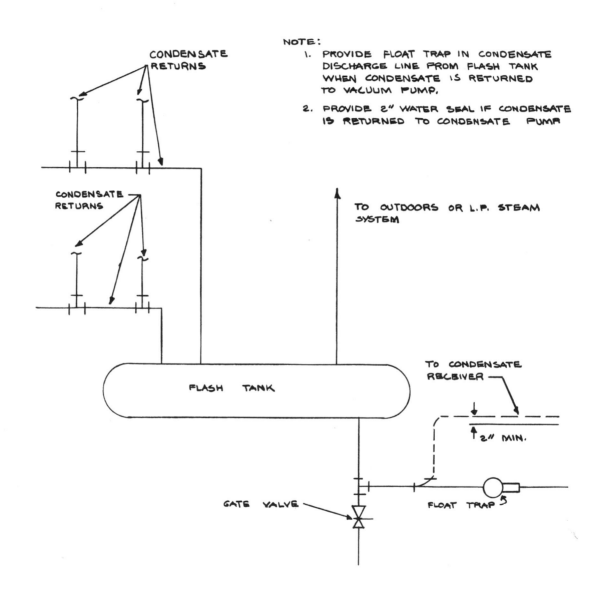

TYPICAL FLASH TANK PIPING

FIGURE 2-50

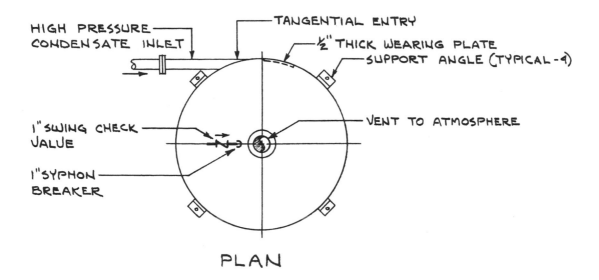

HIGH PRESSURE CONDENSATE INLET

TANGENTIAL ENTRY

½" THICK WEARING PLATE

SUPPORT ANGLE (TYPICAL-4)

VENT TO ATMOSPHERE

1" SWING CHECK VALUE

1" SYPHON BREAKER

PLAN

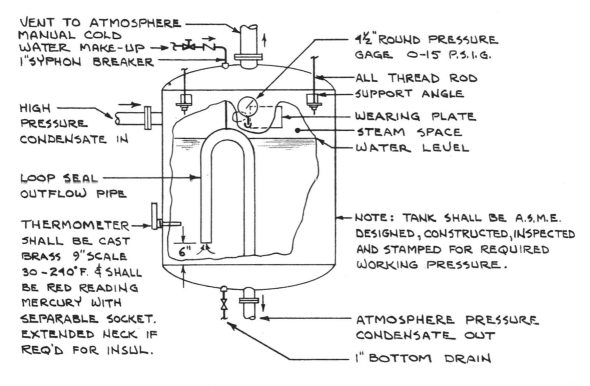

VENT TO ATMOSPHERE
MANUAL COLD WATER MAKE-UP
1" SYPHON BREAKER

4½" ROUND PRESSURE GAGE 0-15 P.S.I.G.

ALL THREAD ROD SUPPORT ANGLE

WEARING PLATE

STEAM SPACE

WATER LEVEL

HIGH PRESSURE CONDENSATE IN

LOOP SEAL OUTFLOW PIPE

THERMOMETER SHALL BE CAST BRASS 9" SCALE 30-240°F. & SHALL BE RED READING MERCURY WITH SEPARABLE SOCKET. EXTENDED NECK IF REQ'D FOR INSUL.

6"

NOTE: TANK SHALL BE A.S.M.E. DESIGNED, CONSTRUCTED, INSPECTED AND STAMPED FOR REQUIRED WORKING PRESSURE.

ATMOSPHERE PRESSURE CONDENSATE OUT

1" BOTTOM DRAIN

ELEVATION

INSTALLATION OF FLASH TANK

FIGURE 2-51

will be loud and violent. Most likely it will crack the low pressure condensate piping, causing numerous leaks.

The flash tank allows for a safe location in which the high and medium pressure condensate can expand and become safe and usable low pressure condensate that can be handled by the low pressure return system on its way back to the deaerator to start the steam cycle all over again.

Frequently, the plans given to the contractor do not completely detail the flash tank, which is a special device and not simply an empty tank. And many times the specifications do not really spell out what is needed. Figures 2-50 and 2-51 very specifically describe the flash tank itself. Note, for example, the gauges, the syphon breaker on a cold water line to provide added cooling if required, and the loop seal overflow piping to ensure that condensate, not steam, comes out of the tank. Finally, the tank is ASME rated. With the detail in Fig. 2-51 you will get a usable, workable flash tank.

Steam Turbine Control: In Fig. 2-52 we illustrate a typical detail of a steam control system for a turbine driven device such as a boiler feed pump. This detail

was alluded to in our previous discussion of boiler feed pumps. It was noted in that discussion that if the load on the steam driven boiler feed pump was relatively constant, no control was actually needed as the steam flow could be manually set with a small amount of overcapacity and manual bypass. But this is frequently not the case.

The controller in Fig. 2-52 is a differential pressure controller selected by the design engineers to maintain a steam flow based on the steam requirements of the turbine and the maintenance of proper pressure on the feedwater discharge line. As the boiler requires more feedwater, the pressure in the feedwater line tends to drop, which through this controller creates an increase in the steam flow to speed up the turbine. The more steam, the more work available at turbine and consequently higher feedwater pressure is available from the turbine driven feedwater pump. The steam line to the turbine should always be trapped as shown, and the steam from the turbine discharge (noted on the detail as excess steam) is normally piped to the flash tank.

Condensate Meter: Figure 2-53 depicts two common versions of the condensate meter installation. The meter is a common type of measuring device that is

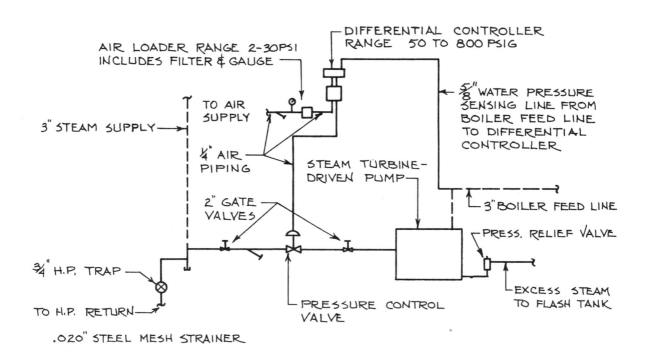

TURBINE PRESSURE CONTROL

NO SCALE

FIGURE 2-52

used when all of the condensate from all sources can be brought to a common point for measuring purposes. Normally this is the point in the building at which the condensate returns to the boiler plant or leaves to flow to a central utility source.

Since we are measuring for billing purposes, the following rules apply:

1. The return piping should be properly pitched and free of leaks.

2. Traps to meters should be continuous flow and not bucket types.

3. A meter on the discharge side of a pump or bucket trap should be preceded by a vented receiver.

4. The discharge from the meter should be full-sized and pitched in the direction of flow.

5. A meter on a vacuum system should have an automatic discharge to the sewer to prevent flooding when the vacuum effect is lost.

6. Strainers must be cleaned regularly.

7. Returns above 150°F should be run through a cooling coil ahead of the meter.

Condensate meters are usually either of the gravity

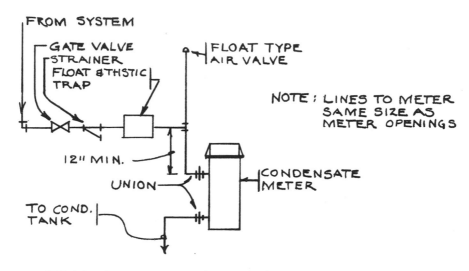

TYPICAL GRAVITY INSTALLATION

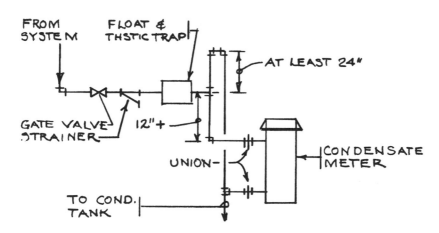

VACUUM INSTALLATION WITH MASTER TRAP

CONDENSATE METERS

NO SCALE

FIGURE 2-53

or vacuum type. The gravity meter is designed to discharge its condensate to a sewer or to an atmospherically vented receiver. It is not airtight. In the upper part of our detail there is a typical gravity installation utilizing a constant flow trap. In the lower part of the detail there is a typical vacuum system installation, again with a constant flow trap. Note that the one major difference in the piping is an air loop rising above the trap at least 24 in. This is to prevent the system from becoming air bound. The 12-in differential in elevation between the trap and the meter is noted as 12-in minimum; it will increase in height as per the manufacturer's recommendations for the required flow rate. The value, for example, of a meter rated at 12,000 lb/hr of steam is 18 in.

THREE

Domestic Hot Water Source Equipment

Frequently overlooked in the design of heating systems is the fact that the supplying of domestic hot water is a separate system. Even in residential work, especially when the heating source is a forced warm air furnace, the domestic hot water supply system is a separate source and requires at least minimal detailing. Gas and electric hot water sources generally require similar details. The gas heater has two additional requirements not needed by the electric heater. These are a smoke pipe and a gas train. We therefore chose to depict the gas heater; the piping depicted, other than the flue pipe and gas train, are the same as for an electric heater. Assuming that the flue pipe is properly shown on the plans, no additional detailing is required of this item.

The details cover gas fired water heating equipment that heats water for washing, pool water, and the like. The details do not show the gas piping to the unit (or oil or electric) since oil and gas piping details are separately shown in Chap. 2; the electric requirements would simply consist of a electric line to the heater unit with a disconnect switch. The details that are presented are not for the small, individual storage and heater unit of a residential or small commercial building but rather for larger commercial, industrial, and institutional applications.

As we previously noted in Chap. 1, solar hot water systems operate in the same configuration and use the same collector plate arrangement as solar heating systems do. The collector plate has the ability to capture an average of 1000 to 1300 Btu of the sun's heat per day. In the summer it captures more heat; in the winter, less; and at night, none at all. Since it takes 833 Btu to raise 1 gal of water from 40°F to 140°F, the general rule of thumb is that 1 sq ft of collector plate will heat, on average, 1 to 1¼ gal of water every day of the year. The heating is obviously done during the daylight hours. Also it should be obvious that the rate is greater in the summer than in December or January.

A heat exchanger can be as simple as the tankless hot water heater inserted below the water line in the residential steam or hot water boiler. This is a water-to-water heat exchanger regardless of whether the boiler is used in a steam or hot water heating system. In the steam system the heater is always located below the boiler water line, keeping it in the water (condensate) side of the steam system. Generally in larger installations the tankless heater may be supplemented with a storage tank whose aquastat controls, operating through the basic boiler-burner controls, use the tankless heater to heat the water in the storage tank, which then becomes the residential source. Our details proceed from that point to the more complex, larger project installations.

Gas Fired Water Heater

Figure 3-1 shows a typical piping detail for a separate heater and related storage tank. Because of the common problem of sudden, large demands for hot water, this system, typical of all similar systems, requires a storage tank. The calculation of demand and storage capacity requirements for water heating is well documented in the *ASHRAE Systems Handbook*. In addition a number of water heating equipment manufacturers provide similar, detailed calculations in their equipment catalogs.

Figure 3-1 is straightforward with all piping, valves, pumps, and controls clearly depicted. Certain observations should be especially noted. Except for very small tanks it is not wise to depend on gravity circulation. When hot water is used in the system, it is pulled out of the top of the storage tank. This permits cold water makeup at building water pressure to enter the system and be pulled through the heater. The heater will react, based on its temperature control, and heat the water. The demand may very well be greater than the capacity of the heater, and consequently the water will not be fully heated when it leaves the heater.

But the above problem is not serious unless the storage tank is undersized; otherwise the tank has a sufficient quantity of properly heated water to accept the demand. While the mixture of improperly heated water and storage hot water will, if the demand lasts long enough, create a problem, in actuality the problem is normally not noticeable. The pump on the line is controlled by an aquastat in the storage tank. Its job is to preclude stratification in the tank by keeping the water circulating between the tank and the heater until all the water in the tank is back to the desired temperature. Note also that both the tank and the

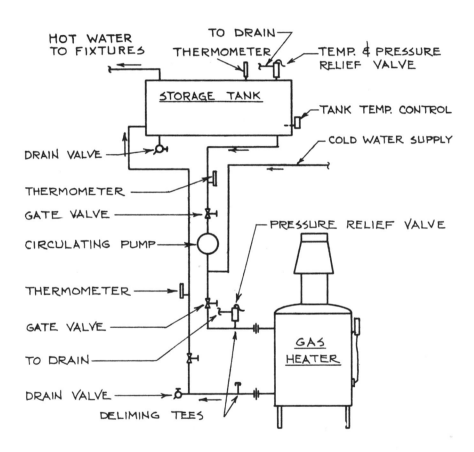

SINGLE TEMPERATURE HOT WATER SYSTEM
GAS FIRED HEATER
NO SCALE

FIGURE 3-1

heater have pressure relief valves and drain valves.

Dual Temperature Water Heater: Figure 3-2 depicts a second common problem of water heating in which a two-temperature requirement exists. Commonly in food establishments the dishwasher and possibly one especially designated sink requires water at 160°F. There are several differences between this detail and Fig. 3-1. The heater is set to deliver water at 160°F and because of the snap action of the dishwasher control cycle a water hammer arrester is placed on the line. Further to preclude stratification in the 160°F water line a gravity circulating line with a flow control clock is installed. Cold water is introduced into the tank when the tank's supply of the hot water is already

above, or tending to be above, the 160°F requirement and, in so doing, the cold water makeup called into play by the demand of the dishwasher is, in effect, preheated prior to entering the heater. There are two temperature control devices. Temperature control 1, located near the cold water entry, will sense colder water and start the circulating pump to keep water flowing from the tank to the heater. Temperature control 2 will also start the heater anytime the 160°F line to the dishwasher drops below its operating limit of 160°F plus a small added value for expected line loss of heat. Again, there are relief valves on the tank and the heater.

Hot Water Circulating System: Figure 3-3 is basically

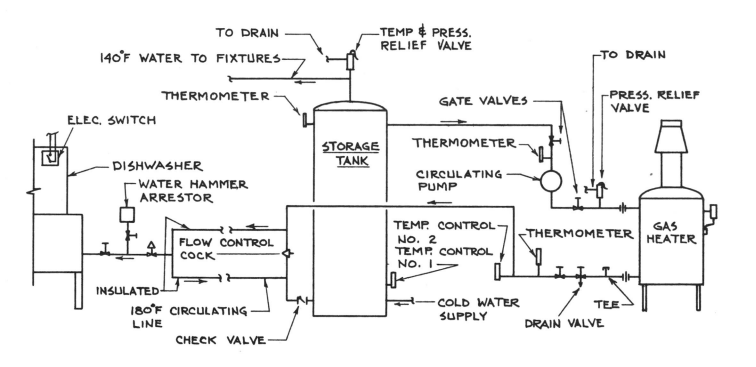

HOT WATER BOOSTER RECOVERY SYSTEM

GAS FIRED HEATER

NO SCALE

FIGURE 3-2

an enlargement and refinement of Fig. 3-1. The enlargement is schematic and the added information would properly apply to any large domestic hot water heating system. Long lines to distant fixtures that are used only occasionally will invariably result in large amounts of lukewarm or cold water being drawn off before normal hot water appears at the top. To guard against this occurrence a circulating (sometimes called recirculating) line is installed with a water temperature controller (aquastat). A partial energy conservation measure would replace or override the aquastat with a time clock which would only allow the pump to run intermittently. Note that the circulating line does not go back to the heater but instead goes back to the storage tank.

This system effectively extends the storage tank to the farthest fixture. Fill system capacity is not required since you are merely trying to avoid line stagnation. Commonly, the circulating line is ¾ in and the pump is also ¾ in. All other devices are similar to those shown in Fig. 3-1. And the size of the storage

tank and heater are the same whether or not the circulating line exists. Making the tank or the heater larger will not in any way solve a problem of stagnation. Only the circulating line can perform that function.

Tempered Water Heating: Instead of the systems depicted in Figs. 3-1 through 3-3 the consultant frequently has problems of dual temperature water for a similar but smaller establishment. Too often, we feel, the problem is treated in too casual a fashion. One of the more common situations is the motel, the school with limited hot food service, or the like that has a need for 160°F water, as noted in Fig. 3-2, plus water for normal fixture use. Since this is a common problem, we show in Fig. 3-4 the detailing required.

The commercial heater shown is most commonly gas fired. It has a recovery rate sufficient to meet both the 160°F water and the normal hot water requirements. It is most likely located somewhere near the kitchen area, but the other use areas are scattered a considerable distance from the heater. To get hot

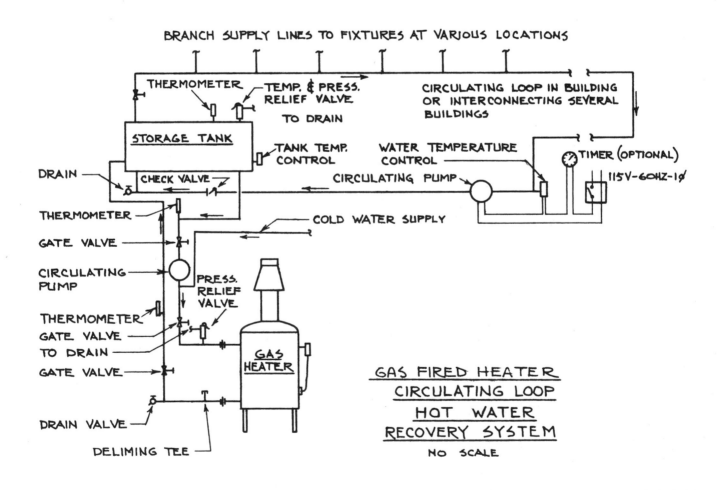

FIGURE 3-3

water to the distant sinks, lavatories, and showers means either that a lot of water is wasted until the water runs warm or there is some kind of circulating system installed on the hot water line. In theory the pump could be omitted as it was on the 160°F water line. But the 160°F using source is, as we previously noted, nearby and that is not true for the lower temperature using sources. Since the recirculation line is small, probably ¾ in or even ½ in, the pump is small and the cost is minimal. This detail can and should be commonly used in small, two-water-temperature situations.

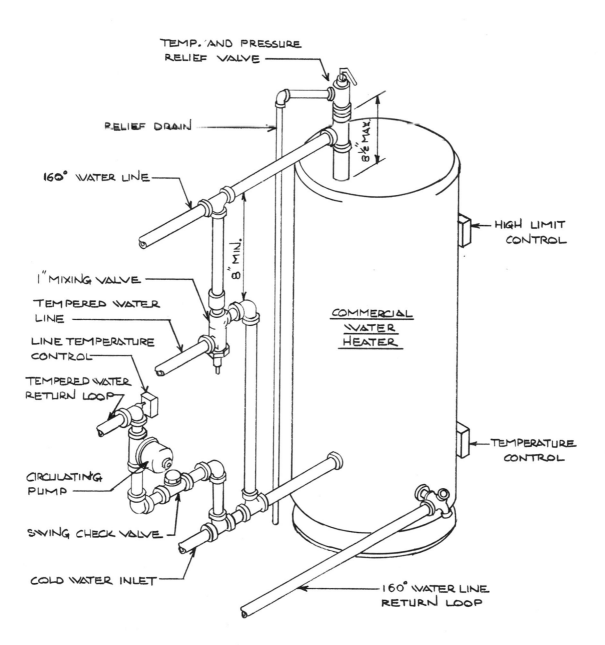

INSTALLATION OF TEMPERED WATER CIRCULATION LINE
— NOT TO SCALE —

FIGURE 3-4

Solar Water Heating

There are a number of items to be noted in Fig. 3-5. First of all the fluid in the collector is not a fluid that you drink or use. Since in the majority of the U.S.A. the solar heater is basically a booster heater, your question might logically be "Why not run cold water through the solar collector, get what heating effect you can get, and then further boost it to the final, desired temperature?" If you do that on cold winter night, you are going to have a frozen, cracked collector and no flow. If on the other hand your collector is installed in an area in which the night temperature does not goes below 32°F, you certainly can do exactly that.

To be safe the collector fluid usually is a nonfreeze glycol solution that goes through a heat exchanger. The heat exchanger can be in a tank or a stand-alone external device depending on what you want to do next. Figure 3-5 shows the heat exchanger in a tank which also has one or more electric booster heaters. There is no rule that requires you to do the same. You could run domestic water into the heat exchanger and out to a gas fired tank heater. If you had a steady demand, you could run domestic water through an external solar heat exchanger and then go directly to any kind of booster. Although it is doing it slowly, the collector is generating a hot water heating effect continuously. You must either store it or use it. Commonly in the northeast in the summer the properly sized collector system is delivering heated domestic water at 180°F.

As we pointed out in general terms in our prior solar heating discussion, the collector at times delivers very little or no heat. Our detail shows a sensor and a thermostatically controlled pump. It may not appear

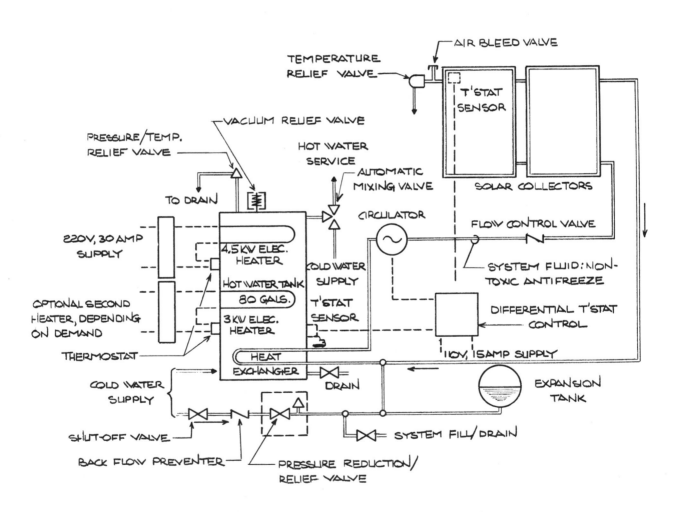

SOLAR HOT WATER HEATER
— NOT TO SCALE —

FIGURE 3-5

logical at first glance, but a collector that is not collecting heat turns out to be a radiator, losing heat. Keep the system going and you will cool, not heat. In that case the circulator stops and the system is shut down.

As in any water heating system the solar glycol fluid expands and contracts; therefore, you need an expansion tank. Also, as in any good system, a properly sized expansion tank and a tight system requires makeup only occasionally.

Collector sizing is available in the *ASHRAE Applications Handbook* and from the collector manufactur-

ers. Some manufacturers also offer free computerized selection program results. In setting the angle for the collector plate the standard rule is 15° plus the latitude of collector location. For example, if your site is at 40° latitude, your plate angle is 15 + 40, or 55°

Heat Exchanger Installation

Figure 3-6 depicts two of the most common methods of getting domestic hot water from a hot water heating system. The boiler itself is not depicted. The basic premise of this system is that the domestic water supply temperature is a fixed value. Inserted in this

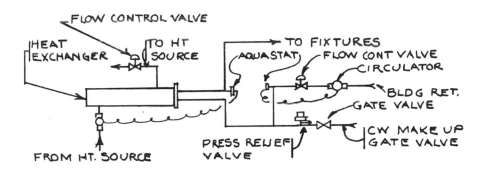

INSTANTANEOUS HEATER PIPING

NO SCALE

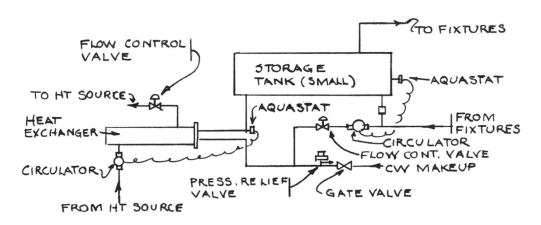

HEAT EXCHANGER WITH STORAGE TANK

NO SCALE

DOMESTIC HOT WATER FROM HOT WATER SOURCE

FIGURE 3-6

supply line at the converter is an aquastat that, as the discharge temperature drops, calls for the line between the boiler and the heat exchanger to circulate 180 to 200°F water from the boiler to the converter. The lines to and from the boiler normally come from tappings on the boiler. These lines could be tapped, one might reason, from the main boiler supply and return headers. This could be done only if at the connecting points the source of supply and return at all times reflects a standard temperature. Frequently, this is not the case, and the heat exchanger performance becomes erratic and below design expectations. Whether the heat exchanger receives 80 or 400°F water or some value in between, the connection to source principal remains the same.

The upper detail in Fig. 3-6 represents the most popular and frequently used source of water heating. It is used both in steam and hot water source systems. When the source system is steam, the connection of the heat exchanger supply and return must be below the boiler water line. The lower detail in Fig. 3-6 is normally used for systems that have heavy rates of demand that periodically exceed the available boiler water heating capability. On demand cold water does not directly enter the tank but rather goes directly to the heat exchanger. Usually the storage tank is sized to handle the difference between peak demand and average demand produced by the heater. In both cases the circulating pump can be used to control the building returns and to control the tank that is also shown in the storage tank installation.

Figure 3-7 is similar in many respects to the upper detail in Fig. 3-6. The basic difference is that the source of energy is steam from a boiler or a steam system. The premise of control is similar except that the aquastat on the domestic supply line controls a motorized steam valve. Again, regardless of exactly where the steam supply source is connected, the supply must be constant and available year round. The steam source normally is varied in quantity by the modulating effect of the steam valve. This modulation is done by a pressure sensing line, as depicted in the detail. And a relief valve is applied as depicted. In the piping of the steam there is an obvious requirement of a trap with a strainer on the heat exchanger. To insure that no condensate builds up in the steam line to the motorized valve a tee connection is commonly employed. It is described as a drip tee. The end of the

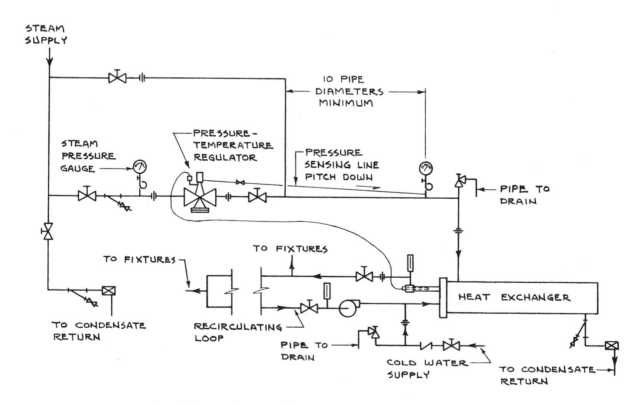

DOMESTIC HOT WATER FROM STEAM

FIGURE 3-7

steam main is dripped (trapped) and any condensate buildup is carefully returned to the condensate system.

Figure 3-7 is basically the same as the lower part of Fig. 3-6. Because the domestic water piping portion is the same as before, we have omitted that portion to concentrate on the steam supply detailing. Commonly, some large-demand water installations, especially the higher temperature source system, and most heat exchangers and storage tank installations that use steam as a source employ the combination storage tank and heater installation shown in Fig. 3-8.

The heat exchanger in this situation is the reverse of the one in our opening discussion on the residual tankless heater. In the residential tankless heater the domestic water circulated in the boiler through the heater piping and absorbed heat from the surrounding hot boiler water. In the storage tank and heater the source steam circulates through the heat exchanger piping and heats the water in the storage tank. The operating control is an aquastat in the tank measuring the tank temperature and opening the steam valve as required. The trapping of condensate at the exchanger and at the end of the steam line is as previously described. There is also the manual bypass around the automatic steam valve which, in both the previously described case and the one depicted in Fig. 3-8, provides for manual, emergency operation in the event of control failure. Note also that there are pressure relief valves on the steam line and on the tank itself.

Heat exchangers, sometimes called converters, can be, and frequently are, used for pure heating, snow melting, and the like. We did not show this sort of piping because we felt it was redundant. We have, instead, concentrated on the source piping to the exchanger. The heat exchanger can, in a high temperature hot water installation, be also used as a reverse type of converter. The 400°F water can be used to turn the exchanger into a medium or low pressure steam source. In any of these situations in which the converter is used as a source for heating purposes the piping from the converter is the same as if the heat exchanger was a boiler.

Process Heating: In many types of industrial processes a form of a heat exchanger, frequently custom-built, is used to maintain process liquid temperature for a specific manufacturing purpose. The variety of this sort of application is endless. Frequently, the temperature requirements are much higher than in nor-

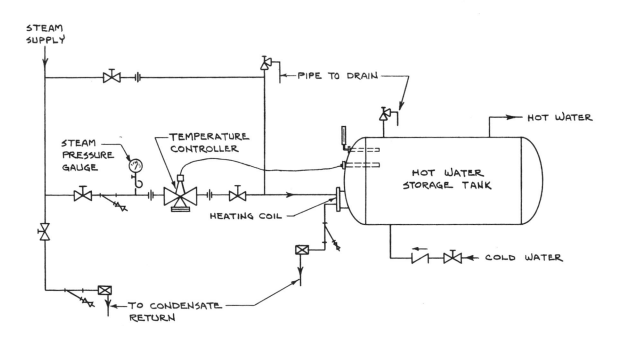

STEAM HEAT TEMPERATURE
CONTROL of HOT WATER HEATER

FIGURE 3-8

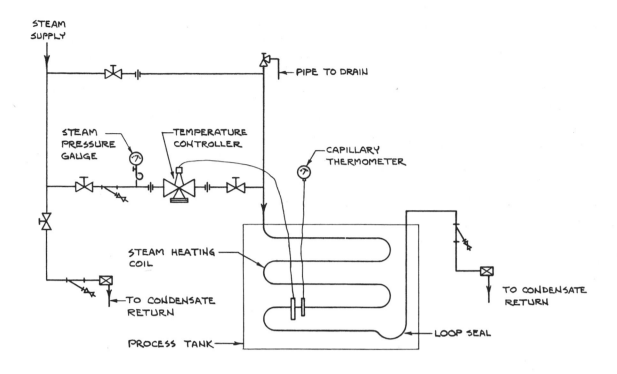

STEAM HEAT TEMPERATURE
CONTROL OF PROCESS TANK

FIGURE 3-9

mal water heating, exceeding even dishwasher temperature requirements. The usual range of required values is 170 to 200°F in the process liquid. Figure 3-9 is a typical example. At first glance it may seem to be simple or partial in its detailing, but that is not the case. Whether the exchanger is used in a closed tank and heater arrangement or an open tank, as depicted, the source is normally low pressure steam, and the steam piping to the exchanger is the same as shown in Fig. 3-8. Generally, a loop seal in the pipe is employed and a separate capillary thermometer is installed to double-check control settings. Otherwise the detail should be totally familiar to you if you compare it to those in our previous discussion on storage tanks and heaters.

FOUR

Refrigeration and Air Conditioning Source Equipment

Although the source for air conditioning is slightly different in arrangement from the source for a lower temperature process or product, the detailing is relatively similar. As in other types of source equipment, today's trend, especially in comfort cooling, is toward the packaged system. Given appropriate roof space the packaged system is hard to beat and detailing is minimal. But there sometimes are drawbacks when it is used on existing roofs.

In a typical installation the rooftop package requires cutting through an old roof. The old roof is often in marginal condition and the weight of the system plus less than perfect cutting, curb, and sealing of ductwork can create a roof leak directly under the unit. Roof framing is a very important item and should be carefully checked *before* you specify any cooling source equipment to be installed on the roof. It is very possible, if not very likely, that you will need an additional roof drain tied to the roof drainage system in the vicinity of your installation. The best detail ever drawn of connections to your roof top unit will not solve roof deflection.

Obviously the total all-in-one-box packaged roof system needs little or no detailing. Our details cover situations in which the parts are separated. We also cover the chiller that has a remote tower, and for low temperature work we cover a skating rink system. Cooling tower details are covered in Chap. 5.

Compressors and Condensers

Air Cooled Condenser: Figure 4-1 is a schematic of the simplest case in which the air conditioning system only operates when the temperature outside is 40°F or higher. A shutter control is all that is needed to generally preclude starting and head pressure problems. Assuming that the components are fairly close together, the condenser acts as the liquid receiver. If necessary an auxiliary receiver could be piped between the condenser's subcooler and the solenoid valves.

Figure 4-2 is a schematic that permits operation of a system below 40°F. The receiver is now a mandatory item that contains the refrigerant used to fill the condenser during lower than 40°F operation. It must be large enough to hold its normal operating charge when the condenser is flooded at its minimum outdoor operating temperature and, to insure condenser subcooling at design conditions during the summer when the receiver runs at full capacity. The check and relief valve trap liquid ahead of the expansion valve to eliminate compressor cycling and to preclude bursting liquid lines that could be caused by expansion in hot weather shutdown.

Compressor Parallel Piping: Refrigerant compressors have certain sensitivities. Paralleling compressors is not recommended if it can possibly be avoided. System pressure equalization is a major problem. Fig. 4-3 illustrates an acceptable piping arrangement if you really have no alternative. Usually the problem involves water cooled condensers with remote cooling towers and therefore our detail is for a water cooled condenser situation. First the oil level in both compressors is equalized. If the machines are different sizes, the height of one of the machine bases must be adjusted so the normal oil level in both machines is the same. If you do not do this, nothing but trouble will occur. Oil will accumulate in the lower level compressor. This level line must not at any point be above the selected level, and the line itself should be level. This

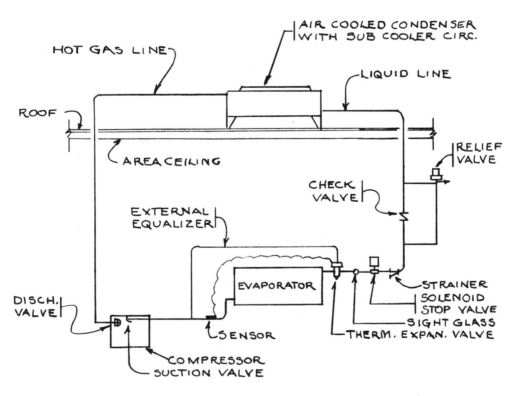

NOTE: NO HEAD PRESSURE CONTROL

AIR COOLED CONDENSER PIPING

NO SCALE

FIGURE 4-1

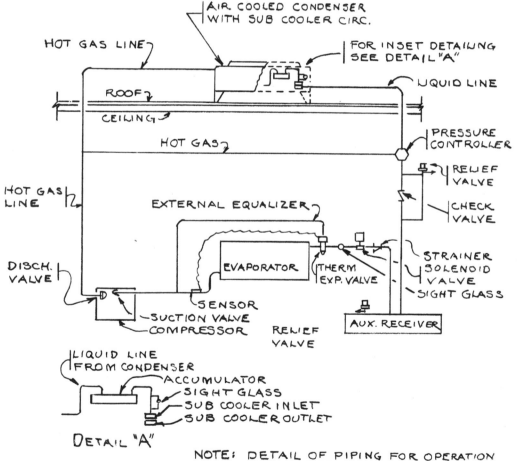

AIR COOLED CONDENSER
WITH SUB COOLER CIRC.

HOT GAS LINE

FOR INSET DETAILING
SEE DETAIL "A"

ROOF

LIQUID LINE

CEILING

HOT GAS

PRESSURE CONTROLLER

RELIEF VALVE

HOT GAS LINE

EXTERNAL EQUALIZER

CHECK VALVE

EVAPORATOR

THERM EXP. VALVE

STRAINER
SOLENOID VALVE
SIGHT GLASS

DISCH. VALVE

SENSOR

SUCTION VALVE

COMPRESSOR

RELIEF VALVE

AUX. RECEIVER

LIQUID LINE FROM CONDENSER

ACCUMULATOR

SIGHT GLASS

SUB COOLER INLET

SUB COOLER OUTLET

DETAIL "A"

NOTE: DETAIL OF PIPING FOR OPERATION
BELOW 70°F AMBIENT

AIR COOLED LOW AMBIENT CONDENSER

NO SCALE

FIGURE 4-2

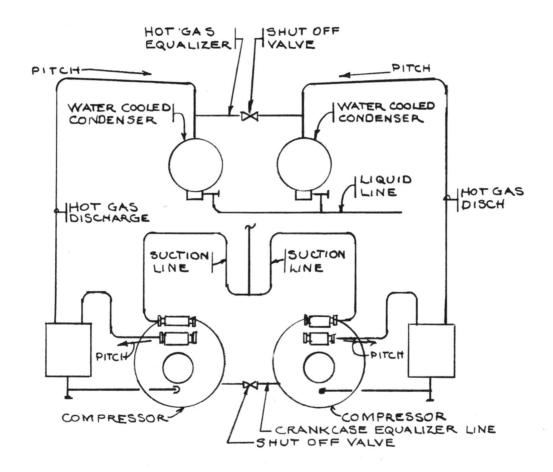

HOT GAS EQUALIZER   SHUT OFF VALVE

PITCH →

WATER COOLED CONDENSER

WATER COOLED CONDENSER

← PITCH

LIQUID LINE

HOT GAS DISCHARGE

HOT GAS DISCH

SUCTION LINE

SUCTION LINE

PITCH →

← PITCH

COMPRESSOR

COMPRESSOR

CRANKCASE EQUALIZER LINE

SHUT OFF VALVE

PARALLEL PIPNG FOR COMPRESSORS
WITH SEPARATE CONDENSERS

NO SCALE

FIGURE 4-3

line is also one place in which there is no danger of making it too big. Note that the discharge lines are also equalized (full size) before they enter the condensers. Finally, note the trapped arrangement in the equalization of the suction return line, which must feed both machines.

The problem can be further compounded if there is only one condenser. Figure 4-4 illustrates the piping in that situation. This system will only work when all of the compressors operate at the same suction pressure. If this is not true, separate condensers are required. The system in general resembles the one shown in Fig. 4-3 except for an elongated suction loop for a condition in which the evaporated return requires one. The suction connection shown on Fig. 4-3 could also apply.

The compressor piping details seem to be filled with more "don'ts" than "dos." Since generally the detailing of compressor refrigerant piping is both limited

and fairly obvious, we felt you would be best served by illustrating the potential problem situations.

Hot Gas Bypass Piping: In Fig. 4-5 we cover the frequently misunderstood subject of hot gas bypass. Again, this could be all furnished in a packaged air cooled chiller and none of the details would be required. We could have shown the hot gas bypass that is used on a chiller in Chap. 4, which covered chillers, but it is a compressor capacity control device and we felt it more logically belonged here.

In Fig. 4-5 when the chilled water load drops toward the freeze point, the hot gas bypass valve opens, allowing gas from the compressor discharge to enter the refrigerant low temperature side between the expansion valve and the chiller, thus mixing discharge gas and evaporator liquid. This raises the superheat in the suction line. The higher temperature actuates the expansion valve through its sensor, which registers

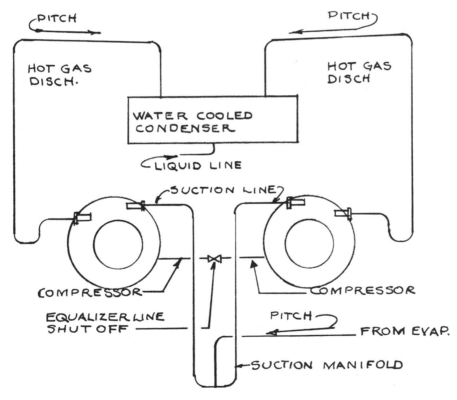

PARALLEL COMPRESSORS— SINGLE CONDENSER

NO SCALE

FIGURE 4-4

this change in temperature. This prevents the compressor from cycling off on the low pressure switch. When the chilled water load rises, the suction pressure will also rise. Sensing this rise in suction pressure, the hot gas bypass will start to close and slowly continue until fully closed. Meanwhile in this cycle of events the water temperature out of the chiller will remain relatively steady.

Chiller Installation

One of the most popular refrigeration systems used in air conditioning applications is the chilled water supply system. The calculations necessary to size chilled water piping are similar to those for hot water heating. If a common 10°F temperature differential is used, the calculation becomes even simpler. And finally if the unit is air cooled and either mounted on grade or on some properly braced and supported roof, the design detailing is further simplified.

Figures 4-6 and 4-7 represent the detailing of a reciprocating water cooled chiller. If the chiller was a larger, centrifugal-type chiller, the piping variations would not be significantly different. The chiller, as detailed, is powered by an electric motor. If the chiller was powered by a steam turbine or was a steam absorption system, the steam piping detailing would be similar to that which we have previously depicted for a turbine powered boiler feed pump in Chap. 2.

In Figs. 4-6 and 4-7 the total detailing could be described as "overkill." Generally, the chillers are very well-balanced and, as such, do not create much in the way of floor vibration. As a consequence the vast majority of floor-mounted chillers do not require a vibration dampening pad and the spring mounting of the pad as detailed. But when they are located in areas in which no vibration transmission can be tolerated, the vibration elimination feature of the pad is required. In this case it should be clearly understood that the enlarged vibration eliminator detail shown in Fig. 4-7 is *typical*; the actual detail may vary and must be calculated and sized by a vibration expert.

There are certain basic caveats which may be added to these details or noted in your specifications. The chiller must be mounted level in both directions. Proper clearances for service and for pulling condenser and chiller piping must be provided. Vibration transmission is an item that must be checked in the start-up procedure. The length of the flexible connec-

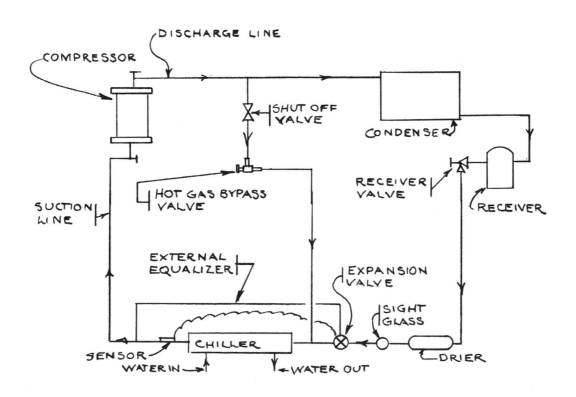

HOT GAS BYPASS CAPACITY CONTROL

NO SCALE

FIGURE 4-5

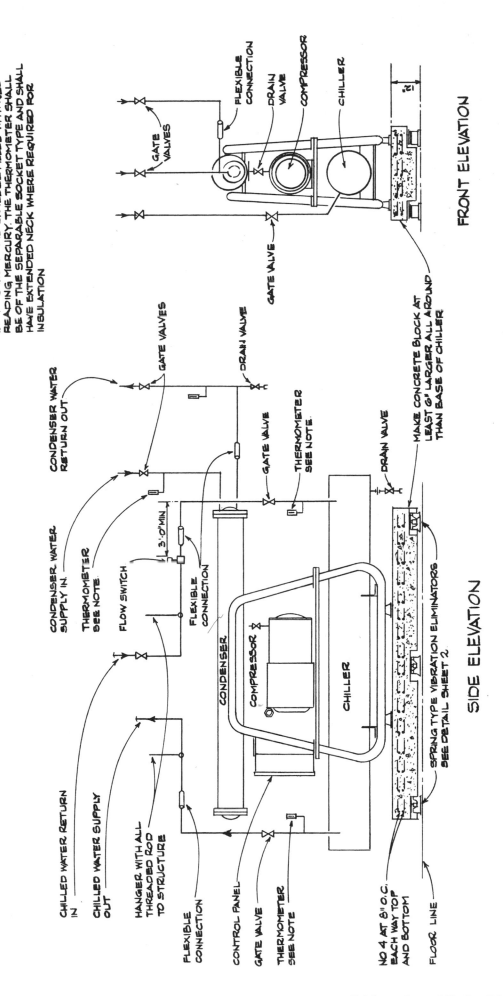

NOTE

THERMOMETER SHALL BE CAST BRASS 9" SCALE, 20°F TO 120°F AND SHALL BE FILLED WITH RED READING MERCURY. THE THERMOMETER SHALL BE OF THE SEPARABLE SOCKET TYPE AND SHALL HAVE EXTENDED NECK WHERE REQUIRED FOR INSULATION

FLEXIBLE CONNECTION

DRAIN VALVE

COMPRESSOR

CHILLER

GATE VALVES

GATE VALVE

FRONT ELEVATION

GATE VALVES

DRAIN VALVE

CONDENSER WATER RETURN OUT

GATE VALVE

THERMOMETER SEE NOTE.

CONDENSER WATER SUPPLY IN.

THERMOMETER SEE NOTE.

FLOW SWITCH

3'-0" MIN.

FLEXIBLE CONNECTION

DRAIN VALVE

MAKE CONCRETE BLOCK AT LEAST 6" LARGER ALL AROUND THAN BASE OF CHILLER

CONDENSER

COMPRESSOR

CHILLER

SPRING TYPE VIBRATION ELIMINATORS SEE DETAIL SHEET 2.

CHILLED WATER RETURN IN

CHILLED WATER SUPPLY OUT

HANGER WITH ALL THREADED ROD TO STRUCTURE

FLEXIBLE CONNECTION

CONTROL PANEL

GATE VALVE

THERMOMETER SEE NOTE

NO. 4 AT 8" O.C. EACH WAY TOP AND BOTTOM

FLOOR LINE

SIDE ELEVATION

RECIPROCATING CHILLER INSTALLATION DETAILS

SHEET 1 OF 2

FIGURE 4-6

Refrigeration and Air Conditioning Source Equipment   97

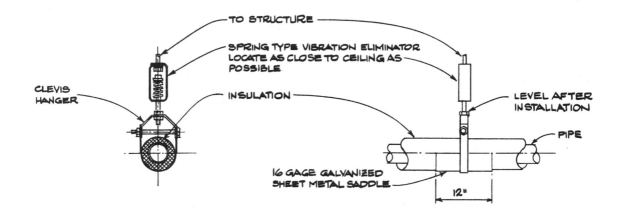

CHILLED WATER PIPE HANGERS

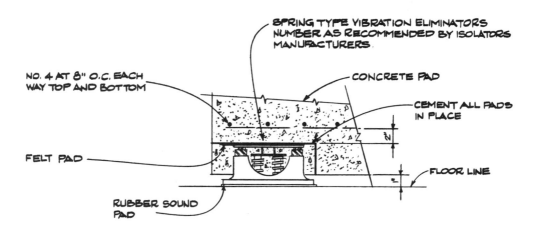

VIBRATION ELIMINATOR DETAIL

RECIPROCATING CHILLER INSTALLATION DETAILS
SHEET 2 OF 2

FIGURE 4-7

tions should be calculated and should normally be at least 24-in flexible connections. All pipes connected to the chiller should be suspended using vibration elimination type pipe hangers. In the usual mechanical room chiller location all pipes passing through walls, ceilings, and floors should do so through pipe sleeves. The space between the pipe and the sleeve should be packed with oakum.

In Fig. 4-6 note that in addition to the obvious locations of gate valves there are thermometers on the supply and return of both the chilled water and the condenser water. This provides a way to measure both chiller performance and the load imposed. Note also that all of the flexible connections are installed horizontally, not vertically. The vertical installation of flexible connections creates a flexible condition that is so limited that it is nearly useless.

Figure 4-7 depicts enlarged details of the vibration mounts and the vibration elimination pipe hangers. Experts in these areas can select and arrange other ways to eliminate vibration besides the springs depicted. However, the springs are the most common device used in this sort of application. Normally the condenser water piping is hung with the same type of hanger as is depicted for the chilled water. In this instance the usual practice is to also insert a strip of felt or rubber between the clevis hanger and the pipe.

The basic chiller control panel is normally part of the packaged machinery. The chiller starter may be part of the control package or for larger machines may be separately located. Most importantly the starter must include a thermal overload protective device for each phase line.

Low Temperature Refrigeration

A common application of low temperature refrigeration is ice making for indoor skating. Ice skating rinks can be indoors or outdoors and are usually operated under a variety of conditions and for a variety of purposes. The normal uses are for amateur and professional sports activities, such as hockey, figure skating, and speed skating, and for recreational skating. Because of the many variables surrounding the ice sheet, as well as the temperature mandated for the ice itself, the calculation for refrigeration capacity, which does not belong in this book of details, is an inexact science at best. The *ASHRAE Applications Handbook* is an excellent starting point for such a design engineering effort. A rough rule of thumb is that it requires 1 ton of refrigeration per 100 to 200 sq ft of ice area. Ice surface temperatures generally start at 22°F for hockey, 26°F for figure skating, and 28°F for recreational skating. Today the primary refrigerant is commonly R-22. Occasionally R-12 and R-502 are used. While the selection of R-22 in a direct expansion system is increasing and direct expansion ammonia has been used, the use of direct expansion ammonia is prohibit-

ed in public buildings by the U.S. ANSI Code B9.1, as well as by the Canadian B52 Mechanical Refrigeration Code. By direct expansion we mean the use of the refrigeration gas directly in the rink piping.

Figures 4-8 and 4-9 comprise the overall piping detail of a system used typically in multiuse public auditoriums in which it may be necessary to melt the rink ice rapidly to make way for a portable basketball floor that is required for the next event.

Figure 4-8 is noted as sheet 1 of 2, and it covers the compressor-condenser, heat exchanger, deice selector valve, brine tank, brine circulating pumps, and city water makeup. In order to show the interrelated piping connections of these items the detail, of necessity, does not show all the piping specialties that may be required or such items as contamination and back pressure precautions that may be required on the cold water makeup. The detail describes the compressors as brine chillers. They are water cooled ammonia units and are sized so that one compressor can maintain the ice sheet under normal conditions. Not shown but included in the specification and control requirements is a sequence controller to alternate the starting of the compressors.

For rapid removal of the ice sheet a low pressure steam-to-water converter was installed. The basic engineering design is based on the heat energy required to turn the total volume of 26°F ice into 40°F water. Selection of the ice or deice operation is controlled by both manual gate valves and automatic selector valves. The brine, as noted on the plans, is a 40 percent minimum glycol solution. The excess system brine is stored in an atmospheric tank (note the mesh screen) which also serves as the point of makeup of city water. If the city water pressure is high, a pressure reducing valve is required on this makeup line. Pumps are duplex with each point mated, and they have the same normal operating requirements as they would for compression.

Figure 4-9, described as sheet 2 of 2, covers the cooling towers, sump, pumps, and heat tracing. It might be well to clear up a seeming ambiguity before we begin our discussion. The terms "supply" and "return" may seem odd at first glance, but these two details were compressed from a very large overall detail. Return here means cooling water from compressor to tower. The opposite of that is supply, which is tower water returned to compressor.

The system described by Figs. 4-8 and 4-9 is a year-round operation. Therefore, the cooling tower must be protected in the winter. This is done by heat tracing the line from the tower to the cooling tower sump and by providing an auxiliary steam-fed tower basin water heater. These devices are shown separately, but they are sometimes combined into one prepackaged device by the tower manufacturer. When energy conservation is to be considered, the substitution of an energy-use application for the cooling tower, such as area heating backup is a common application of conservation technology.

Rink Pipe Installation: Rink piping, as depicted in three small partial details shown in Fig. 4-10, is both simple and to some degree complex. Generally, coil pipe sizes are ¾ in, 1 in, or 1¼ in on 4-in centers. There are other proprietary systems that use smaller pipes on closer center line dimensions. The pipe material is usually mild steel or ultra high molecular weight plastic. It is vital to install the piping so that it is level. Notched iron or welded chair supports are normally used when the pipe is supported above a concrete floor. There are a variety of possible floor, insulation, pipe, and ice arrangements. In Fig. 4-10 we depict a cross section of a fairly common all-purpose floor. This is a floor slab on grade that is insulated and

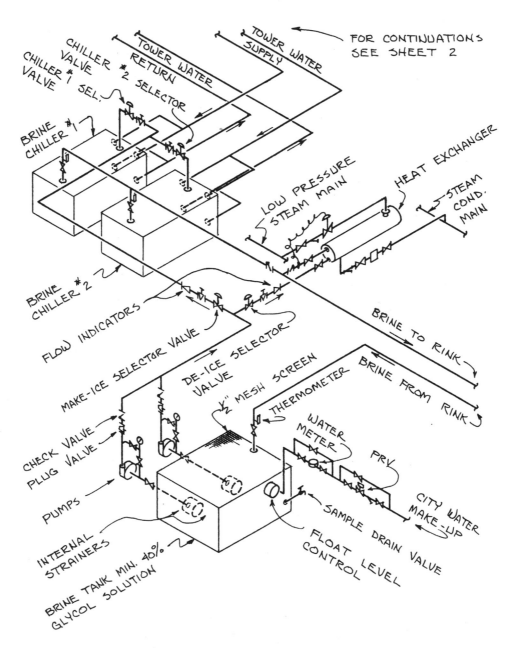

ICE SKATING/HOCKEY RINK BRINE COOLING SYSTEM

NO SCALE

SHEET 1 OF 2

FIGURE 4-8

has piping or tubing buried in concrete. This type of floor is usually designed to steel loading standards and the pipe, as depicted, is normally 1 in or 1¼ in. Since we assume that the floor frequently goes from hot to cold, the refrigeration system is normally designed to freeze a sheet of ice ⅝ in thick in 12 hours, and the insulation shown is mandatory.

In hot water heating system design, one of the best balanced systems is a two-pipe reverse return installation. This is also an ideal rink coil piping arrangement. This is depicted in one portion of Fig. 4-10. However, it is more expensive, and the more common arrangement is the two-header system depicted in the final part of Fig. 4-10. To allow for contraction and expansion the free movement of headers should be arranged in the overall design plan. Generally, circu-

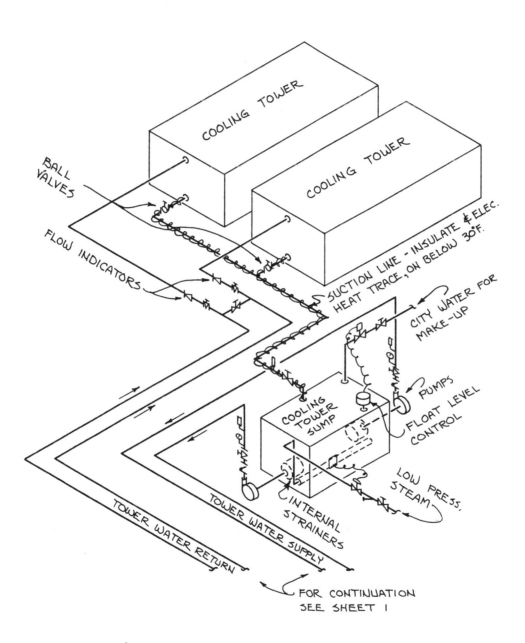

ICE SKATING/HOCKEY RINK BRINE COOLING SYSTEM

NO SCALE

SHEET 2 OF 2

FIGURE 4-9

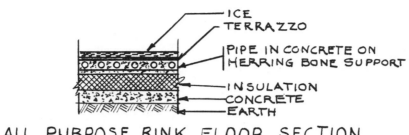

ALL PURPOSE RINK FLOOR SECTION

NO SCALE

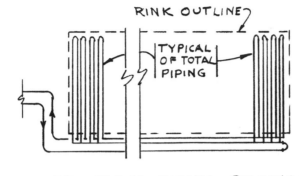

BALANCED FLOW PIPING DIAGRAM

NO SCALE

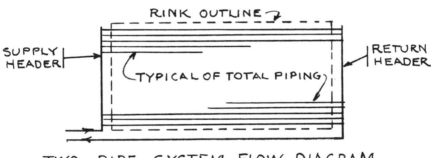

TWO PIPE SYSTEM FLOW DIAGRAM

NO SCALE

SKATING RINK PIPING

FIGURE 4-10

lating brine pumps are designed for 10 to 15 gpm per ton and operate at 20 to 25 psig. Steel piping is welded and plastic piping must use special fittings designed for the temperatures involved. Headers are generally located in specially designed trenches.

Not shown in our details is the customary snow melting pit provided at one end of the rink; it is usually detailed on the structural plans. This pit handles the scraped off rink "snow" and ice. Hot water from the heat exchanger depicted in Fig. 4-8 or from a waste heat recovery system or from some separate source must be provided to melt the snow and ice. Your plumbing plans should require a drain with a screen filter and overflow arrangement for this feature.

Refrigeration and Air Conditioning Support Equipment

In our discussion in this chapter we focus on items that aid in removing the heat created by the refrigeration process. We faced two dilemmas over what to include, and we handled each differently. Certainly the air cooled condenser belongs in this category, but it is not here. The whole subject of refrigerant piping details is a specialty of its own and therefore we include the air cooled condenser and its related refrigerant piping details in Chap. 10.

The air washer as used today might logically have also qualified as an item for Chap. 10, but long before the successful design, development, and fabrication of today's standard finned coil systems, the air washer was the common method of air conditioning. Therefore it is included at the end of this chapter. This rather simple device consisted of a series of spray nozzles which, under pressure, created a water vapor through which the air was drawn. The common descendant of this system is the well-known coil humidifier.

There are two basic types of cooling towers: mechanical draft and natural draft. The mechanical draft tower uses a motor driven fan to draw or push air through the tower. In the natural draft tower air movement depends on atmospheric conditions.

In the usual mechanical draft tower the water to be cooled enters the tower's distribution header at the top, flowing out of holes in the rear piping. The falling water hits a number of splash bars which break the liquid into small droplets. Air is usually pulled upward against this water droplet flow and the resulting interchange heats the air and cools the water. Commonly the falling water collects in a basin from which it is pumped for reuse in the condensing system.

In the natural draft cooling tower the cycle is very similar to the mechanical draft cycle described above except there is no fan. The water enters the top of the tower distribution header as before and descends again in the same pattern. Cooling is accomplished by natural breezes or, in the case of a very large tower, by the air movement created by the slight pressure differential between the top and bottom of the tower.

In the past, when water supply and resource conservation were seldom considered as related subjects, there were a number of air conditioning comfort applications in which spring or well water which could be obtained at temperatures of 50°F and below were used directly in the water coils of the system as the cooling source. This source was also used in certain types of air washer systems. This is generally not true today and is not the subject of this short chapter.

Everyone must be familiar with the various environmental cases involving the power companies' use of river, stream, or lake water in their cooling plants. Generally, this is an argument that revolves around the vast amount of water used, the temperature differential, the resultant rise in stream temperature, and the possible adverse effect this has on marine life and the ecology of the stream. This, too, is not part of this chapter.

There are, however, certain situations in which the use of colder than usual stream water presents an opportunity to eliminate the cooling tower in the heat rejection system of the usual water cooled chiller installation. Therefore, we feel this book would not be complete if it did not at least address the subject and describe how such a system might be detailed.

Cooling Towers

Cooling Tower Piping: As detailers we are, as always, concerned not with the devices themselves but with their related connections to a system. Figure 5-1 is typical. Not all of the specialty piping on the tower connection is shown. The detail implies that the project is for a small commercial application and that the condensing water circulating pump is inside, not outside. This is a somewhat typical 10-ton to 50-ton tower and condenser arrangement. The tower has a makeup water line from the nearest available, properly sized cold water source. This pipe is usually 3/4 in. The most important part of the system is the three-way valve which is controlled by a pump discharge line aquastat and set to maintain 85°F to the condenser. Most systems are designed to utilize a 10°F or 20°F temperature rise through the condenser and the standard leaving water condition is 105°F. The plugged tees are normally used for thermometers. The globe valve is the pump flow control device. Without the three-way valve there is usually an unacceptable unbalance of water temperature in the system in the spring and fall.

Condenser Piping: As your system grows larger, it is usual to have a single larger tower that handles two, three, or more large condensing units. For many design reasons there is a choice of whether the associated pumping and control devices are located indoors or outdoors with the tower. These items should not be duplicated. They are either indoors or outdoors but not in both locations simultaneously. Figure 5-2 is the typical arrangement of a tower located outside and above the condensers or on the same level. Not shown is the tower whose connections are similar to the detail of Fig. 5-1. Here we have shown all the specialties for an indoor pumping setup. There is a common supply, return, and bypass line, and there is a common pump.

The three-way valve maintains 85°F in the pump discharge regardless of flow. Globe valves are provid-

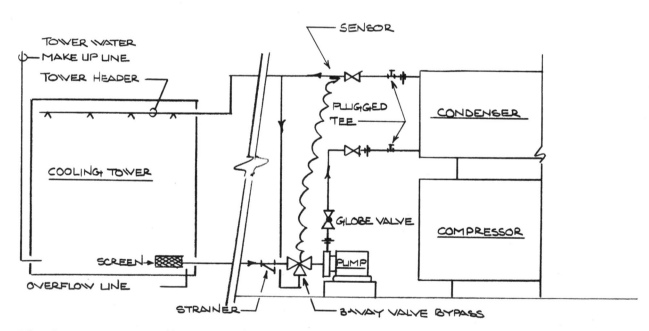

TYPICAL COOLING TOWER PIPING
— NOT TO SCALE —

FIGURE 5-1

ed to adjust the flow through each condenser. The condensers are manually connected and disconnected from the system by gate valves. The supply and return headers must be sized for full flow of all units, as are the pump and tower. The tower can be drained through a drain valve in the suction riser. The system and each condenser can also be drained through individual drain valves. All piping contains appropriate unions and strainers.

Condenser-to-Tower Arrangements: As previously mentioned, the tower and its associated pumping and controls could also be outdoors. In Figs. 5-3 and 5-4 we show versions of this sort of arrangement, one in which the condenser is below the tower and one in

which it is above the tower. In both drawings the pipe at one point continues with arrows and the words "to condenser." If you compare Figs. 5-3 and 5-4 with Fig. 5-2, you will see that the missing piping is the gate and globe valve connections at the condenser—with or without an appropriate header, depending on whether you have a single or a multiple condenser installation.

There are in both figures complete tower piping detailing not previously shown. In Fig. 5-3 in which the tower is above the condenser, first note the float controlled makeup water line. Commonly a bleedoff is specified. This is located and detailed. In the collector pan we not only have the usual suction connection but also an overflow connection with a gate valve drain

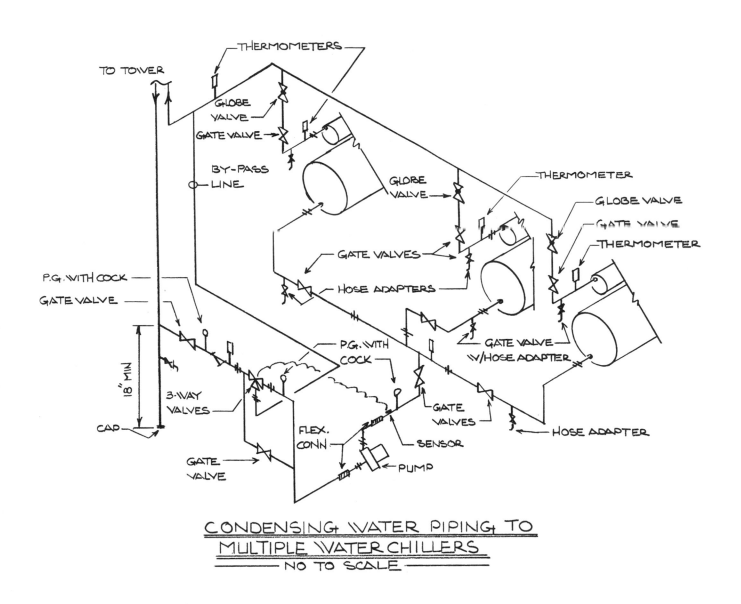

CONDENSING WATER PIPING TO
MULTIPLE WATER CHILLERS
— NO TO SCALE —

FIGURE 5-2

bypass connection. The three-way valve controls the discharge temperature.

Figure 5-4 shows a situation in which the tower is below the condenser. All things, at first glance, seem similar to those in Fig. 5-3. That is generally true; however, there are certain special requirements. The condensing water to the tower discharge must have a positive water seal of a minimum height that is equal to that of the cooling tower, with bleeding valve bypass takeoff ahead of the seal to prevent the drawing of air into the tower. Note the special bypass line connection detail and the connection of the bleedoff valve. Further, the tower basin must have the capacity to hold all the water from the supply line without overflowing. Generally, when the total length of this line is more than 125 ft, other methods must be incorporated to preclude overflow.

Tower with Indoor Storage Tank: As the need for year-round tower operation grows, so also grows the problem of freezing tower heads, especially during

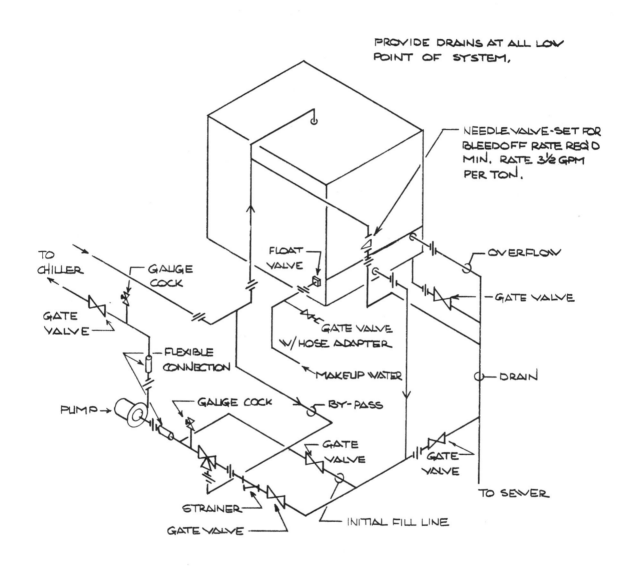

COOLING TOWER PIPING DIAGRAM
CONDENSER BELOW TOWER
—NOT TO SCALE—

FIGURE 5-3

cold weather shutdown. For smaller installations the use of an indoor storage tank is the most economical solution. Figure 5-5 is a typical schematic of this sort of detail. The pump is located indoors and takes its suction from an indoor tank. It pumps through the condenser and up to the tower. The water then flows through the tower and drains by gravity to the indoor tank.

The tank is normally the open type with capacity sufficient to contain the total system. A rule of thumb is that the tank capacity is three to five times the number of gallons circulated per minute by the pump. The tank includes suction, overflow makeup, and drain-cleanout connections. Not that all the specialties required are detailed and for these you should refer to Figs. 5-3 and 5-4. The cooling tower piping consists

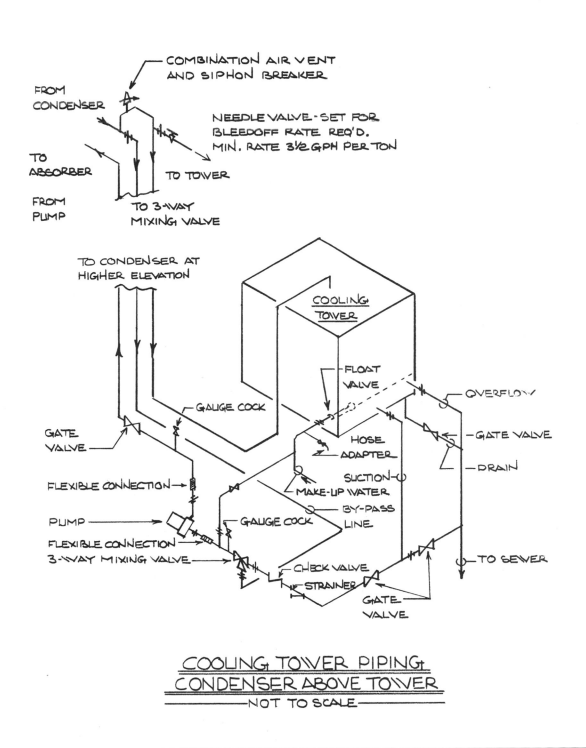

FIGURE 5-4

only of the supply and return connections. When the tower height is more than 10 ft above the pump, a discharge check valve to prevent backflow must be installed.

In this detail, as in all the others, a bleed valve is shown since every tower system should waste a small percentage of water, usually 3½ gph per ton refrigeration to preclude the buildup of dissolved solid. Normally, this valve is at the highest point inside the building. To preclude tower icing and to better control the system, a thermostat cycle that is controlled by a sensing element in the tank is usually provided to cycle the fan.

For large systems indoor storage is normally not feasible. Generally, the tower specification will include an electric immersion basin heater sized to maintain a minimum basin temperature. In addition the exterior lines are usually heat traced in the way we describe under the heat tracing section in Chap. 9.

Tower Bypass Piping: In Fig. 5-6 we show a refinement of the bypass we have depicted as a solution to the cold weather problem. This involves the use of a second bypass line and an instantaneous heater. Through a manual valve bypass on the return line to the tower some of the water is allowed to go directly to the tower basin without going through the normal

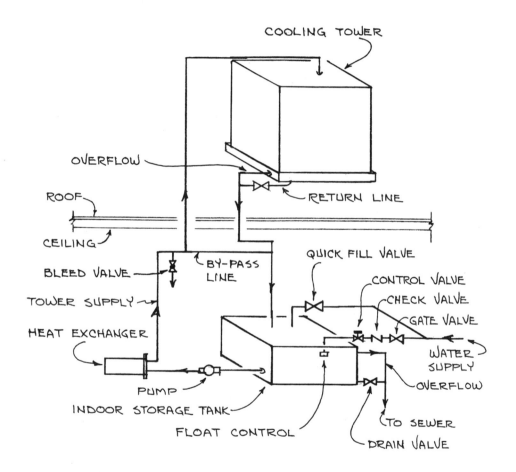

COOLING TOWER WITH INDOOR STORAGE TANK
NO SCALE

FIGURE 5-5

cooling cycle. In the line from the tower this hot water bypasses the condenser. It has its own circulation which passes water through the shell of a steam or hot supplied heat exchanger. This heated and bypassed water system is controlled by a motorized valve on its supply and an aquastat in the tower return line. Normally this system is manually opened in the fall, and the controls are activated by a manual start-stop switch. In the spring it is manually stopped; the two supply and return bypass gate valves are also manually opened and closed when the pump is started or stopped.

In this case the only heat tracing required is for the cold water makeup. There are considerable calculations required to size this system and not all the connections to the specialities are depicted.

Stream Cooling

Condenser Application of Stream Cooling: Figure 5-7 represents the typical arrangement for using a cold water source to cool the condenser of any chiller. Since we know that the cooling tower supplies water at 85°F and that this is totally satisfactory to any chiller provided that the proper quantity of water is available,

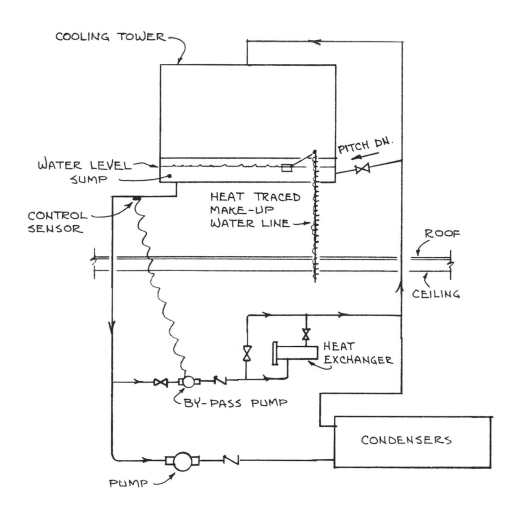

COOLING TOWER BY-PASS CIRCULATION

NO SCALE

FIGURE 5-6

and further since we know that a supply of water at this temperature on the order of 3 gpm/t is adequate, it follows that literally any stream will do. The problem is much more complex than that, however, in terms of the quality and quantity of the water available at all times.

In our detail the water enters the chiller at 85°F and leaves at 95°F. Secondly, the assumption is that before the water gets to the piping shown in our detail it is sufficiently clean to preclude fouling and corrosion problems. Finally, it assumes that the system is environmentally acceptable. Generally, the water as supplied is fairly cool and in the neighborhood of 55 to 60°F. As such it is far below the temperature needed by the water cooled condenser.

The water supply may be of a variable quantity. As such we show it being contained in a storage tank. The storage tank has several purposes. First, as a storage device it has usually twice the makeup rate capacity required to fill the cooling system. Second, the maintained level in the tank assures constant pressure on all parts of the system. Third, the stream may have a considerable static head. The tank assures that the pressure of the three-way valve remains low and does not exceed the pressure close-off rating of the valves. Excess water in the tank or system, as detailed, overflows back to the stream or sewer.

Initially the system is filled through the manual bypass valve, which is then closed. The system has two controllers, T1 and T2. T1 is set to maintain 95°F leaving the chiller. T2 is set to maintain 85°F going into the chiller. The three-way valve blends the mixture of cold water and 95°F water to provide the 85°F water required by T2. At all times except very unusual light load occasions there is an excess of return water at 95°F. This excess is wasted to a stream or sewer

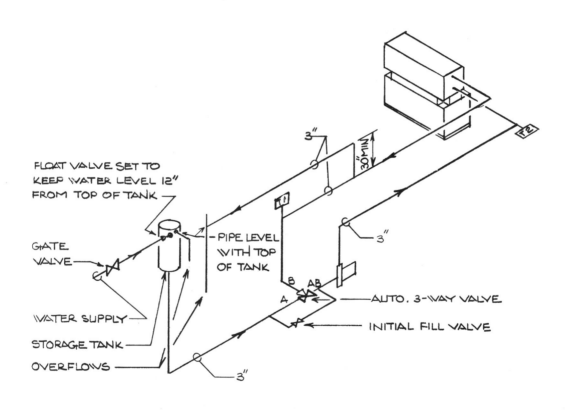

CONDENSING WATER PIPING FOR
STREAM WATER APPLICATION
NOT TO SCALE

FIGURE 5-7

through the overflow line ahead of the T1 control point. The pump is sized for the maximum flow rate required at 85°F.

Air Washers

The full-sized air washer is always located on the suction side of the fan. Nozzles normally deliver about 2 gpm of water per unit. Downstream of these nozzles are eliminator plates which, as deflecting baffles, prevent the droplets of water from being carried along the moving air stream. The eliminator plates are designed to create a collection of zig-zag patterns in the moving air stream. Dust particles are also captured on these plates. As such the washer is also something of a filter, but greasy particles, soot, and smoke can pass through these eliminators without being captured. When used in specialized applications today, the air washers are normally part of an air handling system which includes other filters and cooling and/or heating coils. Properly designed and installed, the air washer can not only do a credible air conditioning job, but it is an ideal way to provide humidification to any system.

Fig. 5-8 is the typical piping connections to the air washer portion of a system in which the function of the

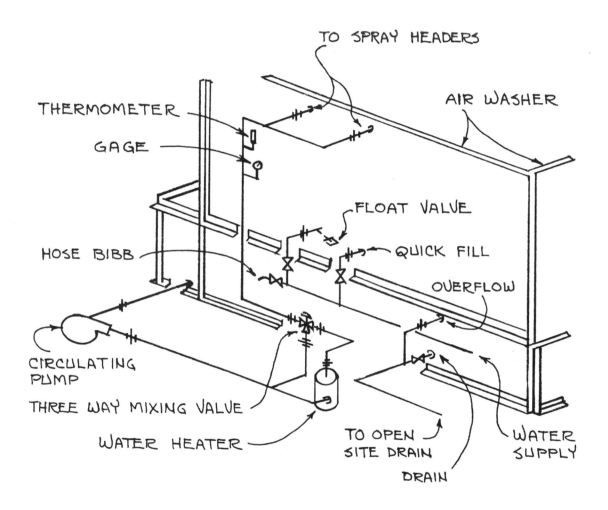

AIR WASHER PIPING FOR HUMIDIFYING SYSTEM

NO SCALE

FIGURE 5-8

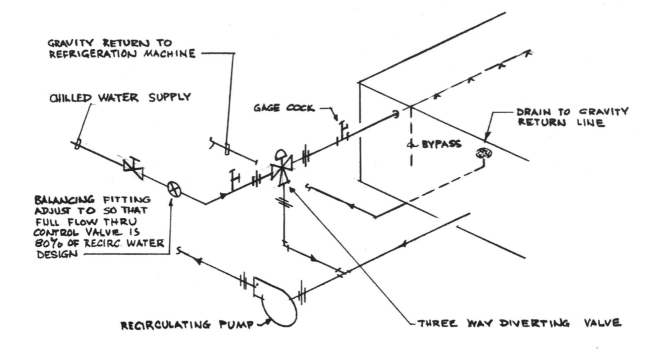

GRAVITY RETURN TO REFRIGERATION MACHINE

CHILLED WATER SUPPLY

GAGE COCK

DRAIN TO GRAVITY RETURN LINE

BYPASS

BALANCING FITTING ADJUST TO SO THAT FULL FLOW THRU CONTROL VALVE IS 80% OF RECIRC. WATER DESIGN

RECIRCULATING PUMP

THREE WAY DIVERTING VALVE

AIR WASHER PIPING USING A THREE WAY CONTROL VALVE

FIGURE 5-9

air washer is to provide for system humidification. At a quick glance the piping is similar both to that used on a cooling tower with remote heating and to a typical heating coil. The pumps and three-way valve control arrangement should be quickly recognized. And as in the cooling tower, water is recirculated from the basin, which also has makeup and overflow arrangements. Normally, to control and maintain the proper quantity of water added to the moving air stream, the heater and three-way valve control, through a discharge line sensor, sets the temperature of the water to be evaporated. Based on calculated rates of evaporation, the humidity can be controlled very accurately. And since the total stream of air, top to bottom and side to side, receives equal exposure, the possibility of stream bypass is avoided.

When the pump and the air washer are mounted on the same floor, there usually is a small suction head on the pump. And a strainer should be on the pump discharge line.

Figure 5-9 is really the same as Fig. 5-8, except the piping shown in Fig. 5-9 is the rearranged portion that is required when chilled water is delivered to the air washer in a cooling application. Commonly the drain in a cooling application does not waste chilled water to a sewer but arranges for the water to flow by gravity back to the chiller.

The use of air washers has become limited, for the most part, to air conditioning installations in the hot, dry southwest areas of the country. When large amounts of controlled moisture additions are required in a humidification process, the washer is an ideal solution.

SIX

Water Source Equipment

In this chapter we are concerned with various types of domestic water supply systems. Insofar as the detailer is concerned, the basic two sources are the public or private utility source and the driven well. When the source is the utility company, our concern lies in detailing the metering of this source. This is a relatively simple process when the service is small, as in a residence or a small commercial, industrial, or institutional building. But when the project is large, the detailer must not only show the correct meter detailing but also depict the meter pit, including clearances, structural capability, and proper access to all of the items in the pit.

Wells are generally described in terms of the method used in their construction or the method of casing and screening used; they are referred to as single-cased or double-cased wells or described as dug, driven, bored, or jetted wells. The American Water Works Association has adopted standard specifications for the construction of deep wells for water works which are not included in our book.

Within the confines of generally accepted areas of engineering coverage, the subject of wells and their associated engineering, design, and detailing is not a mechanical but rather a civil-sanitary engineering subject. However, in the mechanical design of many elementary and secondary schools and small commercial and industrial structures the water supply system, if its source is a well, is not really covered in detail anywhere on the drawings.

Well systems, like on-site sewage treatment systems, should be designed by sanitary engineers or, at the very least, by mechanical engineers with experience and expertise in that area.

Small wells for private water supplies are sometimes dug wells. These are wells excavated using hand tools, with the well digger descending into the pit as the digging progresses. Usually the well lining is porous masonry, which allows for the relatively high water table to fill the well. A driven well is constructed by driving a well point, which is attached to the lower end of a porous pipe, into the earth. The pipe performs both the function of a well casing and of a suction pipe. Driven wells are seldom more than 30 or 40 ft deep.

Larger walls may be core or percussion drilled. The percussion drilled well, under construction, has an external well-drilling rig that is similar in appearance to the familiar oil drilling rig. The applicable detailing is not concerned with how the well is drilled but rather with the parts that are involved in the installation of the water supply system. The two basic parts are the well casing and the pump.

Sprinkler systems are also part of this water source chapter. It should be noted that the design of a sprinkler system has changed considerably over the past two decades. In the not too distant past it was fairly common for sprinkler plans to show little more than the location of sprinkler heads and the location of the incoming main sprinkler service line. Accompanying this brief presentation would be a performance specification. The specification in effect required the contractor to draw up a sprinkler plan, get it approved by the appropriate authority, and then submit the approved plan to the engineer for approval.

To the newly formed engineering firm this sort of arrangement must sound both unbelievable and delightful. Let the contractor do your work and collect a fee for watching him do it. And if the plan created installation problems it was, after all, not the engineer's plan that failed. Many contractors still think that this is the best solution. The contractor is still required by the authorities to provide a fully dimensioned and detailed plan for their approval. Why clutter up the process with an engineer's plan that is not correct?

Today this is becoming a much more rare situation. The client expects the consultant not merely to perform the work for which he is paid but to be certain it is coordinated with all other mechanical and electrical trades.

The design of all types of sprinkler systems is thoroughly covered in Vol. 2 of the *latest* edition of the fire codes published by the National Fire Protection Association, Batterymarch Park, Quincy, Massachusetts 02169. Again, this book is not a book on design, and even if it was, there is little to be detailed in the overall piping layout except for detail clarification of conflicts between the installation of the sprinkler piping and other portions of the mechanical and electrical system installation.

Water Meters

The installation of a water meter is neither a very challenging idea nor a very challenging concept. Therefore, it is small wonder that many completed plans involving a water service show little or no detailing of the water meter. In addition there are questions about how much, if any, detailing is required.

In general the water utility company will require, or install as part of their water service, a shutoff valve in some location outside the structure served. Our first detail, Fig. 6-1, illustrates the fairly common exterior residential shutoff valve application. Depending on the frost line in your locality, the simple residential or other small building service has its line installed 3 to 5 ft below grade. The 4-ft 6-in distance shown in our detail is fairly common in northern and northeastern sections of the United States. What appears on the exterior surface of the building, lawn, or paved area is an 8-in × 8-in cast iron frame, cover, related shaft, and

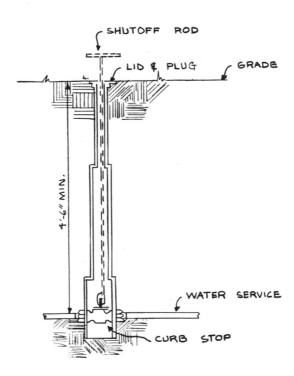

CURB STOP DETAIL

NOT TO SCALE

FIGURE 6-1

valve, all of which are furnished and installed by the utility company, and therefore no detailing is required. When permitted, a common practice is to install shutoff valves around both sides of the meter plus a valved bypass to allow system operation in the event of meter failure or damage. Some utilities do not permit the manual bypass because it provides a simple way to use water that bypasses the meter.

Figures 6-2 and 6-3 show a large meter installation. These two details are taken from the fairly standard practice of utility companies. All of the piping and the meter pit is part of the building and/or plumbing contract. The utility company not only expects that all it will have to do is to install the meters but it also expects that these details, which we have not shown with dimensions, will be precisely dimensioned so that

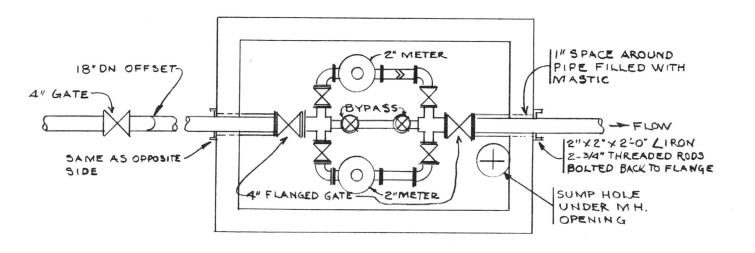

4" SERVICE – DUAL METERS

— NOT TO SCALE —

FIGURE 6-2

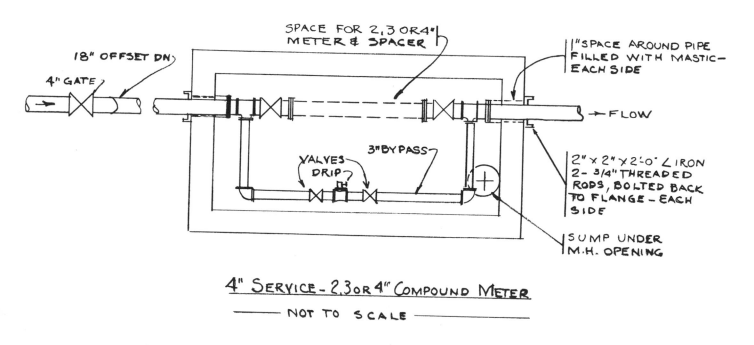

4" SERVICE – 2, 3 OR 4" COMPOUND METER

— NOT TO SCALE —

FIGURE 6-3

their meter installation fits exactly without further piping work on its part.

Since meter sizes and types create variations as to the precise clearance requirements of the utility company, it is very necessary that you not only submit this detail with dimensions for the company's approval but that such items as shutoff valves that are exterior to the meter pit be carefully verified. While these two details are generally accepted standards, they are not necessarily accepted standards, exactly as shown, in your locality.

Meter Pits: Not only are the utility companies concerned with your detail of the meter installations and related service piping, but many companies are equally concerned about the pit itself. Figures 6-4 and 6-5 show, this time partially dimensioned, a standard

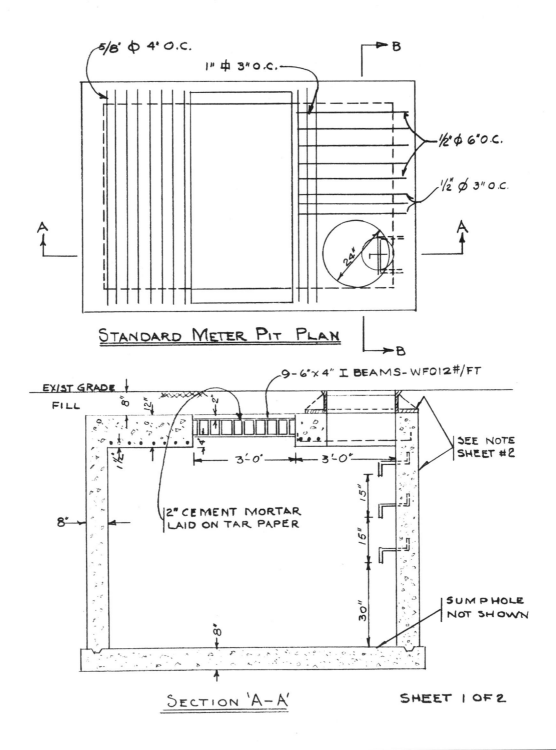

FIGURE 6-4

meter pit plan. The overall size of the pit is not shown although one could reasonably deduce that this was a pit with a surface area 8 ft 0 in long by 6 ft 0 in wide by 6 ft 0 in deep. The implication of the two piping details, is that the pit would fit one of them. In fact it does fit the dual meter installation acceptably for at least one large utility company.

The reinforcing is not arranged to roadway or other heavy-use situation standards. If your pit requires such standards because of its location, the structural reinforcing will need to be redesigned to suit your particular requirements.

Meter pits have differing requirements for meter removal and manhole access. We show both a removable cover and an access manhole. We also show a pit drain, and the detail has notes about other requirements, such as bypass lines. In short the detail tries to cover all the points that may come up in your installation. The proper solution to your final detail is to use this one as a check against local requirements and to

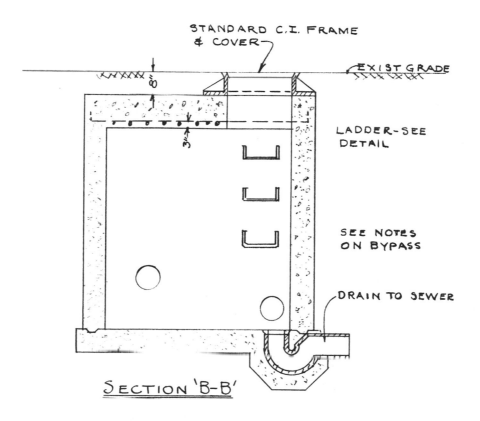

SECTION 'B-B'

NOTES
1. PIT NOT TO BE USED UNDER ROADS OR WHERE HEAVY LOADS ARE IMPOSED.
2. WATER PROOF TOP & SIDES OF PIT WITH 2 PLY TAR PAPER & TAR OR CONSTRUCT OF WATER PROOF CEMENT.
3. BYPASS MAY BE CONSTRUCTED OUTSIDE & AROUND PIT WITH DRIP EXTENDED THRU WALL. BYPASS MUST ALWAYS BRANCH TO RIGHT LOOKING ALONG DIRECTION OF FLOW.

SHEET 2 OF 2

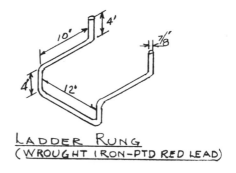

LADDER RUNG
(WROUGHT IRON-PTD RED LEAD)

FIGURE 6-5

make such corrections as are necessary. For large water services a detail such as this is almost always a mandatory item.

There are now available prefabricated concrete meter pits that are usually furnished by the suppliers of prefabricated septic tanks and electric manholes. Sometimes these work well in one area and require considerable modification in another area. The concrete for our detail is the standard 2500-lb concrete.

Wells

Single-Cased Well: Figure 6-6 is a typical detail of a single-cased well and pump. Not shown, but an obvi-

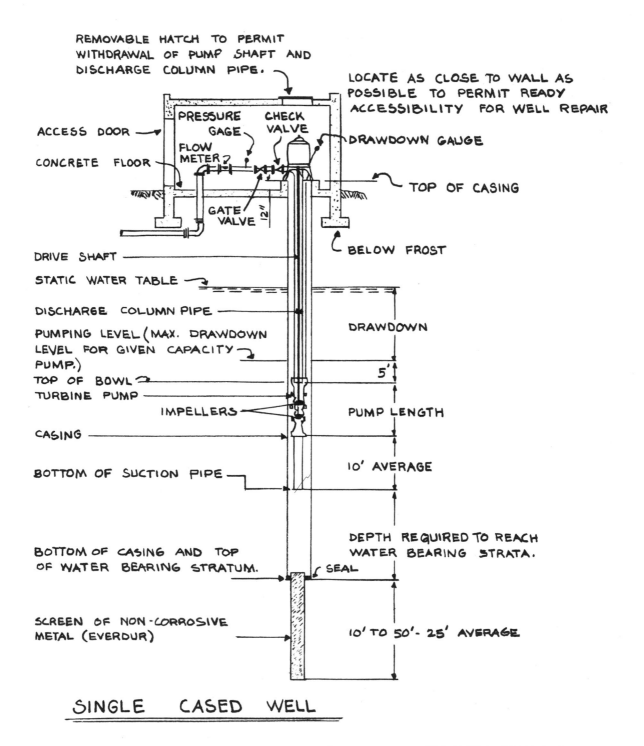

SINGLE CASED WELL

FOR USE IN ORDINARY SANDY & GRAVELLY SOIL

FIGURE 6-6

ous requirement, is the well water storage tank. Single-well casings, such as depicted in this detail, are normally used in wells that have a diameter of less than 12 in. Since a large amount of water can frequently be obtained from a 12-in well casing, it is highly likely that for the vast majority of situations this supply of water will be sufficient. The basic problem in drilling any well is to drill it in a straight line. In the single-cased well, since the force to insert the casing

tends to deform it, the usual requirement is a relatively shallow well depth.

As can be seen in Fig. 6-6, at the bottom end of the casing is a water intake screen. This is from 10 to 50 ft long and slightly less in diameter than that of the casing itself. The clearance area between screen and casing must be sealed. The pump, which is inside the casing, is driven through its shaft by a pump motor on top of the well. The shaft is inside the discharge pipe.

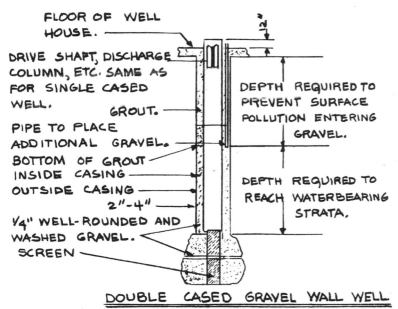

DOUBLE CASED GRAVEL WALL WELL

NOTE:
FOR USE WHERE WATER-BEARING STRATUM IS OF UNIFORM FINE SAND WHICH IS SUBJECT TO FLOWING AT THE VELOCITY THE WATER ENTERS THE SCREEN. THE GRAVEL PROVIDES A MUCH LARGER AREA THAN THE SCREEN AND THE VELOCITY OF WATER AT THE OUTER LINE OF GRAVEL MAY BE REDUCED TO SUCH VELOCITY THAT THE SAND WILL NOT FLOW.

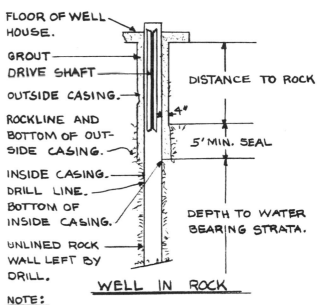

WELL IN ROCK

NOTE:
WELL HOUSE, PUMPING LEVEL, SCREEN AND OTHER DETAILS ARE SAME AS FOR SINGLE CASED WELL.

GENERAL NOTES:

1. WELL SITE SHOULD BE 200 TO 500 FEET FROM POSSIBLE SOURCE OF POLLUTION. LOCATE AT ELEVATED POINT IF POSSIBLE TO PREVENT FLOODING. FENCE IN WELL SITE.

2. WELL BUILDING SHOULD BE FIREPROOF, VENTILATED AND IN COLD CLIMATES, INSULATE THOROUGHLY AND PROVIDE HEAT. ELEVATE ABOVE GRADE TO PROVIDE DRAINAGE AWAY FROM BUILDING.

3. PUMPING LEVEL IS DETERMINED BY CONTINUOUS FLOW TEST AT REQUIRED WELL CAPACITY, AND THE MEASUREMENT OF THE RESULTING DRAWDOWN BELOW THE STATIC WATER TABLE.

4. THE TOP OF THE SCREEN SHOULD BE 50 FEET MINIMUM BELOW GRADE TO AVOID SURFACE POLLUTION UNLESS UNUSUALLY IMPERVIOUS EARTH (10 FEET OF COMPACT CLAY) OCCURS AT THE SURFACE, OR WELL IS FAR REMOVED FROM POSSIBLE SOURCES OF POLLUTION.

FIGURE 6-7

119

The location of the pump is calculated to be below the level of the water table at maximum draw in dry weather conditions. The capacity of the water table and water source, called an aquifer, is part of the well test report data and proper engineering calculations.

Double-Cased well: Figure 6-7 is a combined detail illustrating in generalized terms the application of a double-cased well. Our basic premise in presenting these details is to alert you not only to some of the problems of detailing a double-cased well but also to the basic points of detailing that should be covered.

Submersible Pump: Figure 6-8 is a typical schematic of a small submersible pump installation. The well casing is a single-wall installation. The detail itself is fairly clear and obvious and, surprisingly enough, can be used on some fairly large system applications. In

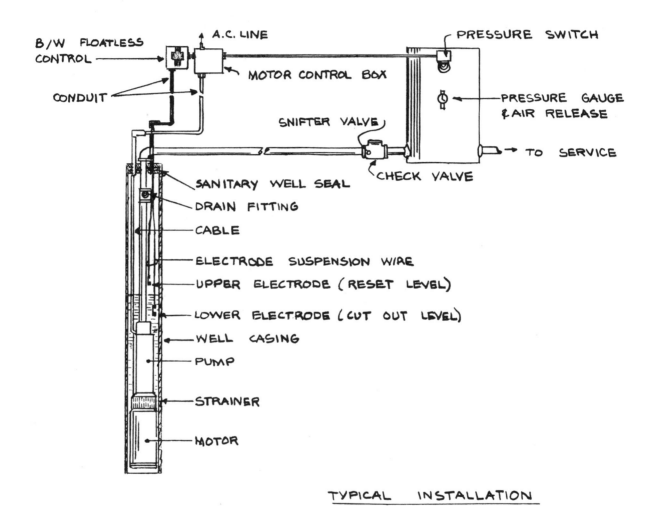

TYPICAL INSTALLATION

SUBMERSIBLE PUMP CONTROL

FIGURE 6-8

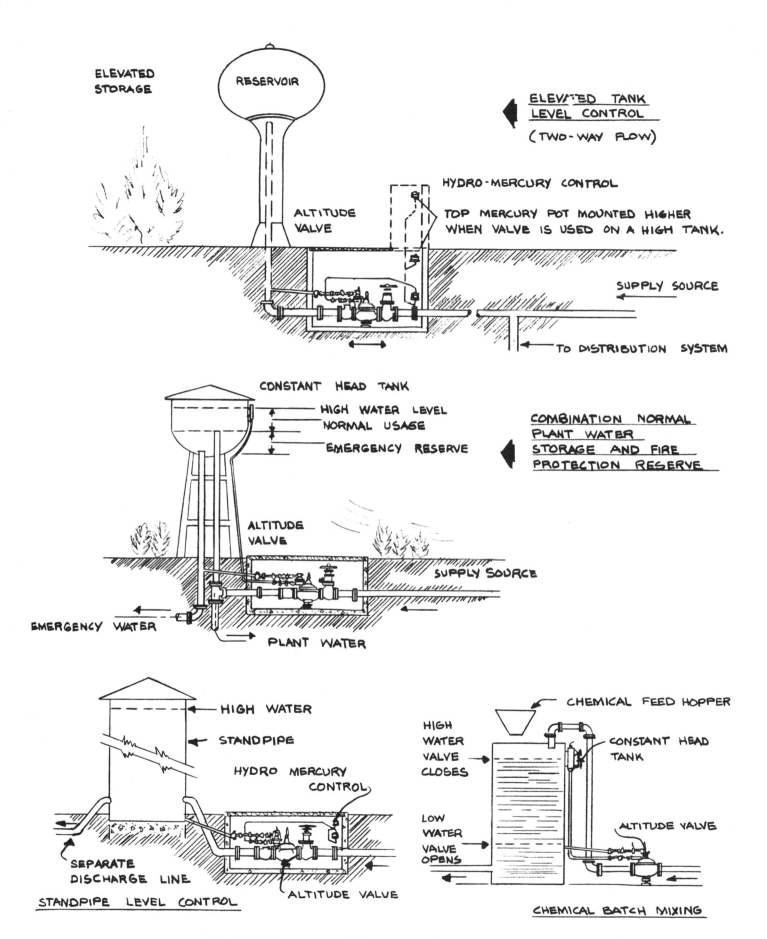

ELEVATED STORAGE

RESERVOIR

ELEVATED TANK LEVEL CONTROL (TWO-WAY FLOW)

ALTITUDE VALVE

HYDRO-MERCURY CONTROL

TOP MERCURY POT MOUNTED HIGHER WHEN VALVE IS USED ON A HIGH TANK.

SUPPLY SOURCE

TO DISTRIBUTION SYSTEM

CONSTANT HEAD TANK

HIGH WATER LEVEL

NORMAL USAGE

EMERGENCY RESERVE

COMBINATION NORMAL PLANT WATER STORAGE AND FIRE PROTECTION RESERVE

ALTITUDE VALVE

SUPPLY SOURCE

EMERGENCY WATER

PLANT WATER

HIGH WATER

STANDPIPE

HYDRO MERCURY CONTROL

SEPARATE DISCHARGE LINE

STANDPIPE LEVEL CONTROL

ALTITUDE VALVE

CHEMICAL FEED HOPPER

HIGH WATER VALVE CLOSES

CONSTANT HEAD TANK

LOW WATER VALVE OPENS

ALTITUDE VALVE

CHEMICAL BATCH MIXING

TYPICAL APPLICATIONS

FIGURE 6-9

121

this system the basic requirements are for the most part not part of the detail depicted. Commonly the owner, architect, or engineer has requested a specific size and depth of well with certain outputs. The driller has installed the casing and screen. The engineering department has selected and specified the submersible pump. The accompanying detail shows all that is required except for an air compressor on the storage tank.

Water Storage Tanks: As we noted in the beginning of this chapter, water supplies, large wells, and their related storage facilities are really civil-sanitary engineering systems and not mechanical systems. But there may be situations in which your firm, in a retrofit project, not only is designing work for the building but also designing some repairs to the existing water supply. Figure 6-9 is a schematic covering some of the typical and common problems. It illustrates in the upper portion the typical detail for controlling the level of water in the tank. In the center detail we show a scheme to be used if the storage supply is to be separated for fire and domestic use. In the lower part of Fig. 6-9 we illustrate another level control and a chemical feeder.

To reiterate, the systems depicted in Fig. 6-9 cross the line between sanitary and mechanical engineering. It is a gray area of practice that occurs in many small application situations. We have not completely detailed the systems because we are reluctant to indicate that these systems really belong in mechanical engi-

neering work. We hope that you will at least consult and verify such details as required with a civil-sanitary engineer or another expert in the water supply field.

Pneumatic Water System

In our previous discussion on wells we took pains to try to mark the line of demarcation between the practice of civil-sanitary and mechanical engineering, pointing out that the well, its pump, and the work pertaining to it, including the exterior storage tank, were not generally construed as mechanical work. In our judgment small storage tanks that are usually indoors and related to small water systems are definitely mechanical items.

Figure 6-10 is a typical storage and pneumatic water supply system for any structure. The source of water is a well pump. As we noted previously, we feel this is basically the logical area of practice for an experienced, mechanical engineering firm that has the necessary expertise to write the specifications for the well driller and to select the proper type of pump, which may be submersible or above ground.

Generally, wells do not have the ability or capacity to handle peak service water demands. Commonly a storage tank of sufficient size to handle this demand is buried underground. It has much the same function as the hot water storage tank in a hot water tank and heater system. Once the water is in storage, whatever discharge pressure the well pump has is lost. A booster

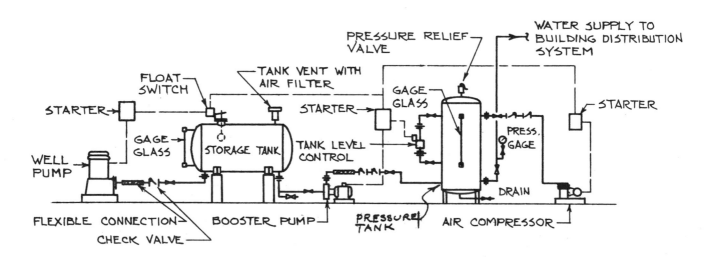

COMBINED STORAGE & HYDRO-PNEUMATIC WELL WATER SYSTEM

NO SCALE

FIGURE 6-10

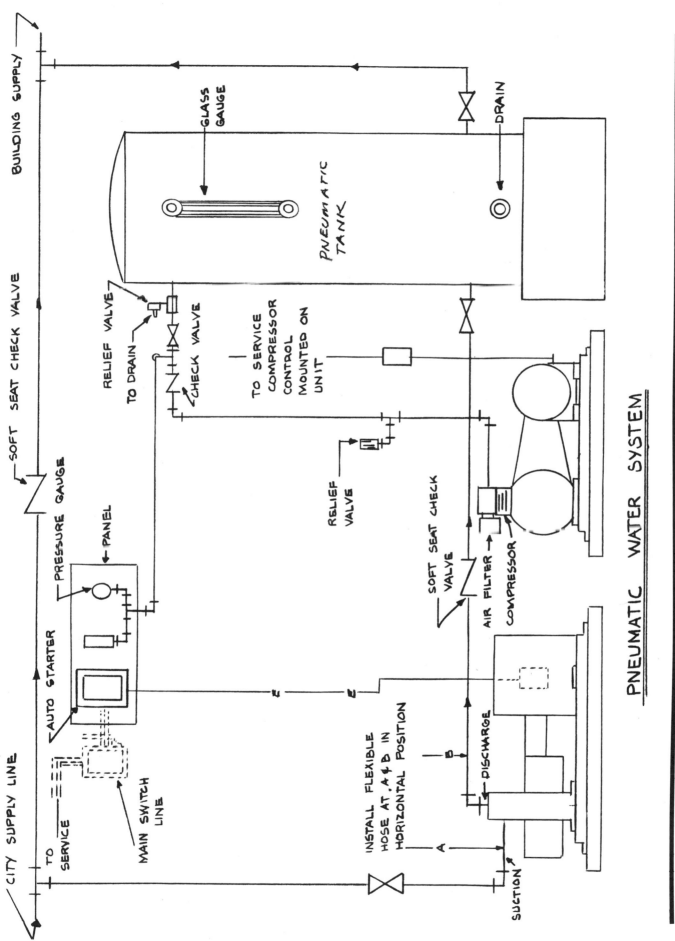

PNEUMATIC WATER SYSTEM

FIGURE 6-11

pump now supplies volume and pressure. What is needed is a pressure balancing device to smooth out the booster pump operation and to provide the system with a steady pressure. This is the purpose of the pressure tank and related air compressor shown in Fig. 6-10.

As can be seen in the detail, there are three interrelated starters. They are operated by float switches in the storage tank and in the pneumatic tank. The booster pump has sufficient pressure to discharge the water from the storage tank into the booster tank. You do not want to have this pump operating every time someone opens a faucet. A head of air pressure is maintained on the supply of water in the pressure tank. System water pressure is provided by this head pressure. If the level of water in the pressure tank drops below a certain point, the booster pump pulls water out of the storage tank. This storage tank water level can also vary, but if it also gets below a certain fixed point, the third starter on the left side of the detail starts the well pump to restock the basic system water supply.

Combined Utility and Pneumatic Systems: There also are domestic water supply situations in which the city or private utility water supply is of sufficient quantity for the use intended, but the pressure may be low at times or is consistently too low to flush toilets, run showers, or handle process requirements. Figure 6-11 is one solution to that situation. Here the city water supply is in the same mode as the well pump in Fig. 6-10 except that it is a constant source. As before, a pneumatic tank provides the final system pressure. Since the supply of water is constant, the booster pump, when called upon, simply draws water from the city supply. The pump, for all practical purposes, has an infinite supply source.

However, there is a very noticeable difference in the piping arrangement. There are shutoff valves on the pump and tank, including the supply to the pump and system supply from the pressure tank. On the bypass line around the booster system there is only a check valve—a very important check valve. Since the system pressure is higher than the city pressure, the check valve precludes back pressurization of the city system. Under normal operation the city water can only go to the booster pump. Also, the booster pump cannot pull water backward out of the pressurized building system.

As we noted at the start of this explanation, there is a water supply, but its pressure is just too low. However, some water is better than no water and if the booster system fails, the pressure will drop and at a certain lower value the check valve will open so that the system will *not* be without any water.

Not shown in our details is the common upper level or rooftop water storage tank. Usually, the height of a very tall building is sufficient to create a water pressure problem in the upper floors. If you mentally view Fig. 6-11 with the air compression and storage tank deleted, the discharge of the booster has no place to go. In the high-rise building the booster raises the water pressure sufficiently to overcome the static and friction pressure of the water consuming fixtures on the top floors and pumps the water into an open or vented storage tank. Water is then pumped or falls by gravity to the fixtures on the floors below.

Sprinkler Systems

Standard Fire Protection System: There are certain required details which relate to the sprinkler service and its system specification. The most common system is a wet pipe sprinkler system. Figures 6-12 and 6-13 cover the service and distribution detailing for that system. Actually, only Fig. 6-12 is really required.

What is shown on Fig. 6-13 is a schematic arrangement of all the components of a wet pipe system. Since it is not uncommon for a wet pipe system to have a few isolated unheated areas that require a dry pipe system, we have made Fig. 6-13 into a composite system. If your system has no dry pipe requirement, convert the tee that feeds the dry pipe system into a straight piece of pipe and do not include the dry pipe component.

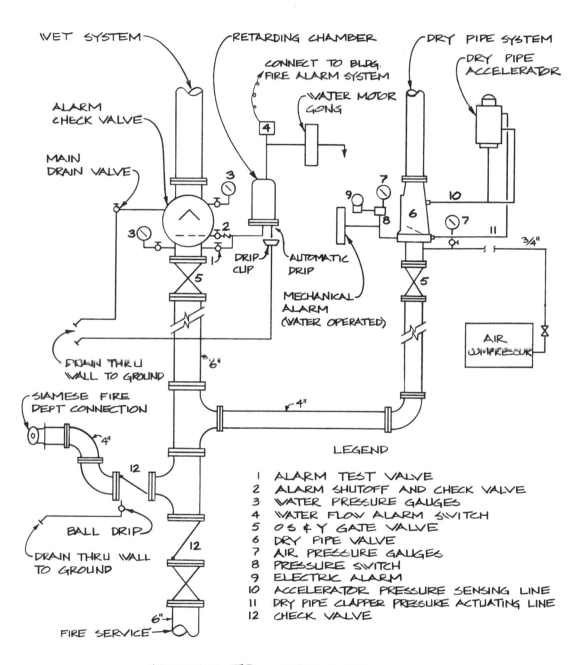

SPRINKLER SERVICE

NO SCALE

FIGURE 6-12

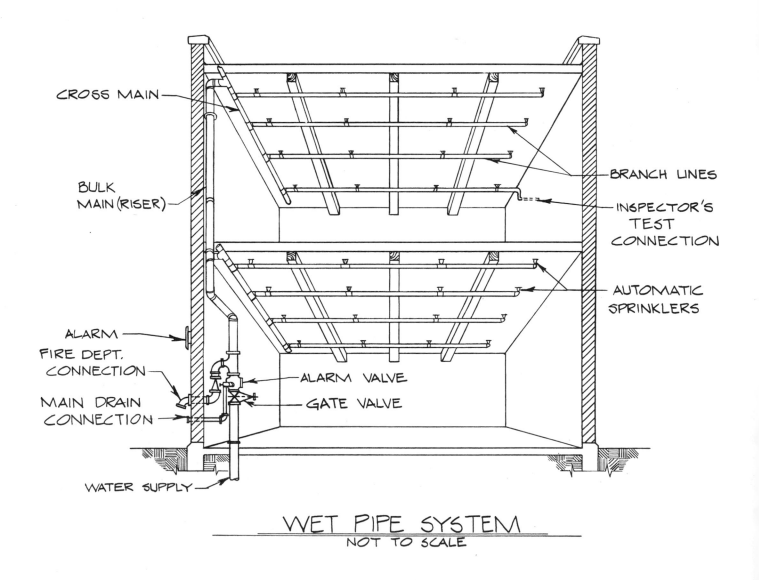

CROSS MAIN

BULK
MAIN (RISER)

BRANCH LINES

INSPECTOR'S
TEST
CONNECTION

AUTOMATIC
SPRINKLERS

ALARM

FIRE DEPT.
CONNECTION

ALARM VALVE

GATE VALVE

MAIN DRAIN
CONNECTION

WATER SUPPLY

WET PIPE SYSTEM
NOT TO SCALE

FIGURE 6-13

Commonly, today's sprinkler systems are connected to the fire alarm system through a water flow alarm switch. Thus, if the sprinkler is activated, not only will the exterior water operated gong sound but all the interior fire horns, bells, and lights will also be activated.

As we previously stated, Fig. 6-13 shows a sectional view of the distribution piping that really is not necessary. However, this plan would clarify the riser piping. Some engineers request that their design department produce a sectional explanatory detail and Fig. 6-13 is offered as a representative example.

Dry Pipe System: When the danger of pipe freezing cannot be avoided, there is no logical alternative, in many cases, to a dry pipe system. In essence water is kept out of the distribution system beyond the dry pipe valve that is near the service entrance. All of the distribution lines are filled with air, not water. If a sprinkler head is activated, the pressure drops and the dry pipe valve, aided by its accelerator, opens very rapidly, releasing air pressure and allowing water to flow to the system. This is what is detailed in Fig. 6-14. Note also that there is a pressure switch to an electric alarm which may also activate the overall fire alarm system.

In Fig. 6-15 we have a distribution schematic whose purpose is similar to that of the wet pipe distribution schematic. There is one added item which we chose to

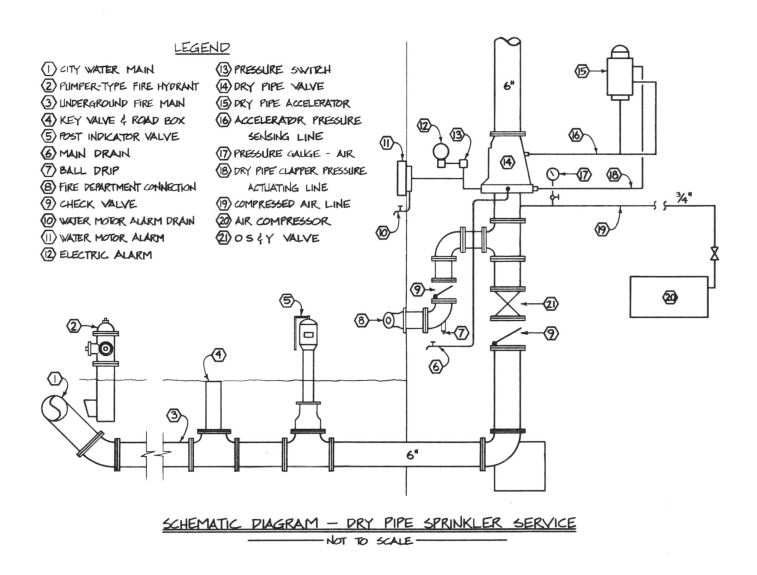

LEGEND

① CITY WATER MAIN
② PUMPER-TYPE FIRE HYDRANT
③ UNDERGROUND FIRE MAIN
④ KEY VALVE & ROAD BOX
⑤ POST INDICATOR VALVE
⑥ MAIN DRAIN
⑦ BALL DRIP
⑧ FIRE DEPARTMENT CONNECTION
⑨ CHECK VALVE
⑩ WATER MOTOR ALARM DRAIN
⑪ WATER MOTOR ALARM
⑫ ELECTRIC ALARM

⑬ PRESSURE SWITCH
⑭ DRY PIPE VALVE
⑮ DRY PIPE ACCELERATOR
⑯ ACCELERATOR PRESSURE SENSING LINE
⑰ PRESSURE GAUGE - AIR
⑱ DRY PIPE CLAPPER PRESSURE ACTUATING LINE
⑲ COMPRESSED AIR LINE
⑳ AIR COMPRESSOR
㉑ O S & Y VALVE

SCHEMATIC DIAGRAM — DRY PIPE SPRINKLER SERVICE
——— NOT TO SCALE ———

FIGURE 6-14

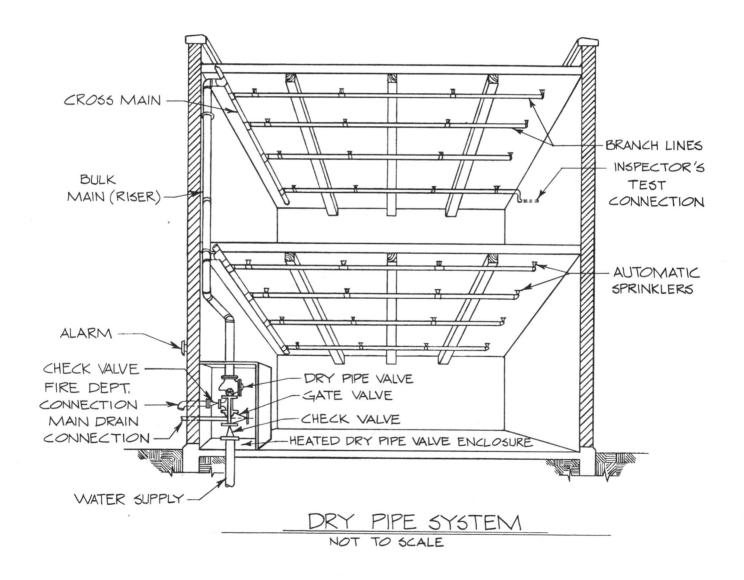

CROSS MAIN

BULK
MAIN (RISER)

ALARM

CHECK VALVE
FIRE DEPT.
CONNECTION
MAIN DRAIN
CONNECTION

WATER SUPPLY

BRANCH LINES

INSPECTOR'S
TEST
CONNECTION

AUTOMATIC
SPRINKLERS

DRY PIPE VALVE
GATE VALVE

CHECK VALVE

HEATED DRY PIPE VALVE ENCLOSURE

DRY PIPE SYSTEM
NOT TO SCALE

FIGURE 6-15

show on this schematic. In a cold area even the service pipe could freeze. Therefore, we have noted that the service is in a heated enclosure. Note that we said "enclosure" and not "room." All that is needed is some kind of small enclosure with a source of heat that is big enough to contain the water filled pipe and to provide access for system test and service functions.

Figure 6-16 is a service detail of a special type of sprinkler system called a deluge system. As can be readily visualized, it requires some heat, usually 160°F or more, to start a sprinkler head. The deluge system has been designed for use with material that creates and promotes the very rapid spread of a fire, a spread too fast, in fact, for the ordinary wet or dry system.

The system detailed in Fig. 6-16 is one that employs open sprinklers attached to a piping system. There is also a separate fire detection system installed in the same area as the sprinklers. The piping system is connected to a water supply system that is activated by the detectors. When this valve opens, water flows into the piping system and discharges from every open sprinkler head that is attached to the system, flooding, or deluging, the area with water.

On-Off System: Figure 6-17 is an on-off system and is described in NFPA standards as a preaction system. It is a system which, like the deluge system, has a separate fire detection system in the same area as the sprinklers. The system contains air, not water, which may or may not be under pressure. The system valve is activated by the fire detectors, permitting water to flow to the area and through the open sprinklers.

Both the deluge and the on-off or preaction systems

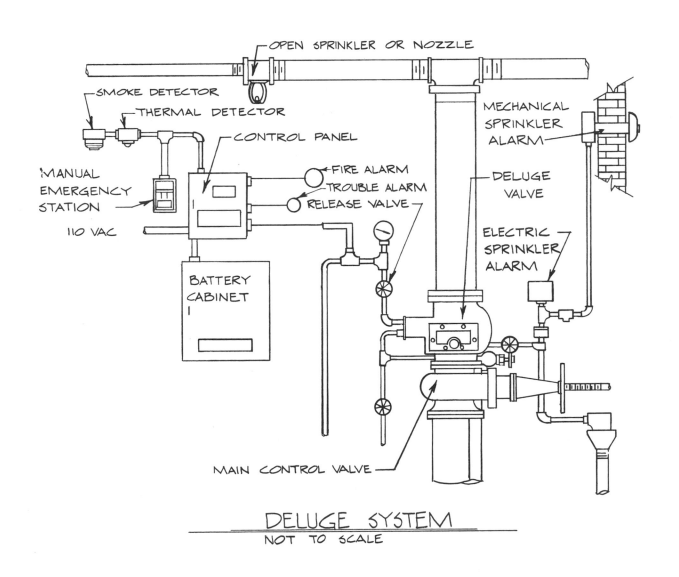

DELUGE SYSTEM
NOT TO SCALE

FIGURE 6-16

are normally dry systems with valved supplies that are operated by heat detectors. They also are arranged for manual valve operation independent of the sprinklers. Sprinkler piping and fire detection devices may or may not be automatically supervised. Open sprinklers may not be automatically supervised. Or there may be a combination of open and automatic sprinkler devices which are or are not automatically supervised. These two special systems require considerable design and engineering expertise in fire safety for proper application.

Water Treatment

Again, as in the subject of wells, we are in an area that is not really part of mechanical engineering and its related details. Primarily the subject of water treat-ment is quite properly in the area of chemical engineering and/or civil-sanitary engineering. In general the mechanical engineer is concerned with water which may be used in a structure or collection of structures, such as a college campus, industrial park, or domestic or process work.

We have only depicted one detail, Fig. 6-18 which shows the application of a chlorinator. Before discussing this detail, a certain amount of explanation of the mechanical engineer's area of concern might be appropriate.

As we have shown in our boiler room equipment details, the mechanical engineer is very much concerned about water quality and treatment in large boiler plants. Other areas of concern would certainly include cooling towers, process control systems, and

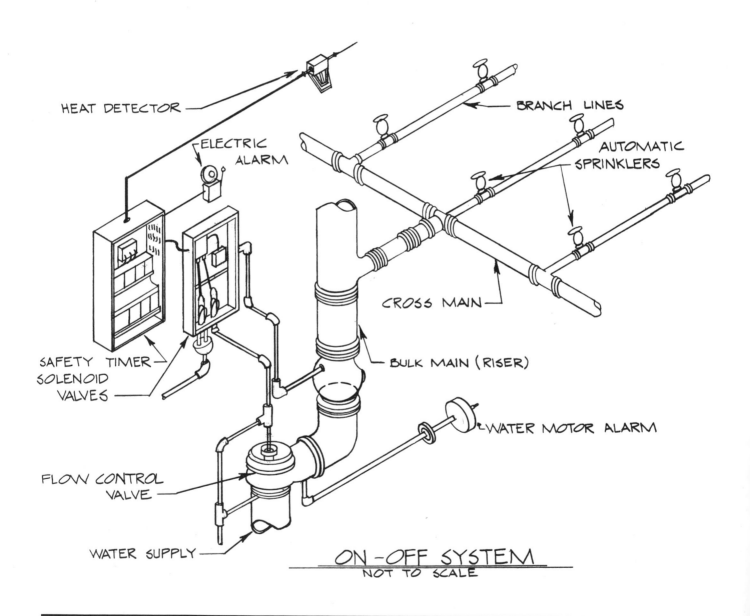

FIGURE 6-17

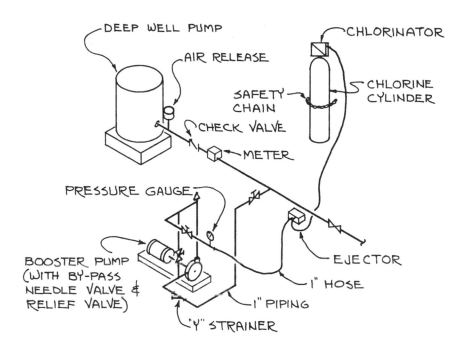

CHLORINATOR INSTALLATION USING
TURBINE TYPE BOOSTER PUMP

NO SCALE

FIGURE 6-18

hot or chilled water distribution systems. There are laboratory, special manufacturing, and chemical process areas in which the treatment of water is a very vital concern of mechanical design.

Generally, the average project uses city water, which is normally safe and potable. Frequently it is believed that there is an unlimited variety of living, disease-causing matter that can be carried in a water system. Actually relatively few pathogenic germs can survive in the environment of a natural water supply. However, we all know of serious, even dangerous, forms of chemical pollutants that can be carried in a water supply, and water can also carry germs that cause hepatitis and gastroenteritis.

We have shown water softeners for boiler plants. The same detail, with different though generally similar equipment, could be used for a water softener for any residential, industrial, commercial, or institutional use. In brief a water softener is a water softener.

Frequently, especially in retrofit work, the mechanical engineer gets involved in a chlorinator installation. To kill the germs in the water the controlled application of chlorine is the common, standard treatment. Most installations are arranged as we show in Fig. 6-18.

The chlorine cylinder, under pressure, has a controlling application device known as an ejector. The line from the ejector can be solidly connected or attached using a flexible hose, as we have illustrated. Normally the system booster pump has a bypass line, and the chlorine injector line is connected beyond that bypass valve and on the discharge line of the pump, as we depict. It is on the pump discharge because we want to treat the system, not the well.

Disinfection by chlorine is not a substitute for other forms of treatment, and not all microscopic organisms are killed by chlorine. Chlorine gas is toxic to humans and animals. However, in the amounts used to kill pathogenic organisms in water it does not do any harm to the humans. Chlorine gas will neither burn nor explode. It is highly corrosive at 300°F and may become corrosive at 195°F. Normally the cylinders can be put in any area below 150°F and should be placed upright as we show in Fig. 6-18. Room ventilation should be arranged to supply air at the top of the side or ceiling of the room with an exhaust at floor level. The chlorine gas is heavier than air, and this arrangement will increase human safety.

SEVEN

Water Source Support Equipment

Normally, other than boiler water treatment, which was covered in Chap. 1, there would have been little or nothing to discuss in this chapter. Smaller systems may have some form of water softener and very large civil works systems would have the treatment facilities at the main water plant. Our book is not involved with civil engineering projects, and the residential water softener really does not require a special detail.

However, the mechanical equipment detailer does have one area in which he or she is often involved and in which there is always considerable controversy over which is the best system. Perhaps it is fortunate that our one subject—the swimming pool filtration system—is the only item covered in this chapter.

Since this is a book on detail application and not on engineering design, we have not covered the design of the systems except to mention in passing in one of our details that its design was based on a 6-hour turnover, which is further explained.

We also decided to include in the details an actual complex working installation which may seem a little peculiar in size and arrangement, but it does depict how at least one very restricted space installation was successfully resolved.

Swimming Pool Filter System

All swimming pools are required to have some sort of filter system. In its essential parts the filter system consists of a filter, a pump, and related water flow balance and control items. In addition some form of chlorination is required to control bacteria. Since a pool filter, like any other filter, requires cleaning to be effective, the pool filter is normally cleaned by reverse flow, called backwash, to remove collected material. The "dirty" liquid from this backwash process is wasted to an appropriate drain. On occasion, for various reasons, the total content of the pool is wasted to a drain. Generally, but not always, the pool and gutter system are part of the general construction and covered by the architectural and structural engineering design phases of the work. Thus this book on mechanical details does not cover the pool and gutter system.

Chlorination: Although the chlorinator is shown on other details in this chapter, our first detail, Fig. 7-1, shows a typical chlorinator installation. Normally, a pump is both required and used, but this is not always the case. Thus our detail shows the location of the injector in both instances. In Chap. 6 we discussed in more detail the function of a chlorinator and the various items to note about its installation. For further elaboration we refer you to that discussion. Suffice it to say that control of bacteria is very important in a pool installation. That is the function of the chlorinator. There is a lot of detailing in connection with a pool filter system and the chlorinator can easily get overlooked. It is for that reason that we provided Fig. 7-1, which is essentially an enlargement of one part of the complex detailing of a pool filter.

High Rate Sandfilter: Figures 7-2 and 7-3 were taken from the actual system of a 183,500-gal, olympic sized pool with a 6-hour turnover. The capability of the pump is 183,500 gal in 6 hours, which means that every 6 hours the entire contents of the pool passes through the filter. In normal rating language the pump has a rating of 510 gpm [$183,500/(6 \times 60)$].

These two figures form a plan and elevation of parts of the system in a sectional format. Both a surge tank and a backwash tank are required by state board of health codes. They are sized according to proper engineering calculations. Because of the point at which

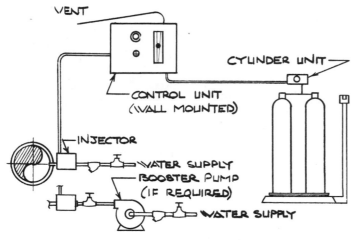

VENT

CYLINDER UNIT

CONTROL UNIT
(WALL MOUNTED)

INJECTOR

WATER SUPPLY

BOOSTER PUMP
(IF REQUIRED)

WATER SUPPLY

EQUIPMENT

1) WALL MOUNTED CONTROL
   UNIT W/ ROTAMETER.
2) INJECTOR UNIT.
3) 25' OF 3/8" O.D. PLASTIC TUBING
4) 10' OF 3/4" FLEXIBLE
   PLASTIC PIPE.
5) PIPE AND HOSE CLAMPS

TYPICAL WALL MOUNTED CHLORINATOR

FOR SWIMMING POOL
— NOT TO SCALE —

FIGURE 7-1

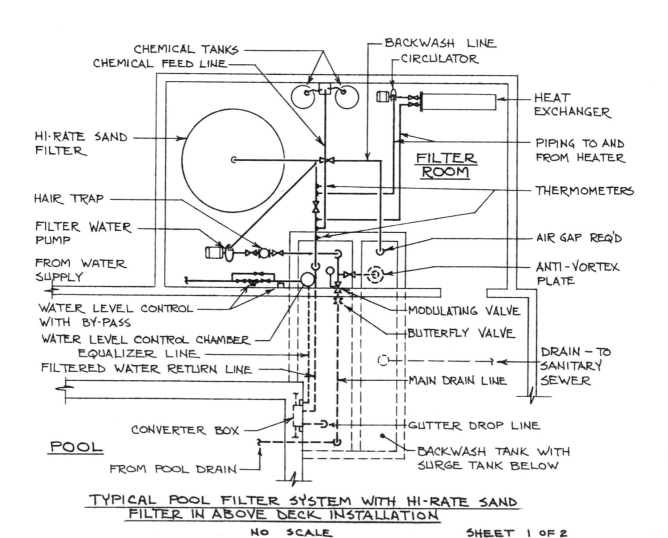

CHEMICAL TANKS
CHEMICAL FEED LINE

BACKWASH LINE
CIRCULATOR

HEAT
EXCHANGER

HI-RATE SAND
FILTER

PIPING TO AND
FROM HEATER

FILTER
ROOM

HAIR TRAP

THERMOMETERS

FILTER WATER
PUMP

AIR GAP REQ'D

FROM WATER
SUPPLY

ANTI-VORTEX
PLATE

WATER LEVEL CONTROL
WITH BY-PASS

MODULATING VALVE

WATER LEVEL CONTROL CHAMBER
EQUALIZER LINE

BUTTERFLY VALVE

DRAIN — TO
SANITARY
SEWER

FILTERED WATER RETURN LINE

MAIN DRAIN LINE

CONVERTER BOX

GUTTER DROP LINE

POOL

BACKWASH TANK WITH
SURGE TANK BELOW

FROM POOL DRAIN

TYPICAL POOL FILTER SYSTEM WITH HI-RATE SAND
FILTER IN ABOVE DECK INSTALLATION

NO SCALE

SHEET 1 OF 2

the elevation cross section was taken, the backwash tank does not appear in the elevation shown on Fig. 7-3. If you visualize the overall surge and backwash tank unit as a rectangular box, looking at the end of the box the backwash tank would be a smaller rectangular box that fits within the larger rectangular box and covers the upper right quadrant. In brief the surge tank is L-shaped and occupies three-quarters of the overall box, and the backwash tank occupies one-quarter of the box.

When the pool has no occupants and is filled to the proper level, the surge tank is nearly empty. The normal pool has outlets back to the pump from the deepest part of the pool and outlets (supplies) properly spaced around the pool some 12 inches below the normal water level line. When bathers enter the pool, this displaces, or unbalances, the normal content of the pool water. The displaced water goes into the gutter and is drained back to the surge, or balance, tank. Not clearly shown but implied in the detail of Figs. 7-2 and 7-3 is that the excess water flows back to the filter, being picked up by the passing return pool line to the circulating pump. This connection is noted as the antivortex plate. The pool level is controlled by an equalizer line.

The various associated filter system components,

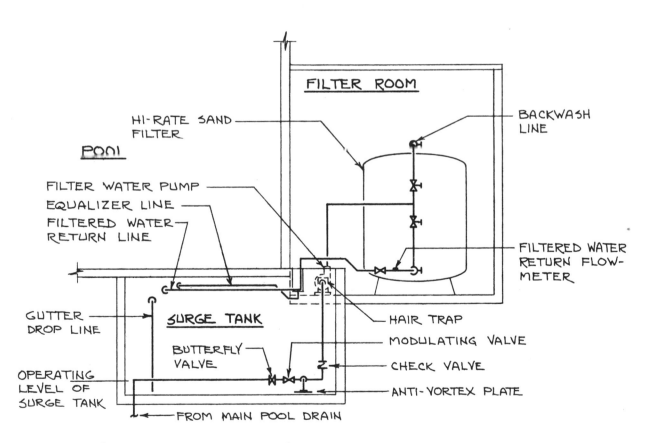

TYPICAL POOL FILTER SYSTEM WITH HI-RATE SAND FILTER IN ABOVE DECK INSTALLATION

NO SCALE   SHEET 2 OF 2

FIGURE 7-3

such as hair traps, chemical feeders, makeup water line, and the like, should be noted. The pool water is heated by a water-to-water heat exchanger that has a separate circulator. It is not necessary to pump 100 percent of the recirculating water through the heater. Usually part of the water is heated and mixed with the circulating water to provide a mixed-water temperature of 80°F for the pool water.

Gutter System: Sometimes trying to include all of the pool filter system piping information can create a complex and confused plan and elevation detail, as depicted in Figs. 7-2 and 7-3. Figures 7-4 and 7-5 show another way of presenting the piping in a schematic format. Figure 7-4 is a three-tank installation which is actually a very large system, and Fig. 7-5 is a one-tank installation of a pool of the same relative size as was depicted in Figs. 7-2 and 7-3.

Figures 7-4 and 7-5 have advantages for the detailer.

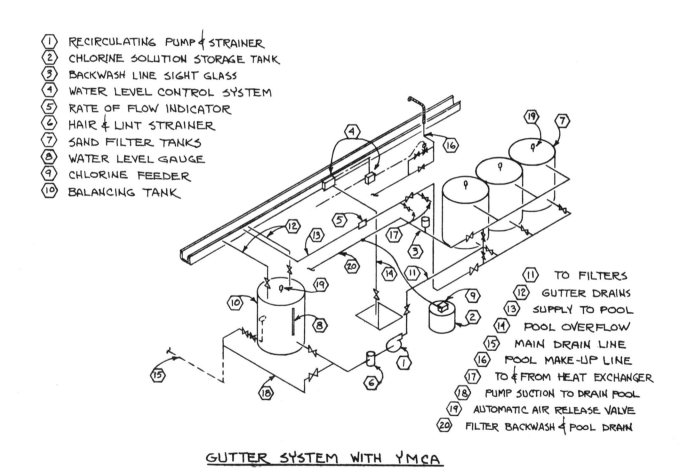

1. RECIRCULATING PUMP & STRAINER
2. CHLORINE SOLUTION STORAGE TANK
3. BACKWASH LINE SIGHT GLASS
4. WATER LEVEL CONTROL SYSTEM
5. RATE OF FLOW INDICATOR
6. HAIR & LINT STRAINER
7. SAND FILTER TANKS
8. WATER LEVEL GAUGE
9. CHLORINE FEEDER
10. BALANCING TANK

11. TO FILTERS
12. GUTTER DRAINS
13. SUPPLY TO POOL
14. POOL OVERFLOW
15. MAIN DRAIN LINE
16. POOL MAKE-UP LINE
17. TO & FROM HEAT EXCHANGER
18. PUMP SUCTION TO DRAIN POOL
19. AUTOMATIC AIR RELEASE VALVE
20. FILTER BACKWASH & POOL DRAIN

GUTTER SYSTEM WITH YMCA
3 TANK SAND FILTER SYSTEM
NO SCALE

FIGURE 7-4

There is no doubt that in both of these figures it is very easy to show every item that is required and exactly how one relates to the other. As such the details are excellent. In addition they create a situation in which the engineers, designers, and specifiers can easily check to be certain that all the designed and specified parts have been noted. It is an ideal way to cross-check plans, specifications, and details.

This sort of detailing requires a very good, properly sized plan and elevation of the actual equipment. In the filter piping the balancing tank is no small item. Nor are the filters and the associated 6-in or 8-in lines, valves, controls, filters, and the like small items. What fits so easily on the schematic detail may not fit in the actual available space. If this sort of detail is accompanied by carefully thought-out and drawn-to-scale plans and elevations of the filter room equipment and piping, we would consider it a superior design detail.

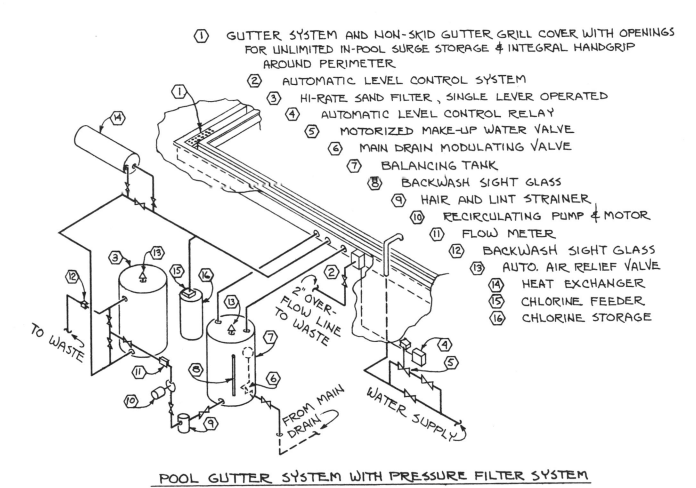

1. GUTTER SYSTEM AND NON-SKID GUTTER GRILL COVER WITH OPENINGS FOR UNLIMITED IN-POOL SURGE STORAGE & INTEGRAL HANDGRIP AROUND PERIMETER
2. AUTOMATIC LEVEL CONTROL SYSTEM
3. HI-RATE SAND FILTER, SINGLE LEVER OPERATED
4. AUTOMATIC LEVEL CONTROL RELAY
5. MOTORIZED MAKE-UP WATER VALVE
6. MAIN DRAIN MODULATING VALVE
7. BALANCING TANK
8. BACKWASH SIGHT GLASS
9. HAIR AND LINT STRAINER
10. RECIRCULATING PUMP & MOTOR
11. FLOW METER
12. BACKWASH SIGHT GLASS
13. AUTO. AIR RELIEF VALVE
14. HEAT EXCHANGER
15. CHLORINE FEEDER
16. CHLORINE STORAGE

POOL GUTTER SYSTEM WITH PRESSURE FILTER SYSTEM

NO SCALE

FIGURE 7-5

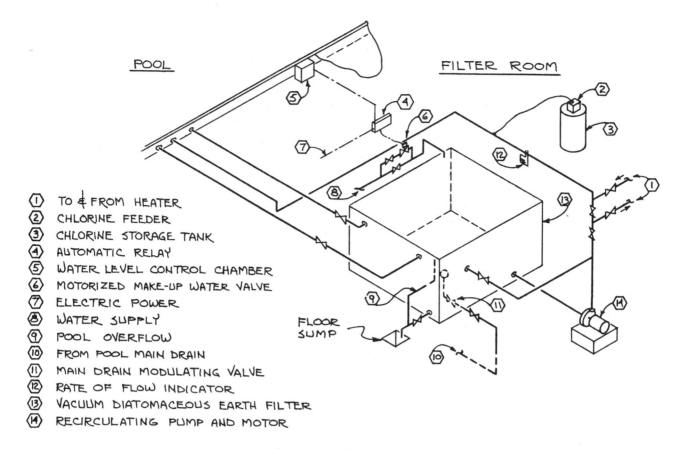

POOL   FILTER ROOM

① TO & FROM HEATER
② CHLORINE FEEDER
③ CHLORINE STORAGE TANK
④ AUTOMATIC RELAY
⑤ WATER LEVEL CONTROL CHAMBER
⑥ MOTORIZED MAKE-UP WATER VALVE
⑦ ELECTRIC POWER
⑧ WATER SUPPLY
⑨ POOL OVERFLOW
⑩ FROM POOL MAIN DRAIN
⑪ MAIN DRAIN MODULATING VALVE
⑫ RATE OF FLOW INDICATOR
⑬ VACUUM DIATOMACEOUS EARTH FILTER
⑭ RECIRCULATING PUMP AND MOTOR

FLOOR SUMP

POOL SYSTEM WITH DIATOMITE FILTER
NO SCALE

FIGURE 7-6

Diatomaceous Earth Filter: Finally, in Fig. 7-6 we show a filter system that uses diatomaceous material instead of sand. There is no rule that requires all pool filter systems to use sand as the filter medium. Actually there are many, many systems that use diatomaceous earth, and we do not want to take sides as to which material is better. To distinguish this detail from the others we made the filter a rectangular box, which of course it actually is not. It uses the same sort of tank that the sand filter utilizes. Again, this is a schematic detail, and the same caveats prevail. It must be accompanied by a very well done set of scaled plans and elevations. In this detail, as in all others, items such as floor sumps, surge tanks, filters, and the like take considerable space.

In general, for all filter systems there are two basic requirements. First, the engineering and design of the system must not only be correct but must conform to all codes. In this area checking with and conforming to your state's board of health codes is mandatory. Second, decide in advance on your detail approach. Either you present plans that are carefully scaled and coordinated with schematic details or your details are sectional, scaled, and detailed. Given a choice, we prefer the carefully drawn plan and elevation with a schematic detail as the composite, explanatory item. Different contractors and suppliers will create variations regardless of what you draw. If your scheme really fits and your schematic detail covers all specified items, you have discharged your engineering and design function properly.

EIGHT

Wastewater Source Support Systems

The areas of work covered in this chapter deal specifically with lift stations, sewage disposal, manholes, catch basins, grease interceptors, roof drains, and vents. The first four items represent outside work and are part of the gray area between mechanical and civil-sanitary engineering. Certainly we would agree that if any of these items—lift stations, sewage disposal, manholes, and catch basins—are part of a full-scale site development plan or are part of a utility system, they belong in the civil-sanitary field. But this is not the situation usually confronting the mechanical engineer. The need is frequently one system or one item to complete the mechanical work. Very commonly the mechanical engineer's client expects him or her to at least know how to perform this task.

The lift station is aptly named and its function is precisely what its name implies—to lift the flow of liquid which is normally sanitary or storm water flow. The purpose of the lift is also fairly obvious. The storm or sanitary line is below, sometimes considerably below, the gravity flow line of the sewer system to which you must make a connection.

There are four lift station details in this chapter. Although the whole subject may seem simple and obvious, we caution you that this is far from true. Our details depict common sizes and will, in many cases, fit or nearly fit your situation. Do not, without proper and careful engineering and design effort, merely copy the details. The seemingly little differences can become considerable, expensive headaches.

Sewage disposal is frequently a specialty of its own. Many of the experts in the field do not have civil-sanitary engineering degrees although their background and experience in this field makes them remarkably qualified. A number of them began in private

or public environmental testing and plan approval areas. Over time they have developed almost a sixth sense of what is good, marginal, or unacceptable design. Assuming that a good design engineer in this area exists in your firm, we believe the details presented in this book will be very helpful to the detailer. The details represent solutions to given problems. Bear in mind that while your solution may be similar, it most likely will not be identical, and that the details will most likely require some modification.

Sewage ejection is a specialized solution to a specialized problem that can occur on any project and generally occurs as a solution to a problem faced by the mechanical engineer who is designing the plumbing system of an industrial or process plant.

Manholes are commonly used in all off-site and utility owned or privately owned sewage systems, and in these situations, there are specific requirements for their installation and spacing. This is definitely an area of sanitary engineering and not in the province of the mechanical engineer or detailer.

In most situations the project design breakdown of tasks, when extensive site engineering is not involved, is to have the mechanical plans include water and sewage services extended to street services as part of that mechanical planning. Frequently, this occurs even though a different trade is involved in exterior work. Commonly this presents no real problem since the lines are merely shown on a site plan connected to sanitary sewers and water mains.

The mechanical detailer's first problem occurs when, after the line leaves the building, the line makes a change in direction. Although some very slight bending or changing of pipe direction could be done by slowly curving the line, the proper way, under practi-

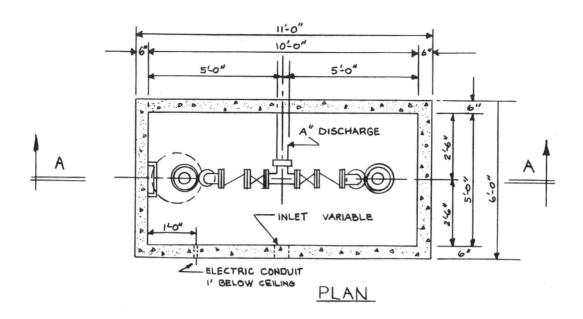

PLAN

NOTES:

1. ELECTRIC CONTROLS TO BE MOUNTED IN ENCLOSURE IN NEARBY BUILDING UNLESS OTHERWISE SPECIFIED.

2. DISCHARGE TO BE PROVIDED WITH DRESSER COUPLING.

A	3"	4"
B	2'-4⅞"	2'-9⅞"

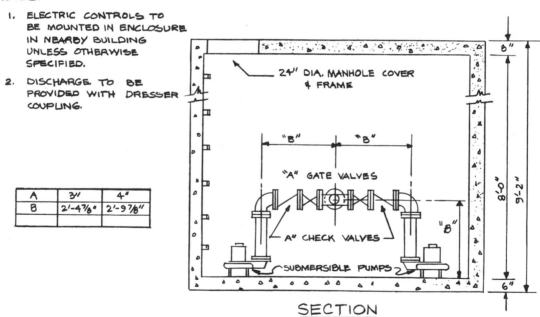

SECTION

SUBMERSIBLE WET PIT SEWAGE LIFT STATION

—NO SCALE—

(RECCOMENDED FOR SEPTIC TANK EFFLUENT USE ONLY.)

FIGURE 8-1

cally all circumstances, is to use a manhole to provide this directional change. Secondly, a line on the property should follow the same good design practice as does the street line, and if practice in the street requires a manhole every 200 ft, lines on the property should also have manholes every 200 ft even if they are in a straight line.

Storm drainage systems for a building frequently involve details covering areas of engineering that may be assumed to belong in the civil-sanitary field of practice but sometimes end up as part of the overall building storm drainage system. We conclude this gray area situation with four details that, by most reasonable definitions, are in the field of civil-sanitary engineering. However, in a limited situation of a drain or two in a parking lot this could end up being the mechanical engineer's job.

Our chapter concludes with three items on which there surely is not debate. Grease interceptors, roof drains, vents through the roof, and thrust blocks are very definitely interior plumbing items and part of the mechanical design.

Submersible Pumps

Figure 8-1 depicts submersible pumps in a sewage lift station. Below the title at the bottom of the page are the words "for septic tank effluent only." This detail is a resolution of the problem that occurs when your client finds there now is a sanitary sewer available and he or she can connect to it provided the effluent of the present sewage tank can be raised to meet the available sanitary sewer elevation requirements. At the top of the detail in the plan view we show an inlet of variable size. The size of the present septic tank outlet will be the size of this inlet.

This lift stage with its duplex pumps is large and relates to a fairly large existing septic tank and field. If it was reduced to a much smaller pump and a much smaller pit, it would be proper for a residential situation. We show, by setting discharge elevations, reasonable depths of liquid with reasonable spacing for 3-in and 4-in cast iron lines. These are guideline values. Proper calculations are required to determine these values. Proper calculations are also required to size the concrete holding chamber and to determine the size, type, and performance of the discharge pumps. Finally, proper sizing and selection of wiring and operating controls is required.

Lift Stations

Figure 8-2 is a typical solution to a similar problem, only this time storm water, not sewage effluent, is involved. In the situation depicted one could make a case that this lift station could solve either or both of two separate problems. First, the combined value of storm water flow during a storm coming from both the 10-in line and the surface drain is just too much for the

leaving 10-in line to handle. Or, second, the system discharges to a storm drainage line whose invert is higher than that of the system served. Note the ball check valves in the pump discharge line. If the main to which you are connecting is overloaded, you do not want water backing up in your system. This is especially true when the main to which you are connecting is above your system. The figure shows an operating and standby pump arrangement. Note the arrangement of operating level controllers. To get at the pump for any reason there is a 36-in access manhole. Do not skimp on this cover. You would be amazed at the size, quantity, and variety of objects that can clog up your pump.

Figures 8-3 and 8-4 are really two versions of the same idea: get the pump into a dry location. Figure 8-3 depicts the motor in a dry location, but the pump is still submerged. Figure 8-4 goes all the way, uses a different type of pump, and gets both the pump and motor into a dry location.

Sanitary engineers and mechanical engineers with sanitation expertise can make compelling arguments over the merit of either approach. Not necessarily totally true but generally true, the submerged pump with the dry motor location creates the deeper pit. Construction costs vary, and one system is not necessarily always less expensive than the other. We have shown concrete pits without reinforcement. This may very well not work in the solution to your problem, especially when any appreciable amount of wheel loading is imposed on your installation.

In both cases you have a different motor and control type than you have with the submersible arrangement. In both pits on Figs. 8-3 and 8-4 we show a separate heater and a separate dehumidifier. Frequently dry pits turn out to be very damp in the spring and fall, and in the winter in the colder areas lines have been known to freeze.

In Fig. 8-4, the totally dry location, we go to the ultimate and depict an exhaust system with piped intake and exhaust connections. Finally, in this figure in which the pumps are in a chamber that is separate from the wet well, the piping connections are made with Dresser couplings.

There are a lot of dimensions shown, which again are arranged to allow for the proper installation and connection of the pipes as sized. These must be carefully checked, engineered, and tailored to fit your exact requirements. They are not to be assumed to necessarily fit your project exactly as shown.

Again, we have not shown reinforcement in the concrete. This may be a requirement for your application. We freely admit that heaters, dehumidifiers, and exhaust systems may not be required for your project; engineering and design judgment is needed. The same caveat applies to which pump and which arrangement you ultimately select. Your tailored ver-

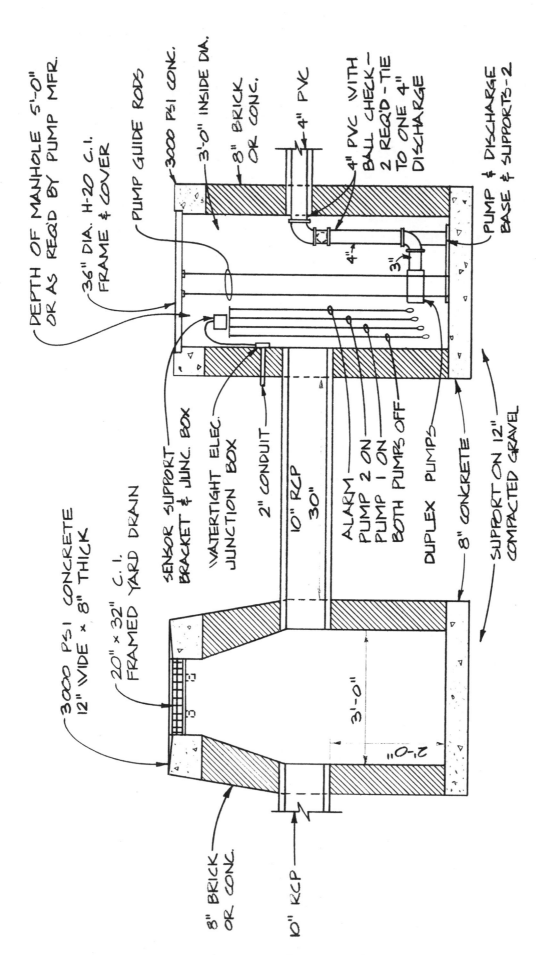

CATCH BASIN & LIFT PUMP STATION

NO SCALE

FIGURE 8-2

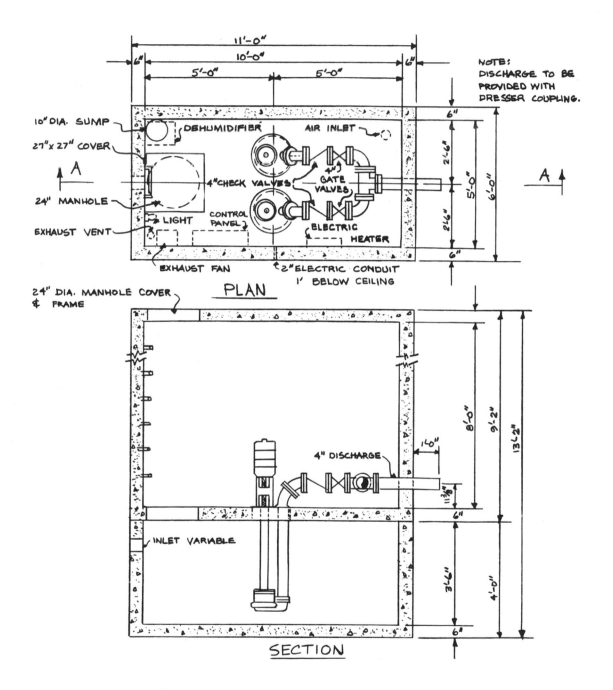

10" DIA. SUMP

27" x 27" COVER

24" MANHOLE

LIGHT

EXHAUST VENT

DEHUMIDIFIER

AIR INLET

4" CHECK VALVES

CONTROL PANEL

GATE VALVES

ELECTRIC HEATER

EXHAUST FAN

2" ELECTRIC CONDUIT 1' BELOW CEILING

NOTE: DISCHARGE TO BE PROVIDED WITH DRESSER COUPLING.

11'-0"

10'-0"

5'-0"   5'-0"

6"   6"

6"

2'-6"

5'-0"

6'-0"

2'-6"

6"

A   A

PLAN

24" DIA. MANHOLE COVER & FRAME

INLET VARIABLE

4" DISCHARGE

8'-0"

9'-2"

13'-2"

1'-6"

11 3/4"

6"

3'-6"

4'-0"

6"

SECTION

SUBMERGED TYPE · WET PIT SEWAGE
LIFT STATION
— NO SCALE —

FIGURE 8-3

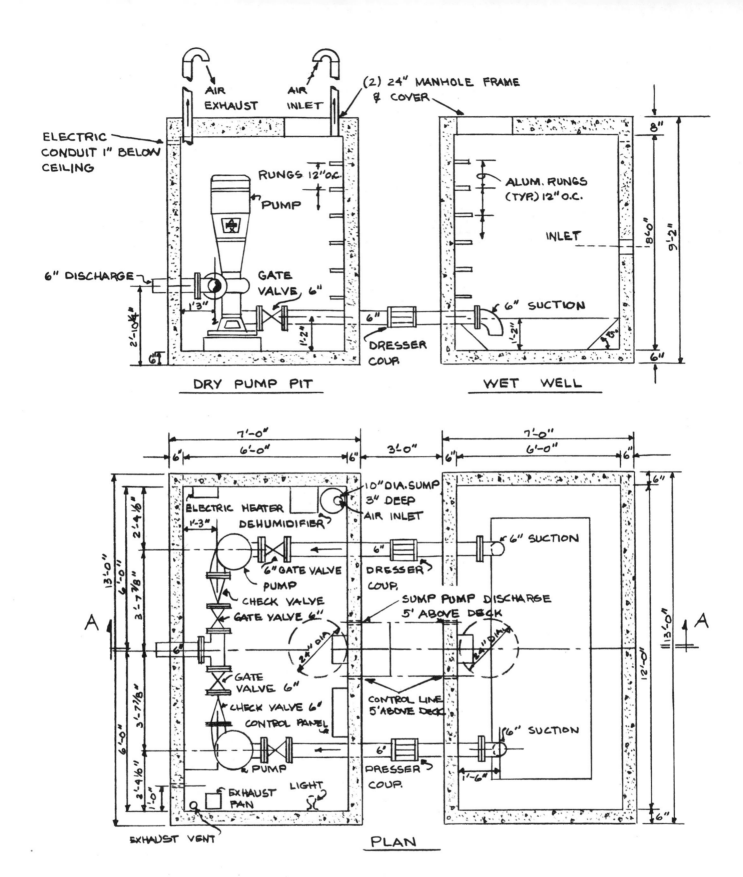

AIR
EXHAUST

AIR
INLET

(2) 24" MANHOLE FRAME
& COVER

ELECTRIC
CONDUIT 1" BELOW
CEILING

RUNGS 12"O.C.

PUMP

ALUM. RUNGS
(TYP.) 12" O.C.

INLET

8"

84'-0"

9'-2"

6" DISCHARGE

GATE
VALVE, 6"

6" SUCTION

1'-3"

6"

1'-2"

6" SUCTION

2'-10¼"

1'-2"

6"

6"

DRESSER
COUP.

6"

DRY PUMP PIT

WET WELL

7'-0"

3'-0"

7'-0"

6"

6'-0"

6"

6"

6'-0"

6"

6"

ELECTRIC HEATER
DEHUMIDIFIER

10" DIA. SUMP
3" DEEP
AIR INLET

2'-4½"

1'-3"

6'-0"

6" GATE VALVE

PUMP

6" SUCTION

6"

DRESSER
COUP.

13'-0"

3'-7⅞"

CHECK VALVE

GATE VALVE 6"

SUMP PUMP DISCHARGE
5' ABOVE DECK

A

24" DIA.

24" DIA.

A

13'-0"

12'-0"

6"

GATE
VALVE 6"

CONTROL LINE
5' ABOVE DECK

3'-7⅞"

CHECK VALVE 6"

CONTROL PANEL

6" SUCTION

6'-0"

2'-4½"

PUMP

6"

PRESSER
COUP.

1'-6"

1'-0"

EXHAUST
FAN

LIGHT

6"

EXHAUST VENT

PLAN

LIFT STATION WITH WET WELL

— NO SCALE —

FIGURE 8-4

sion of one of these two details will be the correct answer. While we frequently prefer not to dimension details, we would recommend dimensioning in this case as the most logical method of detail presentation of a lift station.

Sewage Disposal

When either direct or lift station connections are possible, the problem of the exterior sanitary system is relatively simple. One merely connects the sanitary line to the sewer or lift station and the problem is solved. But when there is no sanitary sewer available, on-site disposal is mandatory.

The design of the septic tank and field begins with two basic calculations—the amount of sewage effluent discharged per day and the amount of effluent per square foot that can be absorbed by the soil at the site. The basic system can be designed based on these facts and proper calculations.

Figure 8-5 shows a typical septic tank and field. Although the scale is very small, the field is actually fairly large, covering an area of 20,000 sq ft, or approximately ½ acre, and the tank as depicted is a 7500-gal tank. Generally, the tank and field are very close to the surface, usually with a foot or two of finished grade.

Depicted with the field are typical perforated tile lines which also can be made of certain types of plastic. Thus our tile field section could vary slightly in construction. Normally, to preclude upward percola-

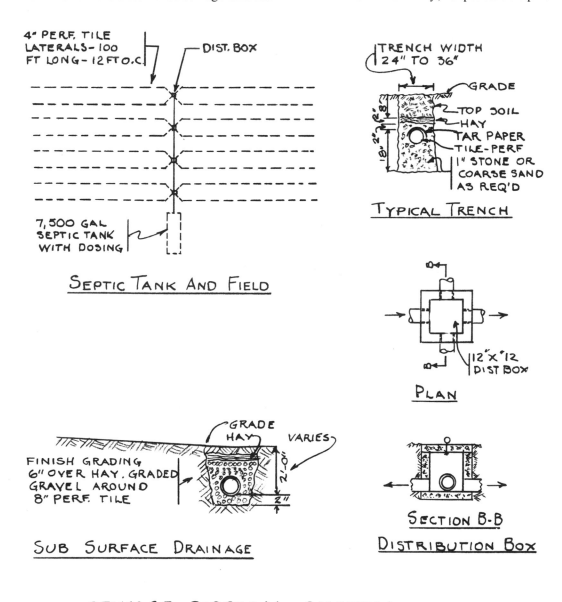

SEWAGE DISPOSAL SYSTEM

NO SCALE

FIGURE 8-5

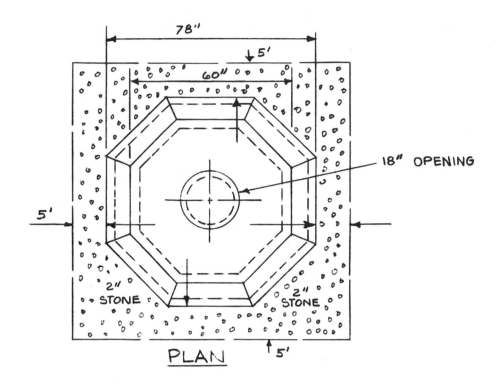

PLAN

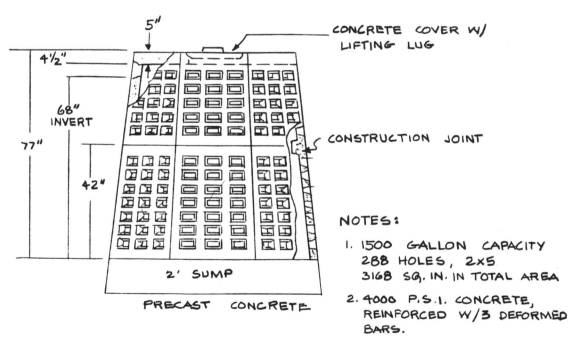

CONCRETE COVER W/
LIFTING LUG

CONSTRUCTION JOINT

2' SUMP

PRECAST CONCRETE

NOTES:

1. 1500 GALLON CAPACITY
   288 HOLES, 2x5
   3168 SQ. IN. IN TOTAL AREA

2. 4000 P.S.I. CONCRETE,
   REINFORCED W/3 DEFORMED
   BARS.

ELEVATION

TYPICAL LEACHING GALLEY

NO SCALE

FIGURE 8-6

tion the pipes in the field are covered with tar-coated paper. The material below the pipe can be either sand or gravel. With good rates of soil absorption it is usually gravel; with poor rates, it is usually sand.

There may be a problem with water on or just under the field. Shown in Fig. 8-5 is a section detail of the typical drainage tile. Considerable care is required in locating this sort of tile to prevent it from acting as a conduit for sewage. And the drainage is of no value if you do not have a proper outlet location. The field indicates diverting boxes and a small detail is shown for them.

Leaching Galley: Figure 8-6 describes a leaching galley, which might be called a solution of last resort. No local or state health department is, most likely, going to accept this sort of solution if it is at all possible to install a leaching tile field. Our detail is representative of a typical installation using prebuilt sections. Frequently there will be limitations on the total depth of the galley. This particular detail came from a situation in which the soil tested acceptably even at 6 ft 0 in below grade. While the detail is correct for our particular situation, capacities and dimensions must be checked with the local and state boards of health having jurisdiction over your project.

Septic Tanks: Figure 8-7 is a typical detail of a 5000-

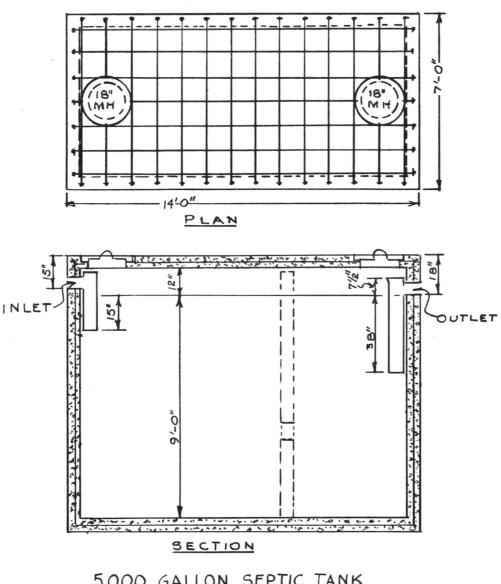

PLAN

SECTION

5,000 GALLON SEPTIC TANK

NO SCALE

FIGURE 8-7

gal septic tank. One word of caution is in order here. Some states and health agencies have specific rules that tanks over 2000 gal must be divided into two compartments with a concrete divider so as to create a space that is two-thirds of the total tank volume in the section nearest to the inlet to the tank. To get flow from the first chamber to the second an opening with a diameter of 6 to 8 in is placed in the partition wall approximately at the midpoint of tank liquid height. In our detail this would be at 4 ft 6 in. The dimensions of our tank are typical. They can vary so long as the liquid depth is 3 ft 0 in or greater. Our tank is reinforced by 6-in × 6-in no. 10 mesh with no. 4 rods, 12 in on center, in both directions. The concrete is 3750 lb.

Standard to all septic tank designs is the 3-in differential between the inlet and outlet piping, which creates the necessary rate of flow. The depth of inlet and outlet pipes can vary from our detail and should be checked against code requirements in your area. The 18-in manholes shown on top of the tank will usually require a masonry access box with frame and cover at grade on tanks over 2000 gals of capacity.

Figure 8-8 repeats the tank shown in Fig. 8-7 and adds a dosing chamber. Generally when the total length of the leaching field exceeds 600 ft, codes will require that there be a means to secure adequate overall field distribution by either pumping or dosing. Usually the dosing device is the Miller siphon, which by the arrangement of liquid pressure and venting will empty at a fixed level, flushing the liquid trap and allowing the contents to be released. There is a certain amount of design in this system. If the dosing period is overly long, the expected good distribution may not occur, as the effluent may seep away too rapidly in one or another portion of the field. Usually dosing tanks are designed to discharge at intervals of 2 or 3 hr at a volume equal to three-quarters of the volume of the tile field. Also, the rate of discharge at minimum head should be at least 25 percent more than the maximum rate of flow of sewage to the septic tank. This excess of flow capacity vents the siphon, which otherwise might discharge continuously.

Sewage Ejection

The problem of the sanitary effluent, at least as far as the mechanical details and the mechanical engineer are concerned, lie normally in solving on-site disposal of effluent or providing a means to make the proper connection into the available sanitary sewer. The

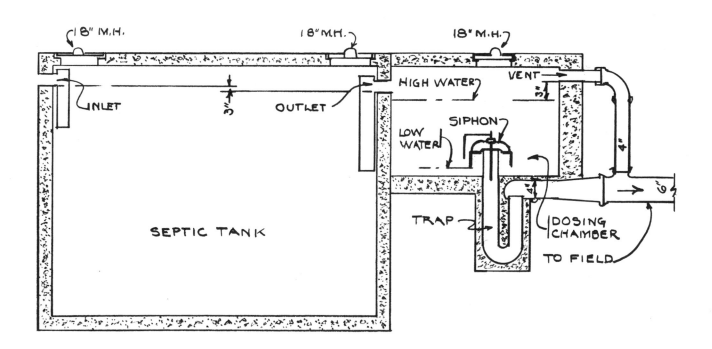

SEPTIC TANK WITH DOSING CHAMBER

NO SCALE

FIGURE 8-8

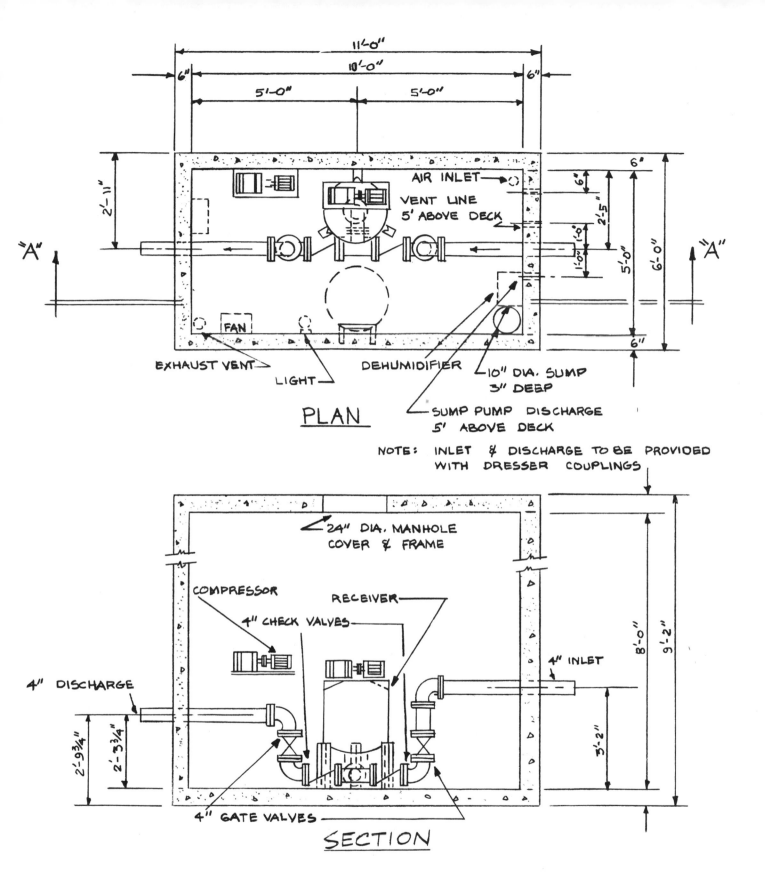

PLAN

NOTE: INLET & DISCHARGE TO BE PROVIDED
WITH DRESSER COUPLINGS

SECTION

PNEUMATIC SEWAGE EJECTOR STATION
NO SCALE

FIGURE 8-9

sewer connection is only a problem for the detailer when a lift station is involved.

There is one other problem that the mechanical designer and detailer may face. Figure 8-9 depicts a typical pneumatic sewage ejector station, and Fig. 8-10 shows the arrangement of compressed air piping to the ejector. As can be seen from the indicated sources of air supply—plant or compressed air—the ejector is used in special situations.

Pneumatic ejectors are frequently used when the

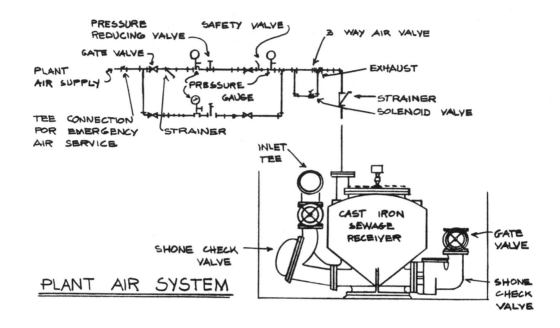

PLANT AIR SYSTEM

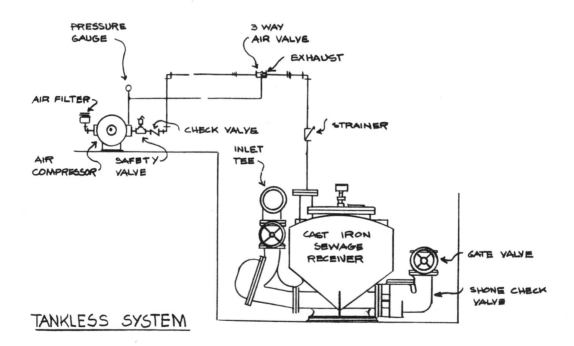

TANKLESS SYSTEM

PNEUMATIC EJECTOR

FIGURE 8-10

rated system capacity is less than 100 gpm. Generally, at the smaller rates of flow the normal centrifugal pump would be oversized since freedom from clogging is usually based on handling solids up to 3 in. The advantage of the ejector is even more striking when larger sewage solids and rags are a problem.

The pneumatic ejector has no moving parts and operates on air pressure. Sewage flows into the receiver by gravity through inlet gate and check valves. Air is vented out of the receiver as it is displaced by the sewage. The end pipe outlet is noted in the upper portion of Fig. 8-9. When sewage fills the receiver, it activates the sensing control of the system, which blocks off the receiver vent and connects the receiver with the compressed air through the three-way valve. The high pressure air forces the sewage out of the receiver and into the discharge piping system. The direction of flow is controlled by the placement of the check valves. When all sewage is ejected, the system is automatically returned to its vented sewage-receiving mode.

The piping detail shown on Fig. 8-10 is powered by plant air or compressed air. The detail is, of course, subject to minor corrections that may relate to your particular situation. These systems as depicted may also in certain sizes be selected as a commercially manufactured and packaged installation.

Since sewage flow usually is not a situation in which related heating is required, none is shown for the pit detailed in Figs. 8-9 and 8-10. The pit does include such items as a sump pump, dehumidifier, fan, and exhaust vent. Whether or not all of these items are required in your situation should be decided by the project engineer.

The discharge of the ejector can be higher than the inlet although the detail does not show it as such. The concrete type and the reinforcing, as well as type of manhole frames and covers, should be based on the ejector station location and its wheel loading.

Manholes

Figure 8-11 is a reinforced concrete manhole that is designed so that it reasonably conforms to a probable prefabricated manhole. Thus, it really is not completely detailed. An argument could be made as to why even show it at all. Primarily this sort of detail is shown to illustrate the features and shape of the manhole in sufficient detail to clarify the specified item. The manhole, for instance, requires an offset frame and cover.

In the detail we have not described the angles of the inlet and outlet. At present unless you so note these angles, you are going to get a straight-through manhole with a spare opening at a 90° angle to this straight-through channel. This spare opening is shown as a circle within a circle. Your situation may not require two incoming connections. If it does not, delete the circle. Secondly, indicate on your detail exactly at what angle the incoming line is relative to the position and direction of the leaving line.

Figure 8-12 is a typical built-up manhole constructed of brick, mortar, and concrete, with a cast iron frame and cover. This is a round manhole with a tapered cone at the top, terminating in a cast iron frame and

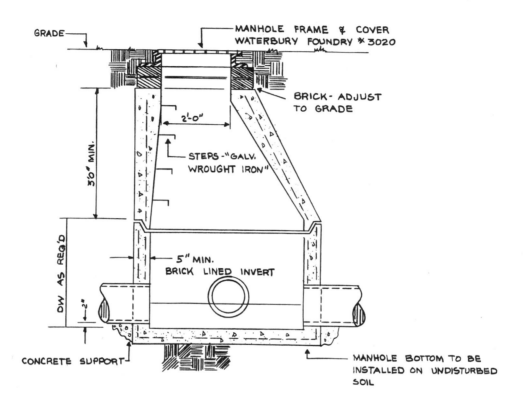

MANHOLE   DETAIL

NOTES:

1. MANHOLE CONE & RISER SECTIONS TO CONFORM TO A.S.T.M. DESIGN C-478-61T.

2. PROVIDE LIFTING HOLES ON ALL UNITS.

3. ABSORPTION NOT TO EXCEED 8% AS PER A.S.T.M. C-76.

4. FIELD BUILT MANHOLE WILL BE ACCEPTABLE - FURNISH SHOP DRAWINGS SHOWING CONST. DETAILS.

FIGURE 8-11

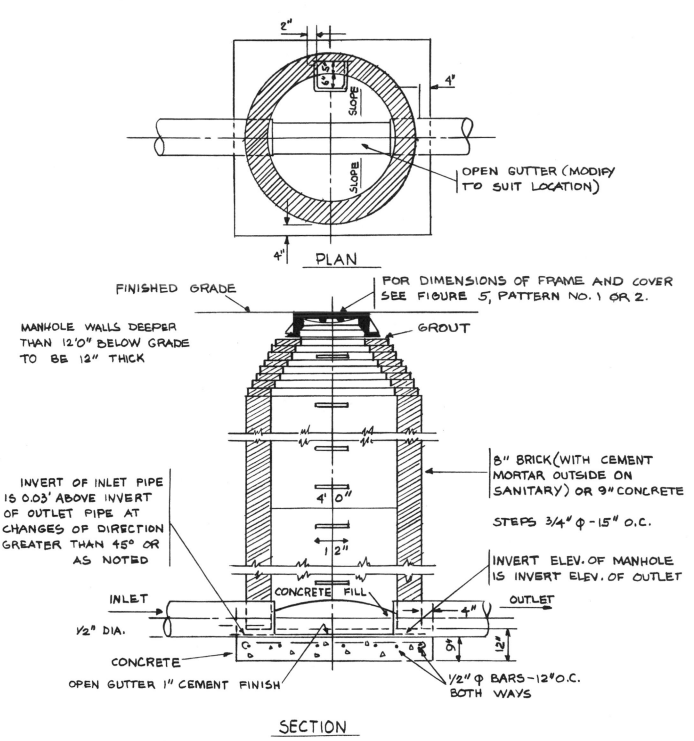

2"

4"

4"

SLOPE

SLOPE

OPEN GUTTER (MODIFY TO SUIT LOCATION)

PLAN

FINISHED GRADE

FOR DIMENSIONS OF FRAME AND COVER SEE FIGURE 5, PATTERN NO. 1 OR 2.

MANHOLE WALLS DEEPER THAN 12'0" BELOW GRADE TO BE 12" THICK

GROUT

8" BRICK (WITH CEMENT MORTAR OUTSIDE ON SANITARY) OR 9" CONCRETE

STEPS 3/4" φ - 15" O.C.

INVERT OF INLET PIPE IS 0.03' ABOVE INVERT OF OUTLET PIPE AT CHANGES OF DIRECTION GREATER THAN 45° OR AS NOTED

4' 0"

1' 2"

INVERT ELEV. OF MANHOLE IS INVERT ELEV. OF OUTLET

CONCRETE FILL

INLET

OUTLET

1/2" DIA.

4"

9"

12"

CONCRETE

OPEN GUTTER 1" CEMENT FINISH

1/2" φ BARS - 12" O.C. BOTH WAYS

SECTION

STANDARD MANHOLE, SANITARY & DRAINAGE

FIGURE 8-12

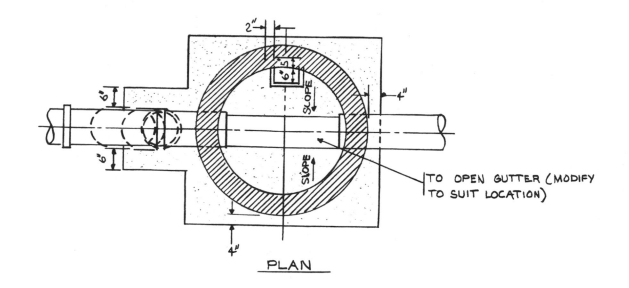

PLAN

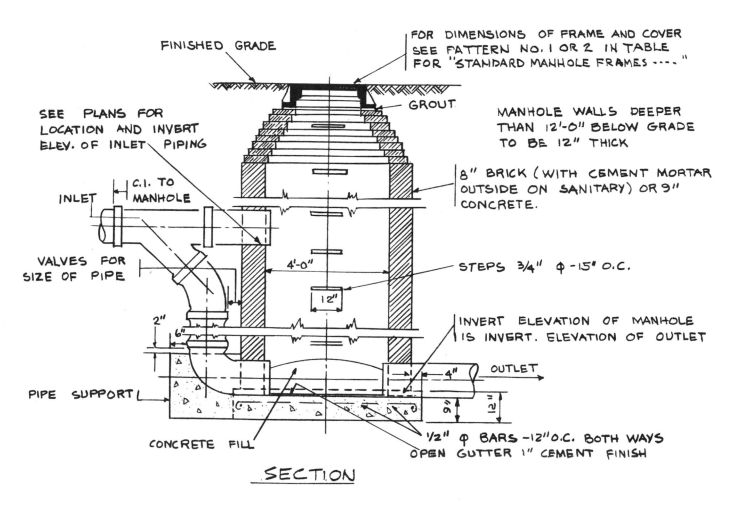

FINISHED GRADE

FOR DIMENSIONS OF FRAME AND COVER
SEE PATTERN NO. 1 OR 2 IN TABLE
FOR "STANDARD MANHOLE FRAMES ----"

GROUT

SEE PLANS FOR
LOCATION AND INVERT
ELEV. OF INLET PIPING

MANHOLE WALLS DEEPER
THAN 12'-0" BELOW GRADE
TO BE 12" THICK

INLET

C.I. TO
MANHOLE

8" BRICK (WITH CEMENT MORTAR
OUTSIDE ON SANITARY) OR 9"
CONCRETE.

VALVES FOR
SIZE OF PIPE

4'-0"

STEPS 3/4" φ -15" O.C.

12"

2"

6"

INVERT ELEVATION OF MANHOLE
IS INVERT. ELEVATION OF OUTLET

OUTLET

4"

PIPE SUPPORT

9"

12"

CONCRETE FILL

1/2" φ BARS -12" O.C. BOTH WAYS
OPEN GUTTER 1" CEMENT FINISH

SECTION

STANDARD DROP MANHOLE SANITARY AND DRAINAGE

FIGURE 8-13

cover. Normally, its diameter is 4 ft 0 in. The base is concrete with a half-round channel, or open gutter, for sewage flow inside the concrete. Sometimes this gutter is created by embedding a half-round piece of sewer tile of the same size as the connecting line.

The detail has a plan view of the type we referred to in Fig. 8-11, the prefabricated manhole. In our detail we depict the manhole as a straight-through manhole. This is where you show on your detail the exact leaving angle and *curve* the channel through the sewer bottom conforms to.

In your design the problem of too high a velocity of sewage flow can be as troublesome as too low a velocity. This is another engineering design problem. However, the manhole can be an ideal place to solve the problem. Figure 8-13 is a detail of a typical "drop" manhole in which at one spot we suddenly lower the sewage line very drastically. This is a fairly tricky thing to do.

Regardless of whether the lines to the manhole are tile or cast iron with bell and spigot joints, the incoming wye is cast iron, as depicted. There is a lot of pressure and velocity here, and we want a tight joint. To save excavation, shoring, etc., we have kept the sewer line up high and now we have got to get it down so that it flows properly to the street sewer.

The manhole is basically the same as before. We really do not want all the splashing and turbulance of a sudden drop, so we have created a sort of high speed entry to the manhole with our wye connection. But there are many times when flow is too great, and we will simply have to accept the splashing with our straight-in connection. This sort of manhole is frequently seen in street sewer piping; the utility company also likes to save on excavation when possible.

Figures 8-12 and 8-13 refer to a detail on manhole frame and cover types. Figure 8-14 depicts that detail. There are basically three types of manhole frames and covers. One is to be used in paved areas, one is to be used in unpaved areas, and one is for situations requiring extra strength. You could use this detail as is or choose the weights that satisfy your requirements.

In our discussion of lift stations we showed details of wet wells. Manholes on occasion are also used as wet wells. This detail is more of a design chart than an actual detail, although it is actually a little of both.

On occasion the "drop" manhole depicted in Fig. 8-13 can be modified as shown in Fig. 8-15 to become a wet well for a pump system. Obviously as the sizes grow larger, you really no longer have a manhole but sort of a hybrid. And the larger sizes are best constructed of reinforced concrete. In one detail we show

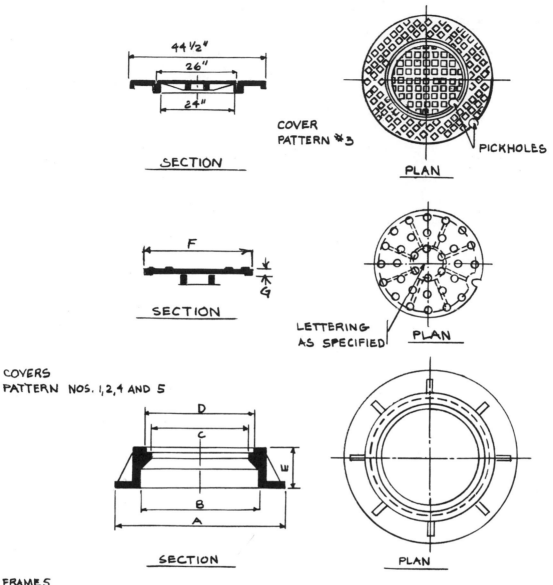

44 ½"

26"

24"

SECTION

COVER
PATTERN #3

PLAN

PICKHOLES

F

G

SECTION

PLAN

LETTERING
AS SPECIFIED

COVERS
PATTERN NOS. 1,2,4 AND 5

D

C

E

B

A

SECTION

PLAN

FRAMES
PATTERN NOS. 1,2,3,4,5

PATTERN	WT., MIN.	A	B	C	D	E	F	G
1	440 LB	36"	26"	20½"	23¼"	9"	1-3/4"	23"
2	330 LB	36"	25½"	20½"	23¼"	9"	1-1/2"	23"
3	1075 LB	56"	46"	42"	44¾"	10"	13/4	44
4	435 LB	38"	27"	23"	25¼"	9"	1½"	25"
5	350 LB	35"	25½"	23"	25¼"	7½"	1½"	24"

NOTE: COVER FOR STORM MANHOLES TO BE SOLID OR PERFORATED
AS SPECIFIED
PATTERN NOS. 1 AND 4 FOR USE IN PAVED AREAS.
PATTERN NOS. 2 AND 5 FOR USE IN UNPAVED AREAS

STANDARD MANHOLE FRAMES & COVERS, PATTERNS #1,2,3,4 & 5

FIGURE 8-14

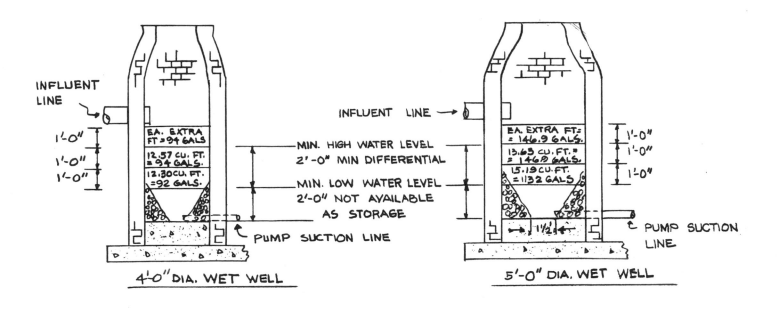

INFLUENT LINE

EA. EXTRA FT = 94 GALS
12.57 CU. FT. = 94 GALS.
12.30 CU. FT. = 92 GALS.

1'-0"
1'-0"
1'-0"

MIN. HIGH WATER LEVEL
2'-0" MIN DIFFERENTIAL
MIN. LOW WATER LEVEL
2'-0" NOT AVAILABLE AS STORAGE

PUMP SUCTION LINE

4'-0" DIA. WET WELL

INFLUENT LINE

EA. EXTRA FT. = 146.9 GALS.
13.65 CU. FT. = 146.9 GALS.
15.19 CU. FT. = 113.2 GALS.

1'-0"
1'-0"
1'-0"

1½"

PUMP SUCTION LINE

5'-0" DIA. WET WELL

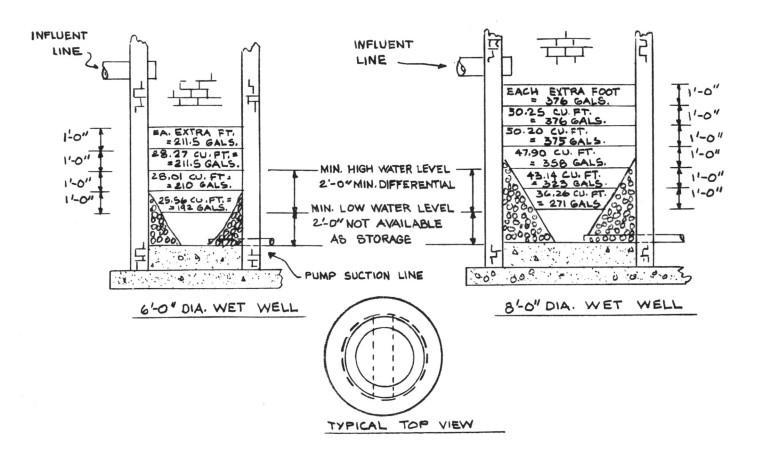

INFLUENT LINE

EA. EXTRA FT. = 211.5 GALS.
28.27 CU. FT. = 211.5 GALS.
28.01 CU. FT. = 210 GALS.
25.56 CU. FT. = 192 GALS.

1'-0"
1'-0"
1'-0"
1'-0"

MIN. HIGH WATER LEVEL
2'-0" MIN. DIFFERENTIAL
MIN. LOW WATER LEVEL
2'-0" NOT AVAILABLE AS STORAGE

PUMP SUCTION LINE

6'-0" DIA. WET WELL

INFLUENT LINE

EACH EXTRA FOOT = 376 GALS.
30.25 CU. FT. = 376 GALS.
50.20 CU. FT. = 375 GALS.
47.90 CU. FT. = 358 GALS.
43.14 CU. FT. = 323 GALS.
36.26 CU. FT. = 271 GALS.

1'-0"
1'-0"
1'-0"
1'-0"
1'-0"
1'-0"

8'-0" DIA. WET WELL

TYPICAL TOP VIEW

TYPICAL WET WELL DESIGNS

FIGURE 8-15

typical capacities in gallons for each foot of height. It is a sort of handy design capacity chart. For reinforcing we suggest you utilize a structural engineer.

Catch Basins

Figure 8-16 is a typical curb-type catch basin that is frequently seen in city streets. However, a shopping center parking lot with sidewalks, curbs, and roads frequently has the same or similar roadway drain requirement. In this detail there appears at grade a 4-ft 4-in × 5-ft 4-in concrete slab with a 2-ft 6-in × 2-ft 0-in drain grating. It is a little larger than the usual residential road drain. Note that the unsized drain line leaves the drain pit some 2 ft 6 in above the bottom.

This frame, cover, pit, and unsized drain line must all be sized according to the amount of surface runoff

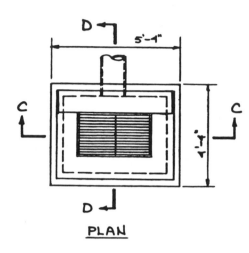

PLAN

WITH CONCRETE MASONRY CORBELLING WILL BE PERMITTED TO A MAX. OF 3"

CATCH BASIN WALL TO BE 12" THICK WHEN DEPTH OF MANHOLE IS GREATER THAN 10' (MASONRY ONLY)

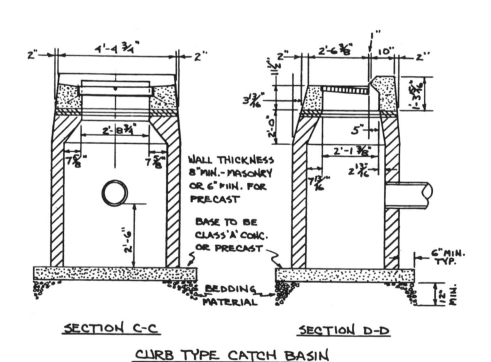

SECTION C-C

SECTION D-D

WALL THICKNESS 8" MIN. - MASONRY OR 6" MIN. FOR PRECAST

BASE TO BE CLASS 'A' CONC. OR PRECAST

BEDDING MATERIAL

6" MIN. TYP.

CURB TYPE CATCH BASIN

NOT TO SCALE

FIGURE 8-16

to the drain. This calculation is required to be certain that the pit is sized to take surges in flow and that there is a surface dirt collector, which must on occasion be cleaned of debris. This sort of drain pit frequently is specified as a precast item. The line sizing is part of the calculations. Without attempting to guess the design, we will only note these lines are normally 8 to 12 in,

depending on runoff, pitch, and other considerations.

Figure 8-17 is very similar, in fact almost identical in pit construction, to Fig. 8-16. All the previously noted caveats apply to this drain. The difference is in the top concrete covering and the cast iron frame and cover. Figure 8-17 is preferred in many instances for parking area drainage because of its two-section cover. In

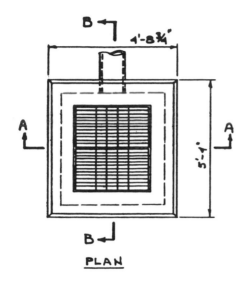

PLAN

WITH CONCRETE MASONRY CORBELLING
WILL BE PERMITTED TO A MAX. OF 3"

CATCH BASIN WALLS TO BE 12" THICK
WHEN DEPTH OF MANHOLE IS GREATER
THAN 10'. (MASONRY ONLY)

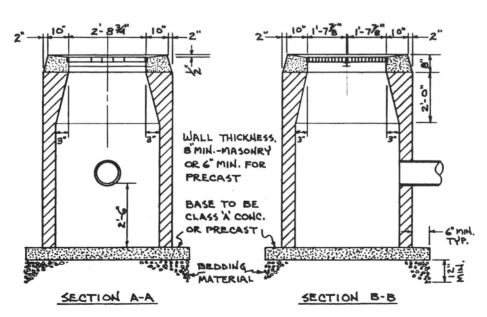

SECTION A-A   SECTION B-B

DOUBLE GRATE CATCH BASIN

NOT TO SCALE

WALL THICKNESS,
8" MIN.-MASONRY
OR 6" MIN. FOR
PRECAST

BASE TO BE
CLASS 'A' CONC.
OR PRECAST

BEDDING
MATERIAL

FIGURE 8-17

relatively "dirty" areas of heavy truck parking and yard work the pit requires frequent cleaning. If you have ever tried to pry up a heavy, semistuck drain cover, you would welcome the chance to merely raise half of it and do your cleaning.

Catch Basin Sections: There are special purpose area drains which may connect to drain lines of fairly shallow drainage systems. Figure 8-18 is one such possible solution. It has a double grate which tends to trap debris more easily at the top. As such it also can plug up at the top more easily and create a small lake in the parking lot, especially after a summer thunderstorm. The drain is on the bottom, and there is a special tapered bottom to the drain to facilitate smooth runoff. While very usable in the proper design situation, this sort of drain requires careful site drainage engineering and analysis.

In Fig. 8-19 we again show the curb drain. This time it is clearly identified as a precast-type. It is obviously part of a pass-through drainage system. Since it is going to arrive on the job with precast holes, the connecting drain lines are a little more clearly noted to finish flush with the drain pit wall. It is a small point, but careless installation somehow seems to be more prevalent with precast items. The detail looks very much like the precast drain in the supplier's catalog. We feel that if you are going to specify a precast drain, your detail should look like the typical available type.

Most mechanical engineers do not, and really should not, get involved with site, street, or road surface drainage. In all honesty if you were involved in a small commercial parking lot of 20,000 to 40,000 sq ft and you researched the subject, you probably could put in a few drainage basins and connect them with reasonably sized piping, probably 10 to 12 in, and do a satisfactory job. Even in this case the hiring a civil-sanitary engineer to check your design is a logical move. Our details are presented more on the premise that a workable design already exists but is being altered or adjusted because of breakage or site repaving, or because you are putting in some underground tanks. You really are not designing a drainage system.

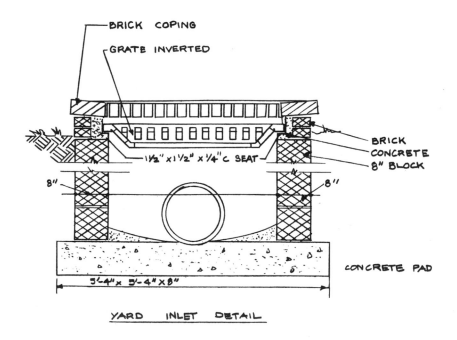

TYPICAL SECTION THRU CATCH BASIN

FIGURE 8-18

Rather, you are rearranging one that is already there with new catch basin locations, and you need a detail of the relocated catch basins.

Grease Interceptors

In a building sanitary wastewater system designed by the mechanical engineer there frequently occurs the problem of grease or acid waste. The two most common sources of these wastes are the kitchen and the laboratory. Resolving the problem usually can be accomplished by referring to standard manufacturer's catalog data. No real detailing is required.

Generally, these standard devices are referred to as grease or acid traps. However, as the size becomes larger, they are generally referred to as grease interceptors or acid neutralizing sumps. The manufacturer's standard packaged items can handle large amounts of grease or, in the case of neutralizing sumps, the discharge of as many as 20 laboratory sinks.

In the case of grease or acid the basin requirement is to contain, or detain, the flow of the liquid long enough to solve the problem. In the case of grease, it will float on top, and the interceptor simply slows the flow enough to allow the grease to coagulate there and the relatively grease-free waste water to be drawn off below and discharged to the sanitary sewer or septic field. Most plumbing codes mandate a grease trap in commercial kitchen installations.

An acid neutralizing sump is somewhat similar to a grease trap in that it provides a large enough capacity to slow down the flow so that some corrective action may be taken. But acid is not grease, and it is usually not going to be removed by mere settling through detention. The acid neutralizing sump usually contains limestone or marble chips which react with the acid to create a neutral salt liquid that is discharged to a sanitary sewer. If it is going to be a septic tank and field, additional chemical treatment and a separate disposal system will most likely be required.

Occasionally the designer may be faced with an existing system which for several reasons cannot

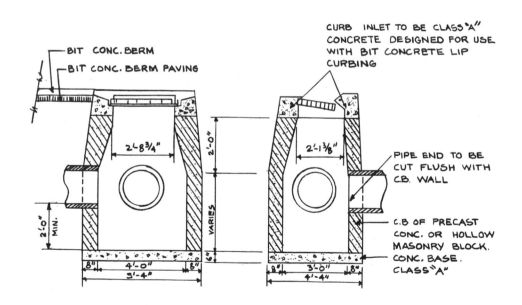

TYPICAL SECTION THRU PRECAST CATCH BASIN

FIGURE 8-19

accommodate an interior grease trap. When such a situation exists, Fig. 8-20 may be the detail to resolve your problem. Generally the volume of liquid in a workable container has a length, in flow direction, that is about twice the width or depth.

Fig. 8-20 shows a more or less typical trap. The grease storage or capture area is some 2 cu ft, which should handle many fairly large kitchens on the basis of twice a year grease removal. The design is really very simple both in design and in premise. The influx of grease-laden water is slowed and forced to make two 90° turns to a larger chamber. This consecutive slowing allows the grease to rise to the top and the relatively clear liquid to be discharged.

Your actual sizing may vary. We prefer the unit to finish slightly above grade, some 4 to 6 in, to preclude surface water from running into the unit since the removable cleaning cover, usually a 12-in × 12-in

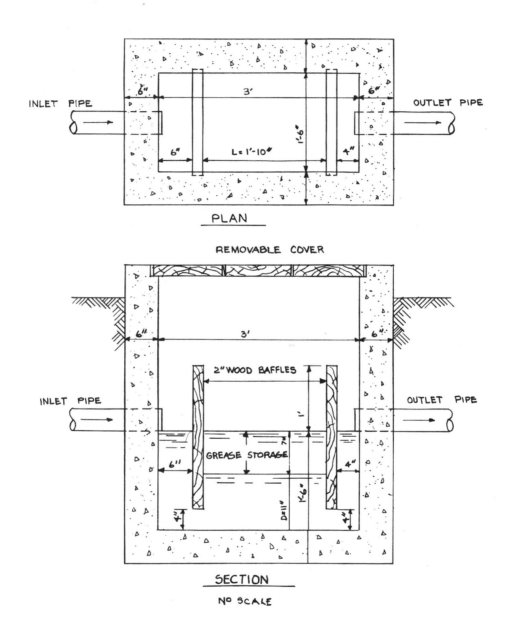

DETAILS OF GREASE INTERCEPTOR

FIGURE 8-20

square is never water tight. Concrete is normally 3000-lb concrete and reinforcement is not required. Intake and discharge pipes, usually cast iron, should be, whenever possible, cast in place during the concrete pour.

Roof Drains

There are various types of roof drains, and there are a variety of roof structures on which the drains may be installed. There are also many sources of a roof drain detail. With all of the above facts the natural question would be "why bother to show any?" Frankly we almost decided to take that course of action.

Figure 8-21 is a typical roof drain. There are other varieties that have slightly different installation requirements. Our detail does illustrate certain common items that should always be part of a roof drain detail.

The basic elements of the drain installation are the drain, its clamping and flashing, the pitch of the roof to the drain, and the division of work among the trades involved. Generally, but not always, the drain is

furnished to the roofer, who sets the drain. Our detail does not specifically cover this point.

Regardless of who installs what, the roof should be pitched toward the drain. Our detail says that this task is the roofer's job, which it usually is. Another area of divided responsibility is the flashing around the drain. Again, our detail says that the roofer does this but that the flashing material is supplied by the plumber.

The roof drain has to be clamped onto the roof structure and onto the flashing. Our detail does not really say who does this, but usually it is done by the roofer. We merely say (in note 3) that the plumber will furnish the sump receiver to the general contractor so that it can be installed by some trade.

In notes 1 and 2 we point out where the drains should be located and how they should be piped. Technically those notes are both design and installation criteria. They do not really belong on the detail, but many offices will put these notes or similar notes on drain details. Although the location of the drains has already been selected, there are certain advan-

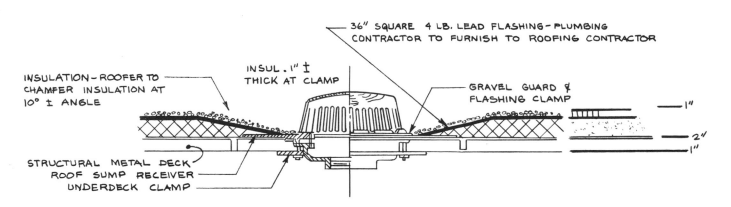

1. LOCATE AT LOW POINTS OF ROOF.
2. PIPE WITH EXPANSION JOINTS OR OFFSETS TO ALLOW FOR EXPANSION.
3. FURNISH SUMP RECEIVER TO GENERAL CONTRACTOR SO RECEIVER CAN BE ATTACHED TO THE DECK.

ROOF DRAIN DETAIL

NO SCALE

FIGURE 8-21

tages in saying they should be installed at low points on the roof. Certainly the note on piping is logical. The note on the furnishing of the sump receiver should be in your specification. If it is not, note 3 may solve many an argument.

Vents

Since all sanitary piping systems require venting, the detail shown in Fig. 8-22 is both obvious and mandatory. This simple detail usually suffers because it is so simple. In far too many plans this detail is treated in true cavalier fashion with the meaningless note, "vent through roof, flash as per code." What it really says is "we don't know, you figure it out."

The installing contractor has no choice but to figure it out and install something. This is exactly what will occur. And if the flashing or the size is wrong, the vent is too tall, or the contractor wants an additional

payment to clear the vent from a window, the problem will be yours alone. Even good contractors occasionally make mistakes in flashing or sizing this simple device.

Flashing, for example, does not have to extend up to the top of the vent and be turned down inside. There are code-accepted flashings that do not do this. If you want what we show and feel is the best job, you are going to have to detail it. If you want an escutcheon on the house-side of the vent as we have shown, it will have to be detailed. The same goes for the vent increases. Frequently lines leading to vents are less than 4 in. Some codes may allow 3-in vents. If you want a 4-in vent, you will have to say so in your detail. Do you want a vent sleeve? If so, detail it.

You may well not want, need, or require all of the items to be as we have detailed them. Obviously there are alternative legal code-acceptable versions which

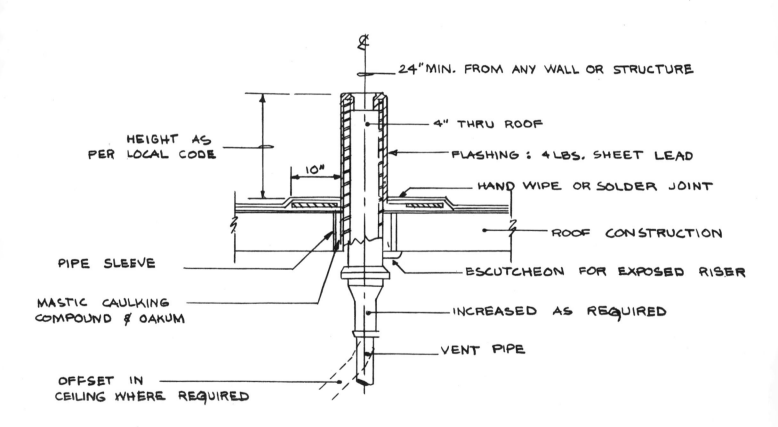

VENT THRU ROOF DETAIL

NO SCALE

FIGURE 8-22

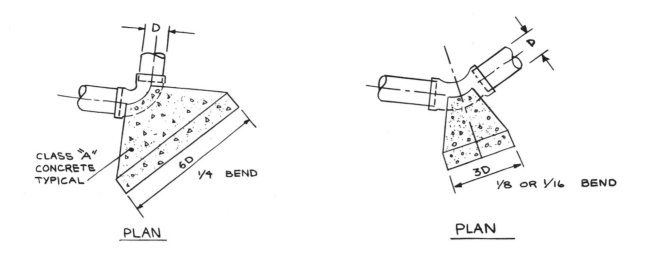

CONCRETE THRUST BLOCK

NO SCALE

FIGURE 8-23

vary from city to town and from state to state. Material also can vary. Although we can easily agree to a modification of our detail, we can never agree to the omission of it.

Thrust Block

One of the common items overlooked on a mechanical engineer's site plan is the thrust block depicted in Fig. 8-23. Our detail is very limited in scope of information for a very good reason. This sort of block is definitely not to be copied as is. It must be part of the engineering calculation of forces created by rapid changes in water pressure.

There are, undoubtedly, completed projects that work perfectly satisfactorily that have no thrust blocks on any domestic water or fire service line to the project's structure. Generally, water lines 4 in and over and all fire lines are cast iron, usually cement lined AWWA class 150. In the vast majority of these installations the direction leaving the building is adjusted inside the building so that the run is a straight line or almost a straight line to the street utility main.

The first mistake commonly made is in the "almost straight line" situation. The cast iron line does not bend. A change of direction must be made with a ¼, ⅛, or ¹⁄₁₆ standard pipe fitting. They do not come in sizes in which the change of direction can require some peculiar angle. Your plan may neatly show such a situation but your plan, in so doing, is wrong. The line cannot be installed as drawn.

Not infrequently the line coming from the street to a domestic or fire service has to make a turn to supply a hydrant or to supply the building. Our detail shows the typical size of block in relationship to the pipe diameter that generally fits your problem. You will note that we do not show the depth of concrete. We only want to show generally how the block will be detailed. The thrust that is created by sudden, rapid changes in water pressure is many times easily capable of blowing apart any pipe-bend fitting. This is not a guess or a copy job situation. It has to be calculated. Commonly, a few more dimensions and notes based on your calculations are all you are going to need to complete the detail depicted in Fig. 8-23.

NINE

Special System Source Equipment

There are five systems that are commonly part of the mechanical designer's assignments that do not merely contain special items of equipment but are specialties of and by themselves. These include the gaseous fire extinguishing systems, which usually use carbon dioxide or Halon as the gas; vacuum systems that are used in cleaning, process, and laboratory work; compressed air systems; heat tracing systems; and snow melting systems.

According to the National Fire Protection Fire Code *Handbook*, Vol. 1, Sec. 12, 12A, and 12B, the gaseous fire extinguishing systems will put out fires caused by flammable gaseous and liquid materials; electrical sources, such as transformers, switches, breakers, and motors; gasoline and other flammable fueled engines; ordinary combustibles, such as paper, wood, and textiles; and hazardous solids. Halon can also be used in computer rooms whereas carbon dioxide cannot.

Vacuum and compressed air systems are in many ways related to each other. Our details on vacuum systems not only cover laboratory and cleaning systems, which are the two most common systems, but also cover compressor and vacuum pump cooling piping. This is a frequently discovered-at-the-last-minute mechanical detail problem. We have not only provided details for the normal 50 and 125 psig compressed air systems but have also included some details of the very special, very high pressure outlets. For pressures up to 125 psig the outlet detailing is usually a simple, valved connection, but at 5000 psig outlets become very, very special design and detail items.

Heat tracing is another not regularly encountered system in which the source may be steam or electricity and the installation and design details are special. Electrical heat tracing is not normally encountered by mechanical designers. The usual mechanical design normally uses steam.

There are three commonly used ways to melt snow from sidewalks, driveways, and parking areas. These are infrared radiant overhead systems, electric resistance cables buried in the slab, and pipe coils that use circulating heated fluids and are also buried in the slab. Of these, the one that concerns the mechanical detailer is the fluid system since the overhead radiant system really requires no detailing and the electric system lies in the area of electrical detailing.

Carbon Dioxide Fire Extinguishing System

In many cases for the detailer the carbon dioxide or Halon system is a nonevent. The relatively simple installation is shown on the plans as a series of cylinders in a designated safe area. From these cylinders a line that connects to all the cylinders is led to a series of outlet nozzles in the room, typically a computer room. The system commonly includes Halon-distributing nozzle outlets on the ceiling and under the raised platform computer room floor. These outlets are energized by a detector which can be arranged to react to smoke, infrared rays, heat, or ultraviolet light.

Kitchen System: The NFPA allows either carbon dioxide or Halon to be used in the categories of protection that we previously noted, however, the following details, which cover specific situations, name only carbon dioxide as the gas. Figure 9-1 is a typical kitchen installation. In this detail, and in the ones that follow, there is very clearly a large amount of extraneous detail. And, most likely, your particular kitchen project is not going to be arranged exactly as the one we depict. But many consultants find it difficult to show precisely what they want in this rather simple situation, and rather than be too brief, we prefer to show a clearly defined situation, even if it means excess detailing. Our detail presents all the items that are related to the fire protection system. When we felt they would be helpful, we added explanatory notes. As far as the real heart of the system is concerned, nozzle locations are of paramount importance. Next in importance are the piping, detector, and tank location. Whatever else that in your judgment is required should be shown as items from this detail or modifications to it. This plus your specification should present a total design package.

Filter Systems: Figure 9-2 is another case in which we try to get what we want. The dust collector, for any number of reasons, needs fire protection. Usually the specified control of the collector operation if a fire occurs is part of the protection system. We want nozzles to emit carbon dioxide on an incoming or outgoing air fire. A simple notation about the heads on the collector and an appropriate specification ought to be all that is needed, and in many cases it is. However, if you are concerned, as we have been, about your message getting across, use our detail, including the notes.

In certain types of industrial air systems the oil bath air filter can be a real hazard. Figure 9-3 grew from a very simple sketch. Obviously the cutaway section

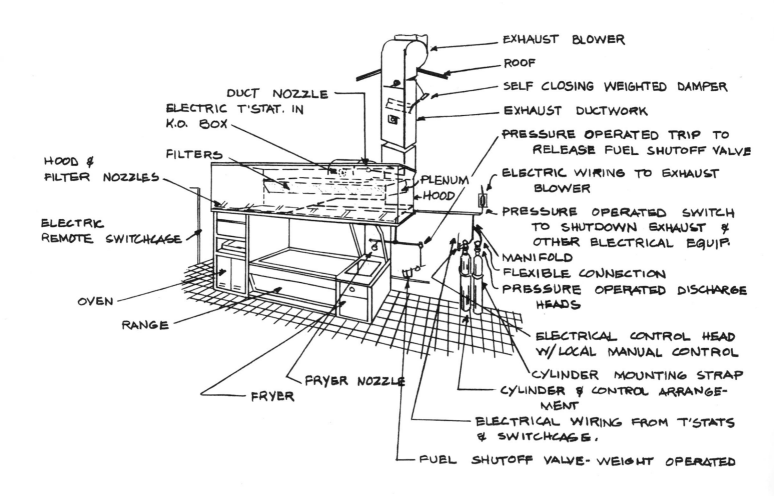

TYPICAL KITCHEN

CARBON DIOXIDE FIRE EXTINGUISHING SYSTEMS

FIGURE 9-1

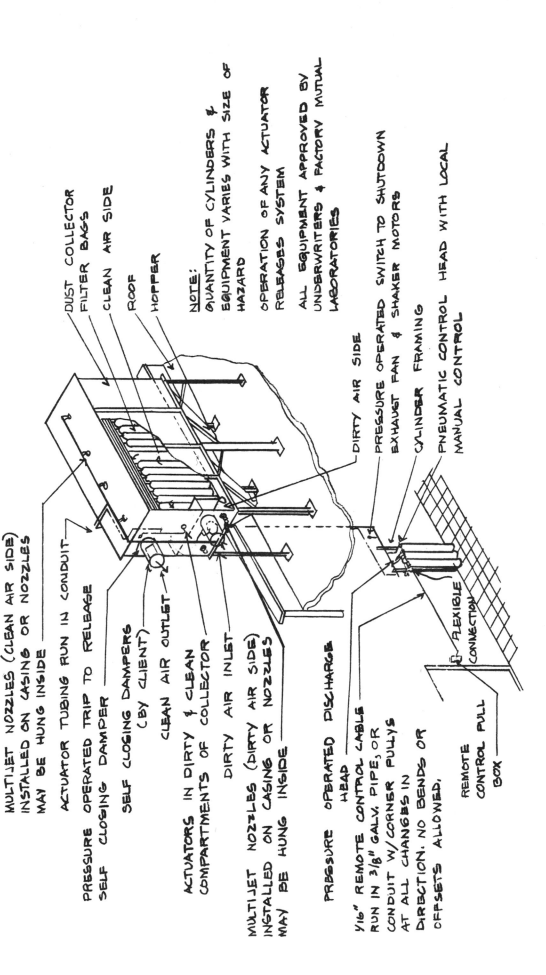

MULTIJET NOZZLES (CLEAN AIR SIDE) INSTALLED ON CASING OR NOZZLES MAY BE HUNG INSIDE

ACTUATOR TUBING RUN IN CONDUIT

PRESSURE OPERATED TRIP TO RELEASE

SELF CLOSING DAMPER

SELF CLOSING DAMPERS (BY CLIENT)

CLEAN AIR OUTLET

ACTUATORS IN DIRTY & CLEAN COMPARTMENTS OF COLLECTOR

DIRTY AIR INLET

MULTIJET NOZZLES (DIRTY AIR SIDE) INSTALLED ON CASING OR NOZZLES MAY BE HUNG INSIDE

PRESSURE OPERATED DISCHARGE HEAD

1/16" REMOTE CONTROL CABLE RUN IN 3/8" GALV. PIPE, OR CONDUIT W/CORNER PULLYS AT ALL CHANGES IN DIRECTION. NO BENDS OR OFFSETS ALLOWED.

REMOTE CONTROL PULL BOX

FLEXIBLE CONNECTION

DUST COLLECTOR FILTER BAGS

CLEAN AIR SIDE

ROOF

HOPPER

NOTE:
QUANTITY OF CYLINDERS & EQUIPMENT VARIES WITH SIZE OF HAZARD

OPERATION OF ANY ACTUATOR RELEASES SYSTEM

ALL EQUIPMENT APPROVED BY UNDERWRITERS & FACTORY MUTUAL LABORATORIES

DIRTY AIR SIDE

PRESSURE OPERATED SWITCH TO SHUTDOWN EXHAUST FAN & SHAKER MOTORS

CYLINDER FRAMING

PNEUMATIC CONTROL HEAD WITH LOCAL MANUAL CONTROL

TYPICAL DUST COLLECTOR

CARBON DIOXIDE FIRE EXTINGUISHING SYSTEMS

FIGURE 9-2

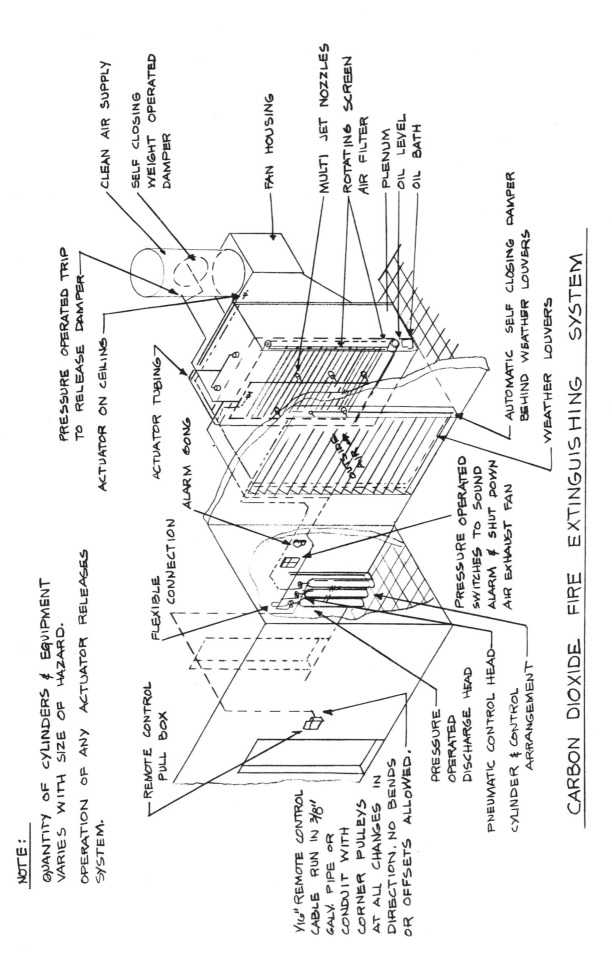

CLEAN AIR SUPPLY

SELF CLOSING WEIGHT OPERATED DAMPER

FAN HOUSING

MULTI JET NOZZLES

ROTATING SCREEN AIR FILTER

PLENUM

OIL LEVEL

OIL BATH

PRESSURE OPERATED TRIP TO RELEASE DAMPER

ACTUATOR ON CEILING

ACTUATOR TUBING

ALARM GONG

FLEXIBLE CONNECTION

REMOTE CONTROL PULL BOX

AUTOMATIC SELF CLOSING DAMPER BEHIND WEATHER LOUVERS

WEATHER LOUVERS

PRESSURE OPERATED SWITCHES TO SOUND ALARM & SHUT DOWN AIR EXHAUST FAN

PRESSURE OPERATED DISCHARGE HEAD

PNEUMATIC CONTROL HEAD

CYLINDER & CONTROL ARRANGEMENT

NOTE:

QUANTITY OF CYLINDERS & EQUIPMENT VARIES WITH SIZE OF HAZARD.

OPERATION OF ANY ACTUATOR RELEASES SYSTEM.

7/16" REMOTE CONTROL CABLE RUN IN 3/8" GALV. PIPE OR CONDUIT WITH CORNER PULLEYS AT ALL CHANGES IN DIRECTION, NO BENDS OR OFFSETS ALLOWED.

CARBON DIOXIDE FIRE EXTINGUISHING SYSTEM

PROTECTION FOR OIL BATH AIR FILTERS

FIGURE 9-3

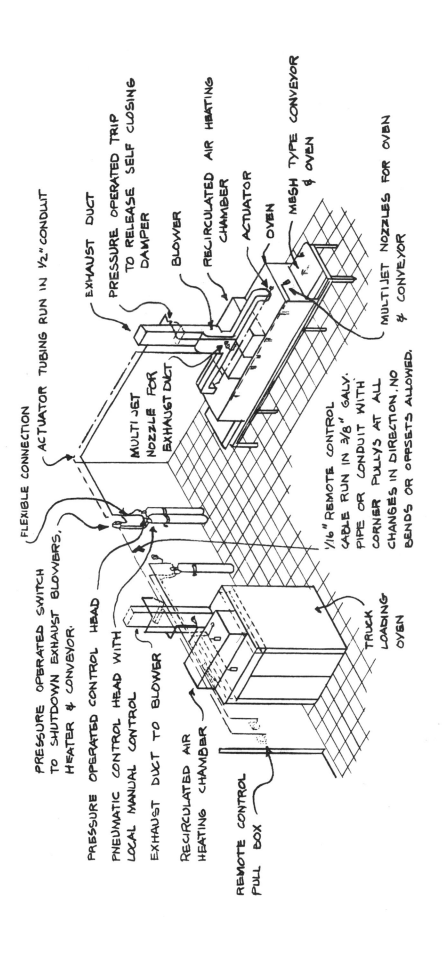

FLEXIBLE CONNECTION

ACTUATOR TUBING RUN IN ½" CONDUIT

PRESSURE OPERATED TRIP TO RELEASE SELF CLOSING DAMPER

EXHAUST DUCT

BLOWER

RECIRCULATED AIR HEATING CHAMBER

ACTUATOR

OVEN

MESH TYPE CONVEYOR & OVEN

MULTIJET NOZZLE FOR EXHAUST DUCT

MULTIJET NOZZLES FOR OVEN & CONVEYOR

PRESSURE OPERATED SWITCH TO SHUTDOWN EXHAUST BLOWERS, HEATER & CONVEYOR.

PRESSURE OPERATED CONTROL HEAD

PNEUMATIC CONTROL HEAD WITH LOCAL MANUAL CONTROL

EXHAUST DUCT TO BLOWER

RECIRCULATED AIR HEATING CHAMBER

REMOTE CONTROL PULL BOX

TRUCK LOADING OVEN

¹/₁₆" REMOTE CONTROL CABLE RUN IN ³/₈" GALV. PIPE OR CONDUIT WITH CORNER PULLYS AT ALL CHANGES IN DIRECTION. NO BENDS OR OFFSETS ALLOWED.

CARBON DIOXIDE FIRE EXTINGUISHING SYSTEM

PROTECTION FOR INDIVIDUAL OVENS

FIGURE 9-4

that shows the cylinder location is not necessary nor is the description of the discharge plenum and duct details. The situation depicted came from an actual job, and we simply repeated the entire detail so that you could see the whole picture. You could stop your detailing at the filter and merely show the cylinders as a diagrammatic representation since your plan will show the actual location of the cylinders.

Process Oven System: Industrial processes use ovens for chemical and heat treating purposes. Because this sort of detail is very small, we put two together in Fig. 9-4 as a sort of composite drawing. These two different oven applications fit two different jobs that we had. You should be careful to check your oven size and arrangement. You may need more nozzles than we have depicted, but for general purposes this detail will be fairly usable, almost as is.

It may seem that we have given the Halon detailing limited attention, but that is not our intent. We have simply shown common applications in which either Halon or carbon dioxide could be used.

Vacuum Systems

While the aerospace industry is well-known for its high altitude simulation chambers, which are created by pulling a very low vacuum in the chamber, the average engineer usually runs into vacuum systems in various hospital operating rooms and in intensive care, and special care rooms. These vacuum systems all require special engineering in their design and special materials in their installation. Apart from the connection details around the vacuum pump, there is really not very much to detail. In our details we present typical piping arrangements.

Clinical Vacuum: Figure 9-5 is a typical clinical vacuum pump connection diagram. Not all of these relative simple systems have a vacuum storage tank. This is a function of system size and design requirements. While you can pull a vacuum with proper pumping and piping, you do not store a vacuum. In most hospital and laboratory systems the vacuum system is used to extract fluids, blood for example, during an operation. The tank may act as a liquid collector. Some of this material may also get into the pump. To resolve this problem water can be used, as shown, in supply lines to the tank and pump, to flush the system. Usually some sort of water-contamination prevention is required. For the tank we have shown an air gap connection and for the pump we have shown a backflow preventer, which is a standard specification item.

1. WATER SUPPLY
2. VACUUM GAUGE
3. FROM SYSTEM
4. VACUUM BREAKER
5. FLUSHING CONNECTION
6. DRAIN CONNECTION
7. LIQUID LEVEL GAUGE
8. VACUUM SWITCH
9. WATER SUPPLY
10. VENT - TO ATMOSPHERE
11. OVERFLOW
12. FUNNEL WITH 2" AIR GAP
13. DRAIN
14. BACKFLOW PREVENTER OR VACUUM BREAKER (IF REQ'D OR AS REQ'D BY LOCAL PLB CODES)

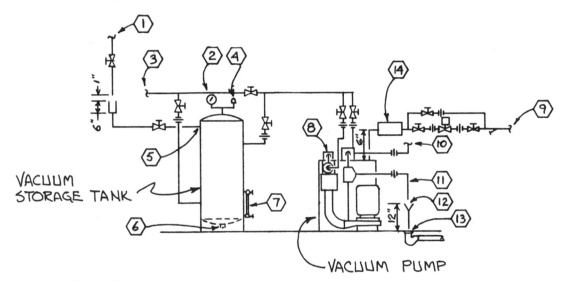

VACUUM STORAGE TANK

VACUUM PUMP

LEGEND

⊣⊢ UNION

⊠ GATE VALVE

⊠ SOLENOID VALVE

⊤ "Y" STRAINER

CLINICAL VACUUM PUMP SYSTEM

NO SCALE

FIGURE 9-5

Laboratory Vacuum: Figure 9-6 really breaks no new ground as far as detailing is concerned. Usually in a laboratory or in an industrial process the vacuum demands are larger and the system is larger. The larger system can also be noisier. Thus we have added two items to the duplex pumping arrangement: a muffler and separator. Essentially the drawing is the same as Fig. 9-5.

Figure 9-7 depicts a central vacuum cleaning system, and Fig. 9-8 depicts an enlarged detail of a vacuum cleaning inlet connection. If you are going to exhaust dust and other dirt particles, you obviously cannot simply blow these dirt particles outside. Thus on the way to and through the vacuum producer or vacuum exhaust pump the air passes through a cyclone separator and a cloth filter. Normally if you show all of the

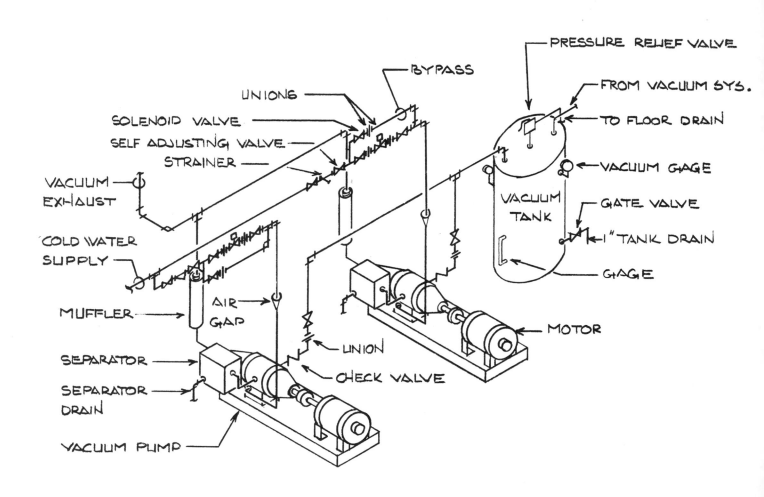

LABORATORY VACUUM SYSTEM
———————— NOT TO SCALE ————————

FIGURE 9-6

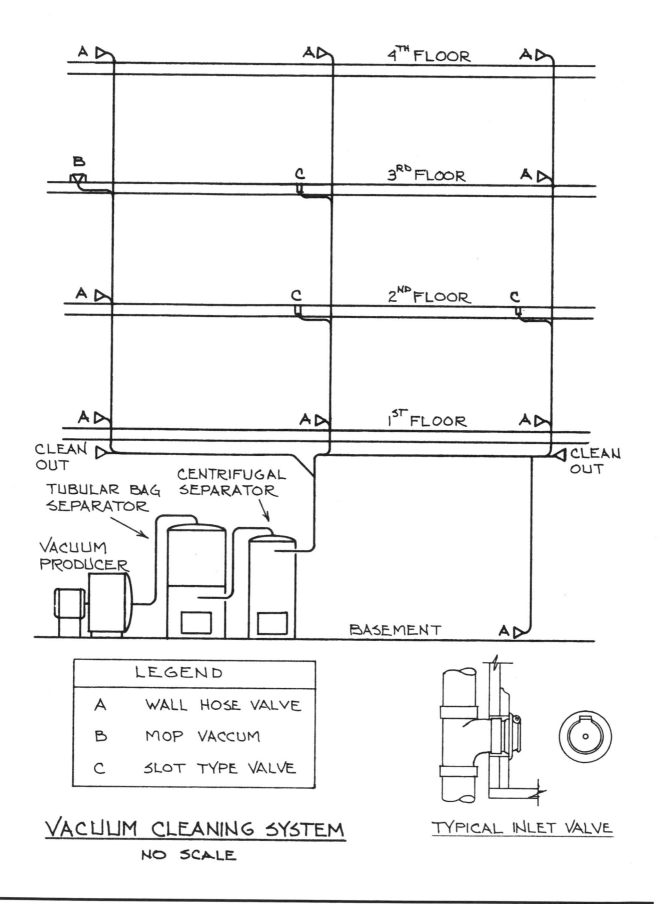

VACUUM CLEANING SYSTEM
NO SCALE

TYPICAL INLET VALVE

LEGEND	
A	WALL HOSE VALVE
B	MOP VACCUM
C	SLOT TYPE VALVE

FIGURE 9-7

risers and cleaning outlets as we do, your detail will have to relate to the correct number of outlets in your particular design project. This part of our schematic detail is merely a representative sample and not an actual situation.

The cleaning head hose is normally snapped into an outlet such as we have depicted in Fig. 9-7. The cost of this system versus the cost a collection of vacuum cleaners and long cords varies. There is no way, short of a thorough investigation of all operating and equip-ment costs, to determine the most cost-effective re-sult.

Vacuum systems and compressed air systems are, in simple terms, the reverse of each other. In most textbooks vacuum and compressed air design is treat-ed under one heading. We are treating compressed air systems in a separate section that follows this one. But compressor and vacuum pump cooling systems are very similar; therefore we are covering the following two pump cooling details together. Vacuum systems

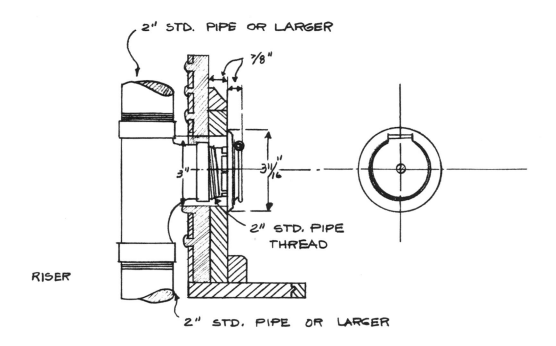

TYPICAL INLET VALVE

CENTRAL VACUUM SYSTEM

FIGURE 9-8

are not usually very large devices and their water requirements are not usually very great. But compressed air systems can be huge and require a large water supply. Cooling towers may well be part of your design requirements. Sometimes chilled water systems are also used. Whenever you design a compressed air system, be certain you have properly allowed for compressor pump cooling water requirements.

Pump and Compressor Cooling: Figures 9-9 and 9-10 show typical water cooling piping to a vacuum pump or air compressor. In both cases the cooling water is wasted to an indirect waste. Figure 9-9 shows a somewhat simpler piping system with solenoid and thermostatic valve control of the supply cooling water. Figure 9-10 shows an air circulating line with a control valve that is activated by the air storage tank.

These two details are both taken from actual jobs, and your compressor or vacuum pump connections may well vary to some degree. To be certain of your actual piping arrangements, do not blindly copy either

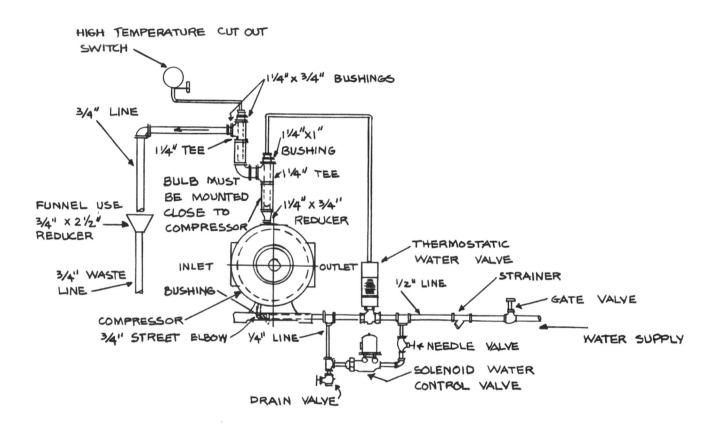

COMPRESSOR & VACUUM PUMP COOLING SYSTEMS

FIGURE 9-9

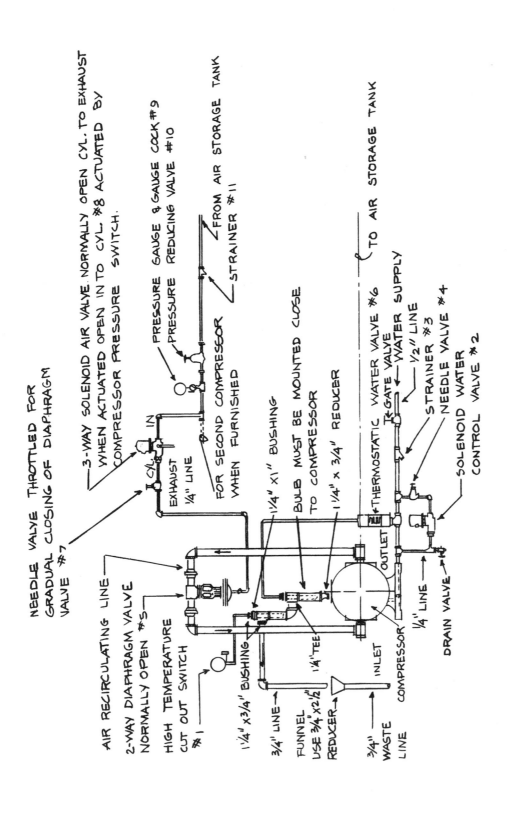

NEEDLE VALVE THROTTLED FOR GRADUAL CLOSING OF DIAPHRAGM VALVE #7

3-WAY SOLENOID AIR VALVE. NORMALLY OPEN CYL. TO EXHAUST WHEN ACTUATED OPEN INTO CYL. #8 ACTUATED BY COMPRESSOR PRESSURE SWITCH.

PRESSURE GAUGE & GAUGE COCK #9

PRESSURE REDUCING VALVE #10

FROM AIR STORAGE TANK

STRAINER #11

FOR SECOND COMPRESSOR WHEN FURNISHED

1 1/4" X 1" BUSHING

BULB MUST BE MOUNTED CLOSE TO COMPRESSOR

1 1/4" x 3/4" REDUCER

TO AIR STORAGE TANK

THERMOSTATIC WATER VALVE #6

GATE VALVE

WATER SUPPLY

1/2" LINE

STRAINER #3

NEEDLE VALVE #4

SOLENOID WATER CONTROL VALVE #2

AIR RECIRCULATING LINE

2-WAY DIAPHRAGM VALVE NORMALLY OPEN #5

HIGH TEMPERATURE CUT OUT SWITCH #1

CYL IN

EXHAUST 1/4" LINE

1 1/4" x 3/4" BUSHING

3/4" LINE

FUNNEL USE 3/4" x 2 1/2" REDUCER

1/4" TEE

OUTLET

1/4" LINE

DRAIN VALVE

INLET

COMPRESSOR

3/4" WASTE LINE

COMPRESSOR & VACUUM PUMP COOLING SYSTEMS

FOR COMPRESSORS ONLY, IN STORED AIR SYSTEMS

FIGURE 9-10

of these two details; they are only typical and representative. The chances are that while your detail connections may differ, the difference is not going to be very substantial. However, it is embarrassing to have a detail that does not totally fit the specified vacuum pump or air compressor.

Compressed Air Systems

As we noted in our previous discussion of vacuum systems and vacuum pumps, the compressed air system and its air compressor are directly related. One does exactly the opposite of the other. There are many types and applications of compressed air, especially in industrial processes.

Figure 9-11 is a typical schematic overview of a compressed air system. The air is supplied at 100 psig. Very commonly a compressed air system operates between 50 and 150 psig. Pipe and fittings are commonly specified to be 150 psig rated. The compressor has a source of air that it compresses. The air is used exactly as described in Fig. 9-11.

Compressed Air Equipment: The detail in Fig. 9-11

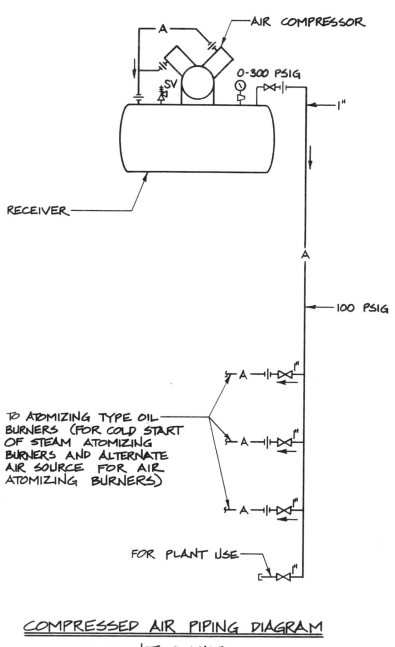

COMPRESSED AIR PIPING DIAGRAM
———— NOT TO SCALE ————

FIGURE 9-11

does not really show the piping at the compressor. In Fig. 9-12 we depict a typical dual 50 psig compressed air supply. The system is not very large. Normally in large compression systems the compressor is a two, or more, stage device. The compressors are water cooled and the cooling water is wasted to a floor drain through an indirect connection. To maintain a system balance and to preclude constant cycling of the compressors the system supply goes through a receiver tank that has been properly designed, engineered, and sized. Compressor discharge air goes through a separator to eliminate any liquid in the compressed air line. The air intake is through an air filter.

Figure 9-13 is a large two-stage compressor installation. From the air intake filter, air is delivered to the first stage in volume and compressed to about two-thirds of its final value. It is therefore warm and is cooled by the water cooled intercooler. It is then compressed further to its final delivered air pressure and passes through a water cooled after-cooler. To be certain no liquids are in the air, the air goes through a separator and then goes to an air receiver, as did the

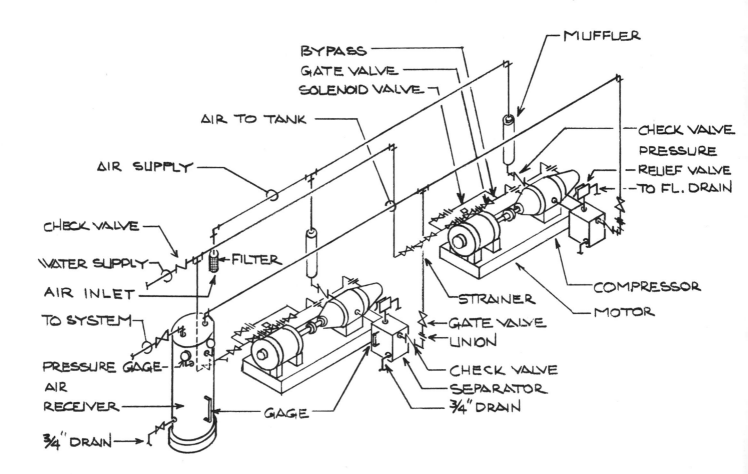

50-PSI COMPRESSED AIR SUPPLY
NOT TO SCALE

FIGURE 9-12

air in the system shown in Fig. 9-12. Finally, the air goes to the compressed air distribution system. Our detail is deliberately not very clear about the disposal of the cooling water except to note that it is wasted to the nearest drain. A system of this size may well be one, such as we have previously discussed, which has a source of refrigerated chilled water or is connected to a cooling tower.

Compressors of the size shown in Fig. 9-13 usually have serious, sometimes severe, vibration problems. They have both pounding vibration problems from the machine itself and pulsed air problems created by the compression of large volumes of air. Pipe sizes of 4 in and 6 in to the system are common. Not only will you require a spring-mounted base, but you will also require horizontal and vertical flexible pipe connections to solve pipe vibration. Finally your system will also require some type of specialized air dryer.

Very High Pressure Outlets: The five details that follow this discussion are all related to a very special type of very high pressure compressed air system. The system pressure is 5000 psig. This special type of compressor is a totally self-contained five-stage unit. Except for the specially rated pipe, its connections are no more complex than any detail we have previously shown. To be certain that the air is dry, air is discharged through a specially designed and installed refrigerated dryer. The distribution pipe is welded double-extra-strong stainless steel. The real detail problem is in the pipe supports and outlets. If you

LEGEND

① WATER-TEMP. REGULATING VALVE - INTEGRAL BULB
② FLOAT TYPE DRAIN TRAP
③ PRESSURE SAFETY SWITCH
④ TEMP. SAFETY SWITCH

⑤ WATER-TEMP. REGULATING VALVE REMOTE BULB
⑥ SIGHT GLASS
⑦ PRESSURE RELIEF VALVE
⑧ TO NEAREST DRAIN

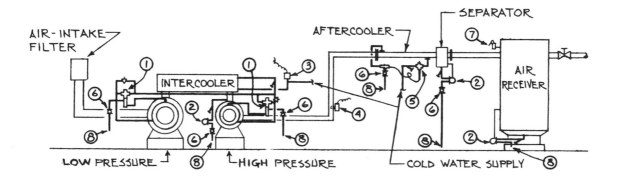

COMPRESSED AIR COOLING & DRAINING DETAILS

NO SCALE

FIGURE 9-13

VALVE OPERATING PROCEDURE
TO REMOVE MOISTURE CLOSE GLOBE
VALVE. VENT DRIP LEG TO ATMOSPHERE
OPENING CAP. REPLACE CAP AND MAKE
AIR TIGHT BEFORE REOPENING VALVE.

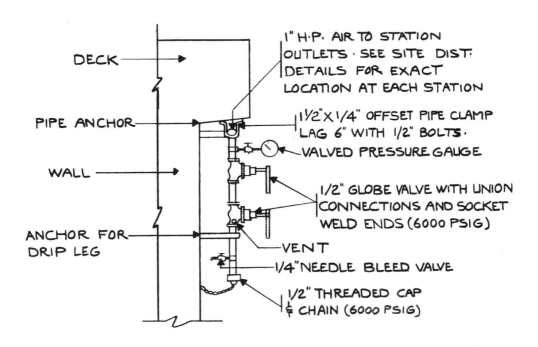

DECK

PIPE ANCHOR

WALL

ANCHOR FOR
DRIP LEG

1" H.P. AIR TO STATION
OUTLETS · SEE SITE DIST.
DETAILS FOR EXACT
LOCATION AT EACH STATION

1½" X ¼" OFFSET PIPE CLAMP
LAG 6" WITH ½" BOLTS.
VALVED PRESSURE GAUGE

½" GLOBE VALVE WITH UNION
CONNECTIONS AND SOCKET
WELD ENDS (6000 PSIG)

VENT

¼" NEEDLE BLEED VALVE

½" THREADED CAP
& CHAIN (6000 PSIG)

SERVICE CONNECTION – DOCK #1
——— NO SCALE ———

FIGURE 9-14

stood in the air path at this pressure, it would blow all the flesh from your frame and you would become an instant skeleton.

Figures 9-14 and 9-15 are typical air supply connections on a shipyard dock. In Fig. 9-14 the air line supply runs just below the edge of the deck, and in Fig. 9-15 the supply runs behind a concrete curb at the edge of the deck. All outlets are redouble-valved, and the service pipe is anchored to the deck wall. A gauge records the pressure.

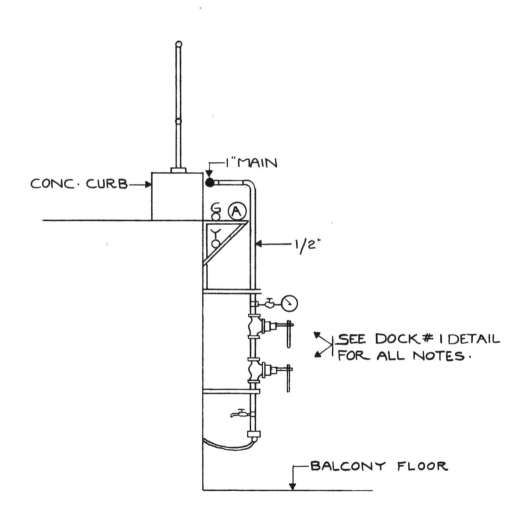

SERVICE CONNECTION — DOCK #3
———— NO SCALE ————

FIGURE 9-15

In Fig. 9-16 there is a typical resolution of the service connection at deck level. The service connection seen in Figs. 9-16 and 9-17 is also the same as in Fig. 9-14 except that it is mounted in a horizontal rather than a vertical direction. The plate is through-bolted to the wood curb. Figure 9-17 is a face view of the plate and the service connection detail is the same as before except that it is mounted horizontally. Pipe supports, such as the one shown in Fig. 9-18 are also substantial, and the high pressure air line is securely clamped to the support.

These five details are from a very special and unusual design problem and its solution. However, they will serve you in any 1500 psig and up situation. For simple lower pressure lines such as 150 psig a valved service opening, most likely without a gauge, would be a standard solution. A similar but not so strong support and clamping arrangement can also serve as part of your detailing.

Heat Tracing

Frequently, in the exterior distribution of fluids, expecially oils, the piping is a run above ground. Generally, the reason for this above ground installation is that the soil is corrosive or that it is in an area in which the existing below ground utilities create a myriad of obstacles to an underground installation. It takes a fairly low temperature to freeze oil. Freezing is not, however, the problem. As oil is cooled, its viscosity is increased and pumping or using the oil becomes a problem. The usual method of resolving the problem is to heat trace the pipe using a steam source.

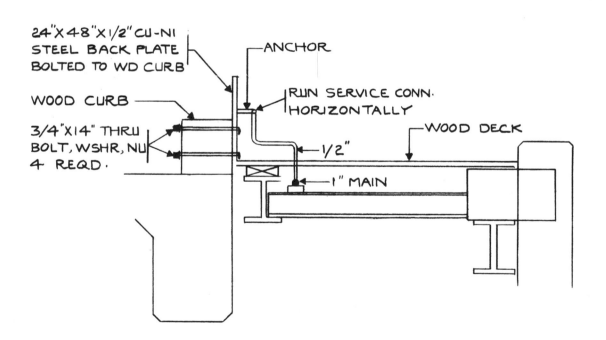

SERVICE CONNECTION – PIER #11,13
NO SCALE

FIGURE 9-16

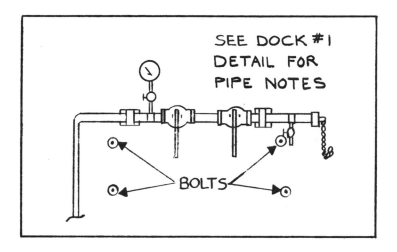

SEE DOCK #1
DETAIL FOR
PIPE NOTES

BOLTS

PLATE DETAIL
— NO SCALE —

FIGURE 9-17

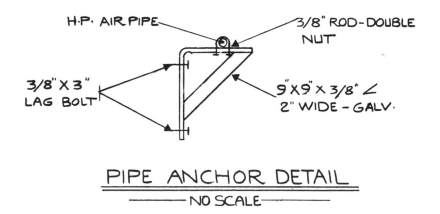

H.P. AIR PIPE

3/8" ROD-DOUBLE NUT

3/8" X 3" LAG BOLT

9" X 9" X 3/8" ∠
2" WIDE – GALV.

PIPE ANCHOR DETAIL
— NO SCALE —

FIGURE 9-18

Figure 9-19 is a typical installation of a heat traced oil line. The installation is common to an oil line supported from a concrete wall or other suitable place with an angle bracket and a double clamp hanger for the oil line and the steam line. The clamp around the steam line has to be free enough from the steam line so that the steam line can expand and contract. Sometimes an aluminum or galvanized iron sleeve is inserted in the clamp that holds the steam line. The condensate line is at the end of the steam main since the system is arranged so that the steam line pitches steadily downward in the direction of flow.

Figure 9-20 illustrates another method of supporting the heat traced oil line when there is no supporting wall or pier available. In this case a series of concrete pads are constructed; the detail illustrates a section through one of these pads. Also shown in this figure is the condensate return line. To get the correct flow pitch relationship for both steam and condensate, the pad heights can be adjusted, or the proper pitch can be obtained through a combination of steam hanger adjustments and shimming the condensate line clamp supports. Insulation around the steam and condensate line is not shown in this detail, but there *must* be an

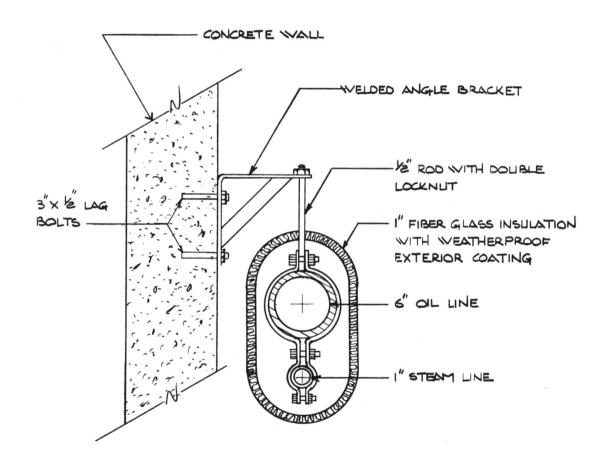

CONCRETE WALL

WELDED ANGLE BRACKET

½" ROD WITH DOUBLE LOCKNUT

1" FIBER GLASS INSULATION WITH WEATHERPROOF EXTERIOR COATING

6" OIL LINE

1" STEAM LINE

3" X ½" LAG BOLTS

WALL SUPPORT SYSTEM
FOR HEAT TRACED PIPING
NOT TO SCALE

FIGURE 9-19

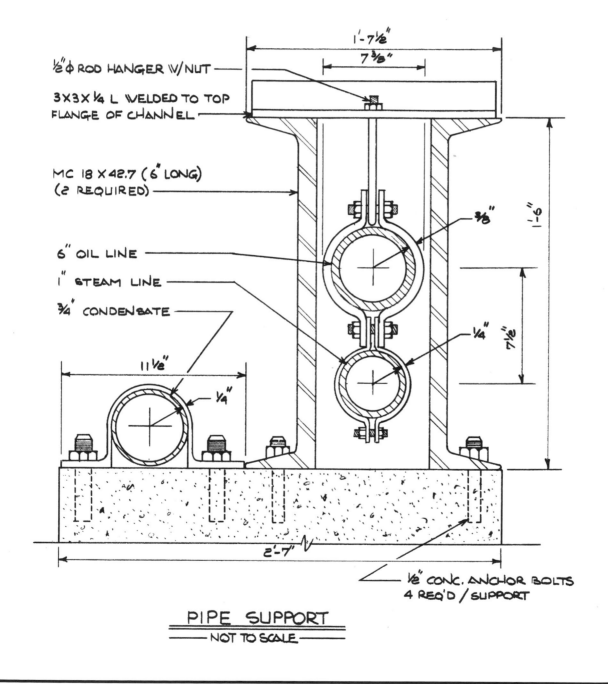

1'-7½"

7⅜"

½"∅ ROD HANGER W/NUT

3X3X¼ L WELDED TO TOP FLANGE OF CHANNEL

MC 18 X 42.7 (6" LONG) (2 REQUIRED)

6" OIL LINE

1" STEAM LINE

¾" CONDENSATE

⅜"

¼"

1'-6"

7½"

11½"

¼"

2'-7"

½" CONC. ANCHOR BOLTS
4 REQ'D / SUPPORT

PIPE SUPPORT
— NOT TO SCALE —

FIGURE 9-20

insulation envelope around the steam and oil lines as depicted in Fig. 9-20 and the condensate line *must* be insulated.

Figures 9-21 and 9-22 illustrate another common form of pipe protection. This is an electrical system either at 220 or 440 v. Commonly the 440-v system has a separate power source for the thermostat which is normally rated up to 220 v. Depicted in both of these figures is a line from one sort of container to another.

In the usual installation this is the sort of protection used on outdoor cooling towers that run year-round. It is not limited to this type of installation but is also a very common protector of hot or chilled water lines in HVAC systems and in domestic hot and cold water piping. There also may be situations in which sections of roof drain piping are subject to possible freezing which can also very successfully and economically employ this type of protection.

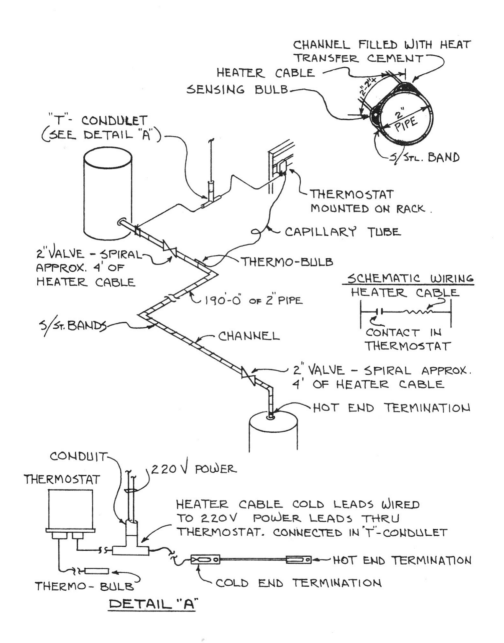

ONE PARALLEL HEATER CABLE — NO CONTACTOR

NO SCALE

FIGURE 9-21

Both details are pipe protection installation details, as well as overall schematic diagrams. The actual installation of the heat tracing wire is usually one or more lines in parallel, as shown on the two figures, or a single cable spirally wrapped and properly insulated around the pipe. In all cases the overall installation of wire on pipe is insulated and the insulation is properly treated for outdoor conditions. Note also that any valves should have their valve bodies specially covered. In a 2-in valve the valve is wrapped with a total of 4-ft 0-in of heater cable. Valves are especially sensitive to freezing.

Whether you protect your pipe by steam or electricity, the amount of protection required should be carefully calculated using methods in the applicable section of the *ASHRAE Fundamentals Handbook* or specific manufacturer's literature, preferably from both sources.

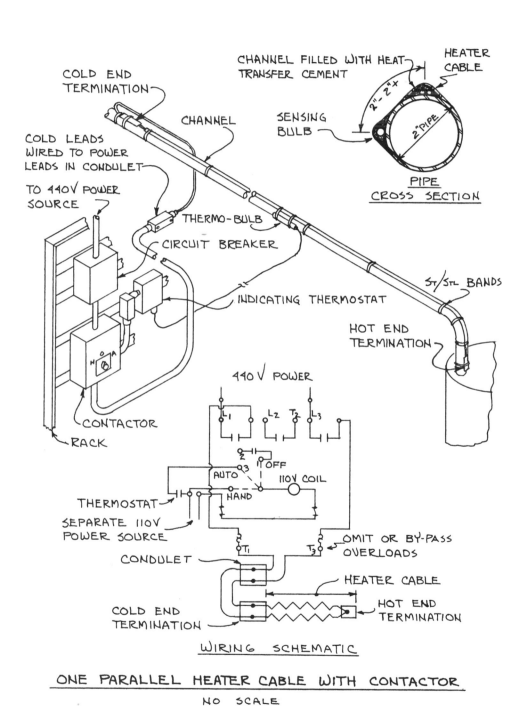

ONE PARALLEL HEATER CABLE WITH CONTACTOR

NO SCALE

FIGURE 9-22

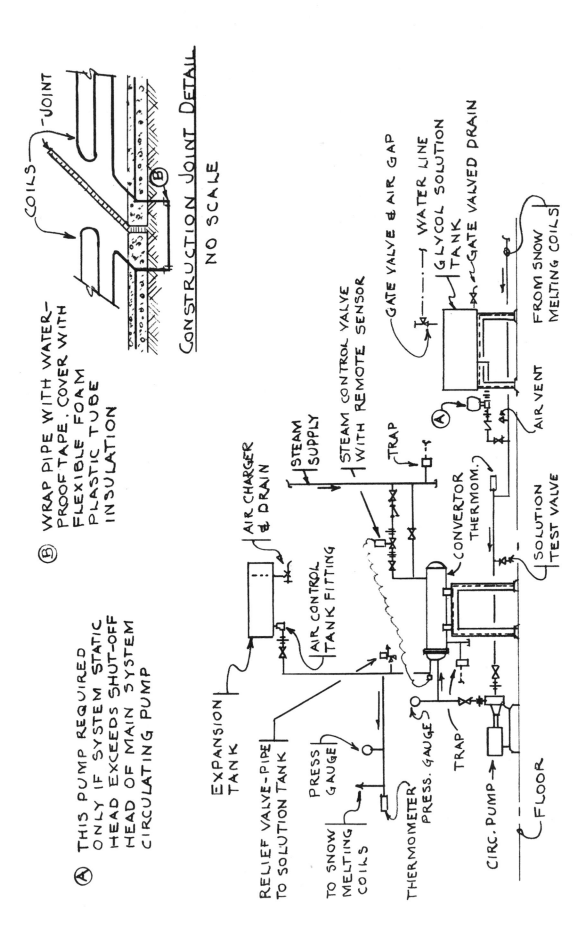

CONSTRUCTION JOINT DETAIL

NO SCALE

(B) WRAP PIPE WITH WATER-PROOF TAPE. COVER WITH FLEXIBLE FOAM PLASTIC TUBE INSULATION

(A) THIS PUMP REQUIRED ONLY IF SYSTEM STATIC HEAD EXCEEDS SHUT-OFF HEAD OF MAIN SYSTEM CIRCULATING PUMP

SNOW MELTING SYSTEM PIPING

NO SCALE

FIGURE 9-23

We have shown the electric protection detail as a fairly complete overall installation because this is the one that can be the most confusing. The steam traced system is really nothing more than a properly sized bare steam main with a trap at the end and a condensate return. Commonly its control is a pipe sensor that operates a two-position motorized steam control valve.

Snow Melting

The source of the radiant coil snow melting system is normally a boiler with a converter. The boiler can be steam or hot water and fired by oil, gas, or coal. Since we have depicted boiler and converter installations on many previous details and we wanted to get all the pertinent points on one detail, we have depicted in Fig. 9-23 only the steam converter portion of the heating source. Bear in mind that the converter could just as easily have been a hot water converter.

Basically the snow melting fluid must be one that will not freeze when the pavement is dry and fluid is not moving. Some quick answers on the percentage of glycol can be made based on the fact that a 35-percent glycol solution will not freeze until the temperature is 0°F, and a 50-percent glycol solution will withstand −30°F.

The snow melting system, like any hot water system, requires an expansion tank and a circulating pump. And it requires a makeup source. Note that under certain conditions the makeup source requires a pump. Your detail should contain the note that the system static head exceeds the pump shutoff head.

The snow melting system, normally being a closed system, requires little in the way of makeup. When makeup is required, it must come from a special source to avoid contamination of the domestic water system. Thus, in our detail water is introduced in a solution tank via an air gap.

Like all radiant coils, snow melting coils should be held level and should be properly supported. Usually the slabs containing the snow melting coils has a construction expansion joint, and the slab piping must pass this joint. Since you do not want to heat the earth and create slab movement problems, the bypass loop should be insulated as we have depicted. For snow melting calculations we refer you to the *ASHRAE Systems Handbook* and to available literature from pump, converter, and pipe manufacturers.

TEN

HVAC and Refrigeration Distribution Equipment

All the plumbing, heating, ventilating, air conditioning, and refrigeration equipment details shown in this chapter may very well be described as the details between here and there. In our breakdown of material we view the source and source support equipment as the supplier or manufacturer, the user of the source as the end-user, and in between these two the middleman as the distributor or transporter.

In the plumbing area there is practically no detailing required since we have the equipment at one end and the service at the other, with only the pipe in between the two. And at the equipment itself we usually have shutoff valves and traps as the final connections to the equipment. But for all the other trades there are a variety of connections. In steam systems we frequently have pressure reducing requirements and converters to go from steam to water and vice versa. In refrigerant piping the user and the supplier are frequently very near each other; thus the distribution detailing for refrigerant piping frequently covers the detailing of the compressor and related evaporator and condenser. Because of this fairly intimate arrangement, we have put all the refrigerant piping detailing in one place in this chapter.

Another detailing specialty is the support of pipes regardless of what system they serve. This sometimes involves specialized types of piping arrangements along the way, and therefore we grouped these details together. Since there is no rule that says that all the piping must run inside the building, we have followed the specialized piping details with details that cover the piping that runs outside the building. These exterior piping details include seismic piping details just in case earthquake protection is part of your detailing requirements.

As far as air systems are concerned, the distribution ductwork is the total area of detailing. At one end of this system is some type of fan with flexible connections; at the other end is the grille, louver, diffuser, collector, or just an open end. The detailing is always concerned with what happens along the way. In exhaust systems we have tried to cover the more specialized situations, including composite details for smoke and kitchen exhaust systems and specialized exhaust hoods.

Converters

Converters and heat exchangers are commonly interchangeable terms. Whenever a liquid or a gas at a given temperature is used to heat or cool another liquid or gas in a tank and interior coil arrangement, the process is a conversion process and the device is usually called a heat exchanger or converter. All converters are constructed using a shell and tube fabrication arrangement. Normally the vapor, if it is a vapor-to-liquid converter, is in the shell, which has the greater volume of space, and the liquid is in the tubes. When both fluids are liquid, the hotter or colder of the two, depending on the requirements (heating or cooling), is in the tubes. In the following four details we present some of the basic details in converter piping.

Swimming Pool: Figure 10-1 is described as a swimming pool heater although this detail could be used as a typical detail for a number of different applications. In this figure the general circulation arrangement of the pool water is briefly and schematically depicted. Once the pool water is heated, the hourly heat load to maintain the pool temperature is fairly small. The source of heat for the pool heat exchanger can be either steam or hot water. By means of a three-way valve a portion of the pool water is diverted through the heat exchanger. Since the heat source, especially if it is steam, will heat the diverted pool water to a temperature that is far higher than the usual 80°F required by the pool, a discharge sensing element controls the opening of the three-way valve and maintains the 80°F discharge temperature.

High Temperature Hot Water to Building: Figure 10-2 is the typical installation of a building high temperature hot water converter which accepts approximately 400°F high temperature hot water and converts it to low temperature hot water for building heating purposes. Here a three-way valve is not used on the incoming high temperature hot water main. The heat load of the building which is imposed on the high temperature hot water system is maintained by a modulating flow control valve that is similar in operation to a modulating steam valve. While one may assume that a main high temperature hot water system is free of air, there is no absolute guarantee that it is. At 400°F the corresponding pressure is approximately 250 psig. Water released to the atmosphere at this temperature and pressure will flash into steam. Venting is usually effected through 300-lb-test air bottles and a double set of manually controlled globe valves with a safety cap on the end of the ½-in vent line. Venting should only be performed by experienced maintenance personnel.

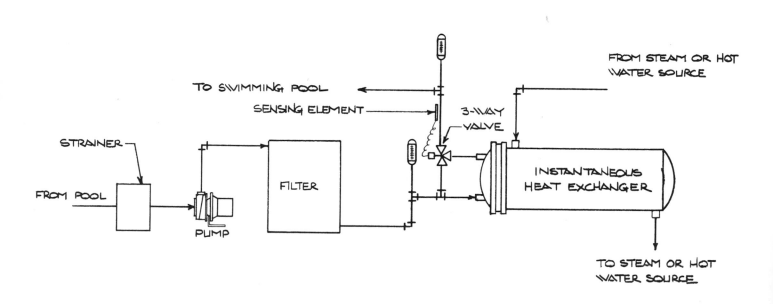

SWIMMING POOL HEATER
—————NOT TO SCALE—————

FIGURE 10-1

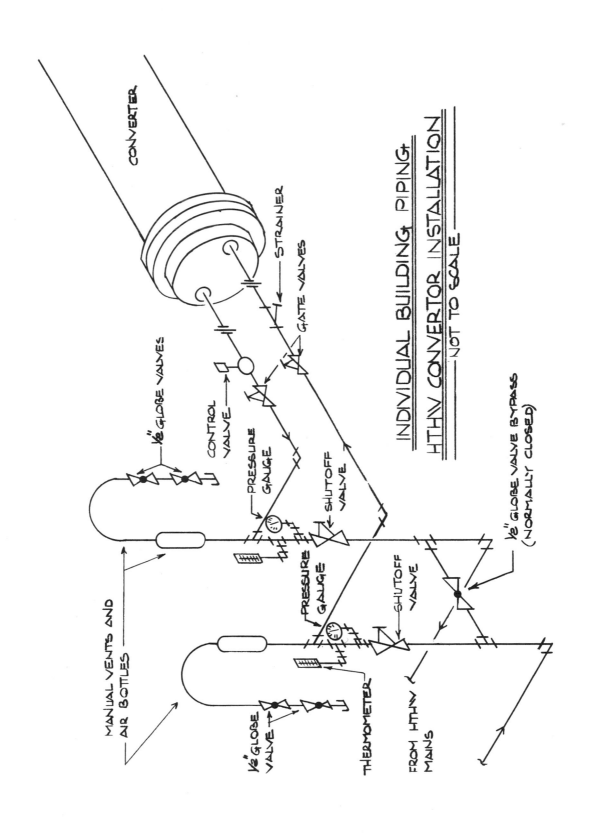

INDIVIDUAL BUILDING PIPING

HTHW CONVERTOR INSTALLATION

— NOT TO SCALE —

CONVERTER

STRAINER

GATE VALVES

½" GLOBE VALVES

CONTROL VALVE

PRESSURE GAUGE

SHUTOFF VALVE

PRESSURE GAUGE

SHUTOFF VALVE

½" GLOBE VALVE BYPASS
(NORMALLY CLOSED)

MANUAL VENTS AND AIR BOTTLES

½" GLOBE VALVE

THERMOMETER

FROM HTHW MAINS

FIGURE 10-2

Hot Water: Figure 10-3, which is a high temperature hot water converter and source of 150 psig steam, bears a striking resemblance to the detail for a steam boiler piping installation which we have shown in Chap. 1. Actually the converter and storage tank is, in effect, the same as an oil or gas fired boiler. In place of the burner and combustion chamber we have a tubular heat exchanger that is supplied by 400°F water. Its controller is actuated by a steam pressure controller that is similar to the steam boiler's pressure controller which activates an oil or gas burner to maintain constant pressure. The tank has a combination feeder and low water cutoff and a pump controller for the steam system that is similar to that on the boiler-burner unit. Also the pressure relief and blowoff piping are the same as for any 150 psig boiler. In essence this converter is a boiler.

High Temperature Hot Water Special Uses: Since Figs. 10-3 and 10-4 are portions of a high temperature hot water system, we thought we would show in Fig.

10-4 a sort of schematic diagram of probably the most common reductions of high temperature hot water to lower temperatures for other uses. The converter, domestic water heater, associated pumps, piping, and controls first reduce the high temperature hot water to a medium temperature of about 290°F. This corresponds to approximately 60 psig, or medium pressure steam, and at this temperature may have many direct uses; it is also readily convertable to other temperatures for indirect uses. The medium temperature hot water system has its own pumps, expansion tanks, and piping that are similar to any high temperature hot water system. One such use of medium temperature hot water is illustrated in the installation of a domestic hot water heater. Three-way valves are sometimes used on medium temperature hot water systems primarily to facilitate better pump operation and performance at low flow. In a larger heating or power and heating system this schematic is, of course, just part of the overall plan. Another common use of medium

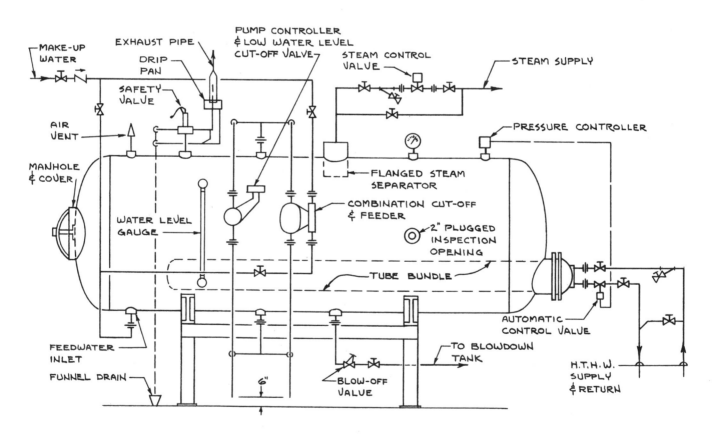

TYPICAL HIGH TEMPERATURE HOT WATER TO STEAM CONVERTOR PIPING DETAIL

FIGURE 10-3

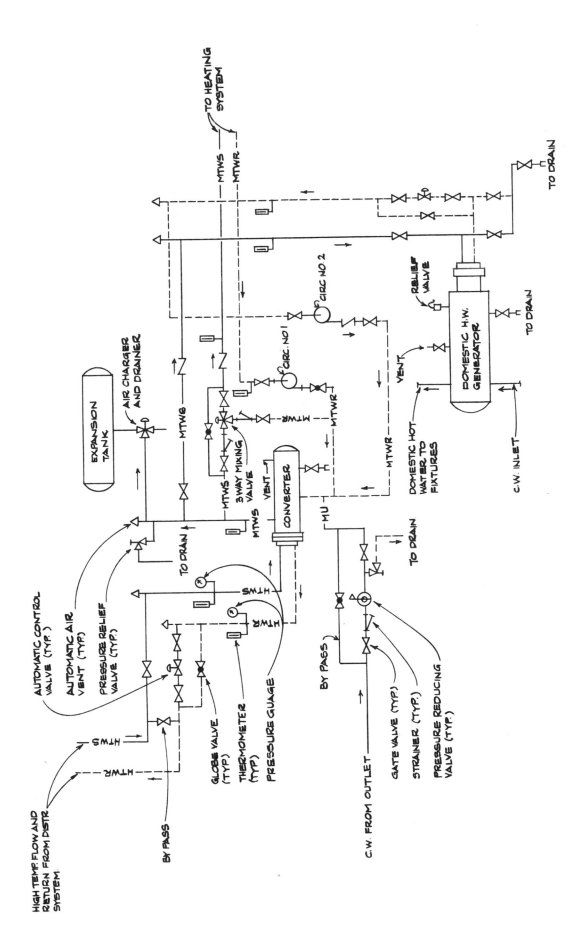

TYPICAL CONVERTER CONNECTION FOR HTW SYSTEM

FIGURE 10-4

temperature hot water is an oil pump and heater set installation in which the main plant is supplied with no. 6 (bunker C) fuel oil that requires preheating.

Pressure Reducing Valves

There are a number of requirements governing the installation of pressure reducing valves, as well as a number of ways in which the proper installation of these valves can be made. Explanations of the various design approaches can be found in the *ASHRAE* *Systems Handbook* and in the literature of the major pressure reducing valve manufacturers.

Figure 10-5 is a fairly standard two-stage installation. It consists of a primary reduction stage and a secondary or regulation stage. The primary control valve would normally reduce, for example, an incoming pressure of 100 to 150 psig to a valve of 40 to 80 psig. It is controlled by a pressure controller that senses the 40 to 80 psig setting and sends its control signal to the primary reducing valve to maintain that

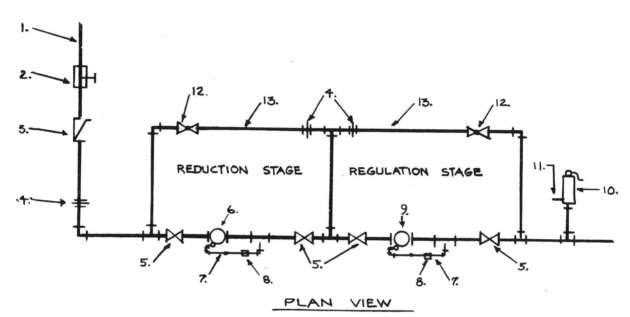

PLAN VIEW

1. SERVICE PIPE
2. SERVICE VALVE
3. STRAINER
4. FLANGED UNION
5. GATE VALVE
6. PILOT-OPERATED PRESSURE-REDUCING VALVE
7. BALANCE LINE
8. PRESSURE GAGE
9. PILOT-OPERATED PRESSURE-REGULATING VALVE
10. PRESSURE RELIEF VALVE (OPTIONAL)
11. VENT
12. GLOBE VALVE
13. BY-PASS LINE

TWO STAGE PRESSURE-REDUCING & PRESSURE-REGULATING STATION EQUIPMENT. USED WHERE HIGH-PRESSURE STEAM IS SUPPLIED FOR LOW-PRESSURE REQUIREMENTS.

FIGURE 10-5

pressure. The secondary controller senses the leaving low pressure steam of 1 to 15 psig and sends its signal to the secondary or regulation stage valve to maintain the 1 to 15 psig final pressure. Ample distance between the two reducing valves must be provided. It is usually recommended by reducing valve manufacturers that this distance should be at least 20 ft of pipe that is sized to take the required flow at the reduced pressure. Both reducing valves are sized to pass the required capacity at the given pressure conditions and both valves must be capable of throttling and shutting off tight against full primary pressure.

In theory a reduction of pressure from 150 to 15 psig could be accomplished with one valve. In practice there are serious problems of noise and control at low flow (called wire drawing) that make this application of a single valve impractical. Generally about 75 lbs of the overall drop is taken in the primary stage and 40 to 50 lbs of drop in the secondary or regulation stage. The control then becomes very precise and effective.

Standard Application: Figure 10-6 is a typical version of one of the common specialty applications of pressure reducing valves. Usually this is a situation in which the necessary reduction can be made in one stage and in which there is also a wide variation in load. The top valve in the diagram is sometimes called the high load valve and is sized for 100 percent of required capacity. The lower valve, called the low load valve, is sized for 15 to 25 percent of the required capacity. Both valves are operated independently, but a transfer relay at the control point permits only one to operate at a given time. As the load changes from a very low load of 15 to 25 percent of full load, the smaller valve operates. Above this value the smaller valve is cut out and the larger valve is used. This, as noted in the title of Fig. 10-6, is parallel operation.

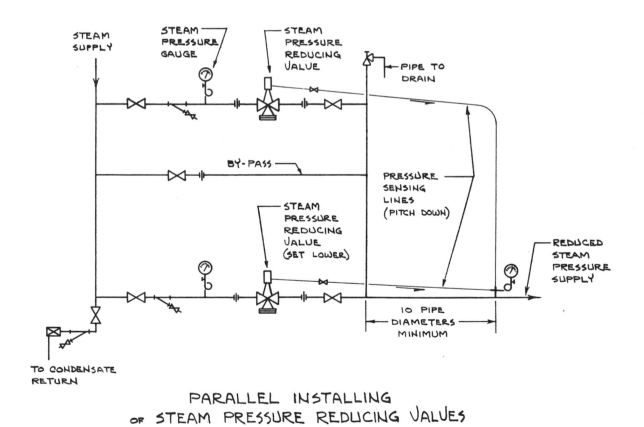

PARALLEL INSTALLING
of STEAM PRESSURE REDUCING VALVES

FIGURE 10-6

Batch Processing: In Fig. 10-7 we illustrate another typical example of the application of a pressure reducing valve. The valve operator can be electric or pneumatic. In our detail it is a direct acting electric actuator or operator which controls steam flow to an industrial batch processor. Some of the commonly installed related indicator and process vents are shown.

Note that in this application and in Fig. 10-6 an end use of steam is implied or shown and a pressure relief valve is indicated. In the case of Fig. 10-5, which is a two-stage reduction, there is also a relief valve at the point of end use which, in straight reduction for an overall purpose, is at the point of final reduced pressure.

Bypasses around reducing valves are mandatory. Generally they are manual bypasses which are only used in cases of valve malfunction in which steam flow must be maintained. They do not provide very accurate control and should never be used as a substitute for the reducing valve except for very short emergency situations.

Refrigerant Piping:

In our refrigerant piping details that follow we have included some relatively simple explanatory elements,

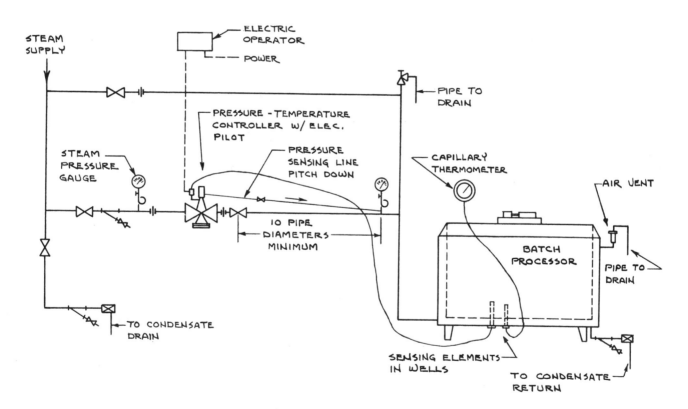

PRESSURE REDUCING VALVE
FOR BATCH PROCESSING APPLICATION

FIGURE 10-7

as well as complete specific equipment details. We are using this sort of presentation because of the peculiarities that occur in piping a fluid which can change its state along the way.

As with steam or water a refrigerant piping system must be designed for minimum pressure drop to avoid decreases in capacity or increases in refrigeration power. However, the refrigerant system is much more sensitive to pressure drop. Steam and water generally do not change in state as they circulate. A refrigerant does. Compressor oil is miscible with refrigerants and travels with them, which does not usually occur with steam or water. Therefore, all details must be concerned with minimizing the accumulation of the refrigerant in the compressor crankcase and with returning oil to the compressor at the same rate as it is leaving it.

One of the basic problems in any system occurs when the coil is considerably above the condenser. This creates a high static head on the liquid line. Liquid can flash into gas before it gets to the evaporator expansion valve. An arrangement as shown in Fig. 10-8 will not preclude the flashing problem, but properly sized and engineered, it can get rid of the flashed gas and insure liquid at the expansion valve.

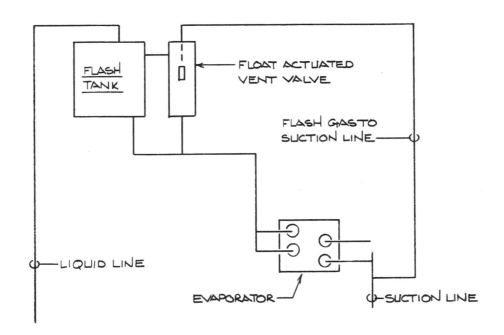

METHOD OF OVERCOMING EFFECTS OF SYSTEM
HIGH STATIC HEAD
—NOT TO SCALE—

FIGURE 10-8

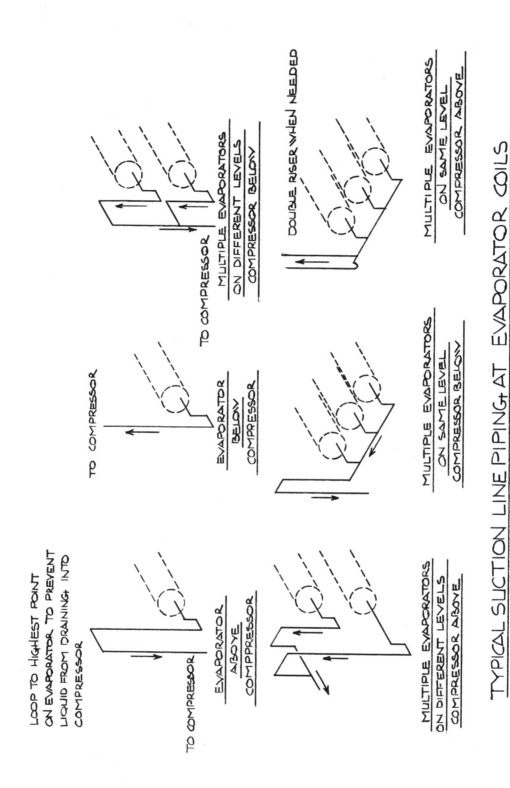

TYPICAL SUCTION LINE PIPING AT EVAPORATOR COILS
NOT TO SCALE

FIGURE 10-9

Evaporator Coil Suction Piping: Figure 10-9 illustrates some typical ways to connect suction piping at evaporator coils. This detail is not a direct plan detail but rather shows a way to illustrate how two very important elements of suction pipe detailing must be treated. All suction piping has two basic requirements. The first is to preclude liquid from draining into the compressor during shutdown. The second is to prevent oil from draining from an active to an idle evaporator. Both requirements are resolved by suction traps and loops into suction headers. Two of the six details show multiple evaporators which seem, at least partially, not to comply with the stated rules. Actually they do comply. A trap is not required when evapora-

tors are on the same level and there is either a looped connection or a pump-down arrangement. A double riser always has the larger line trapped. These six illustrations are not intended to illustrate the only ways of making the sort of connections that are required for suction piping but rather to illustrate the common basic principles.

Suction Header Below Coil: Figure 10-10 illustrates another common piping principle that often is not clearly shown. Common to all evaporator connections is a liquid and a suction line. The liquid expansion valve is modulated by the signal it gets from its sensor on the suction line. To properly balance this signal a second source of information comes from the equaliz-

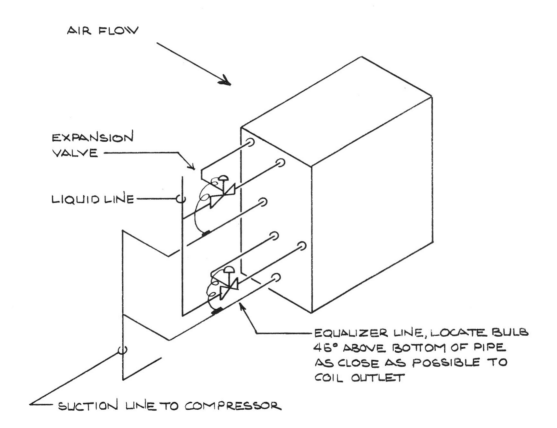

DX COIL USING SUCTION CONNECTIONS TO
DRAIN COIL—SUCTION HEADER BELOW COIL
———————— NOT TO SCALE ————————

FIGURE 10-10

ing line. This is connected to the top of the suction header at the end that is opposite the suction connection.

Direct Expansion Coil: Figure 10-11 illustrates two more principles of liquid and suction line piping. The upper detail of Fig. 10-11 is a continuation of the piping shown in Fig. 10-10. There are two differences in the figures. First the compressor is above the coil, and the question of what to do if two suction risers are the same size as the main is resolved by the dotted lines. Looping risers into the mains is the best solution, regardless of sizes. Second, in the lower half of the detail is a typical simple connection to a direct expansion (DX) cooler.

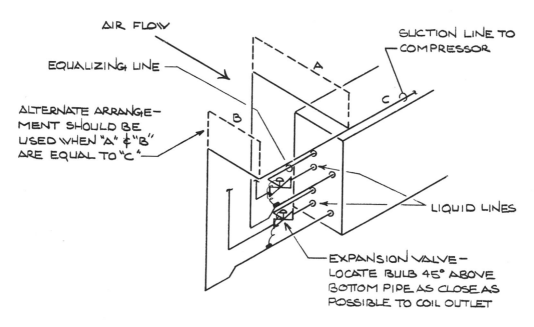

AIR FLOW

EQUALIZING LINE

ALTERNATE ARRANGEMENT SHOULD BE USED WHEN "A" & "B" ARE EQUAL TO "C"

SUCTION LINE TO COMPRESSOR

A

B

C

LIQUID LINES

EXPANSION VALVE— LOCATE BULB 45° ABOVE BOTTOM PIPE AS CLOSE AS POSSIBLE TO COIL OUTLET

DX COIL USING SUCTION CONNECTIONS TO DRAIN COIL SUCTION HEADER ABOVE COIL
—NOT TO SCALE—

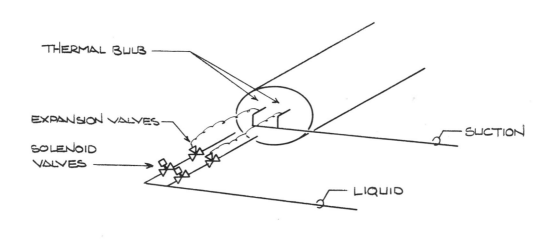

THERMAL BULB

EXPANSION VALVES

SOLENOID VALVES

SUCTION

LIQUID

DRY EXPANSION COOLER
—NOT TO SCALE—

FIGURE 10-11

Figure 10-12 resolves certain special problems for both direct expansion coils and larger direct expansion coolers. In the top detail a resolution of the coil that has connections at the top or middle is illustrated. The piping is arranged with oil drain lines so that this connection does not drain the evaporator. The drain lines connect at the bottom end of the suction connection to a common suction header below the coil. In the lower detail a pilot operated refrigerant feed valve is connected to a small expansion valve. The expansion valve is the pilot device. This cooler is like a typical steam-to-water heat exchanger except that refrigerant, rather than steam, is in the tubes. Generally there are oil return problems, as well as expansion control problems, when the capacity of the compressor is below 50 percent. The detail shown is one solution. A

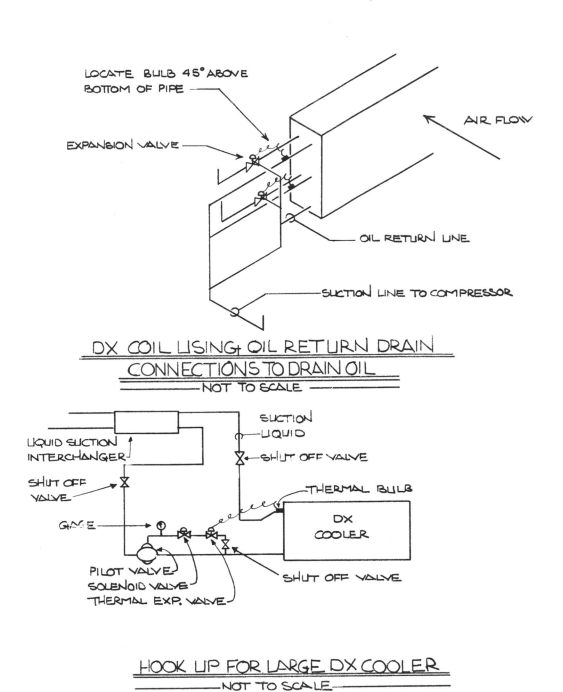

LOCATE BULB 45° ABOVE
BOTTOM OF PIPE

AIR FLOW

EXPANSION VALVE

OIL RETURN LINE

SUCTION LINE TO COMPRESSOR

DX COIL USING OIL RETURN DRAIN
CONNECTIONS TO DRAIN OIL
— NOT TO SCALE —

SUCTION
LIQUID

LIQUID SUCTION
INTERCHANGER

SHUT OFF VALVE

SHUT OFF
VALVE

THERMAL BULB

GAGE

DX
COOLER

PILOT VALVE
SOLENOID VALVE
THERMAL EXP. VALVE

SHUT OFF VALVE

HOOK UP FOR LARGE DX COOLER
— NOT TO SCALE —

FIGURE 10-12

liquid-suction heat exchanger is also part of this solution. This particular detail requires careful engineering for proper sizing of the components illustrated.

Flooded Cooler: Figure 10-13 depicts the refrigerant piping connections for a flooded cooler. In this instance the refrigerant is in the shell and the liquid to be cooled is in the tubes. To assure that the proper quantity of oil is returned to the compressor a continuous liquid bleed line is connected below the liquid level in the shell and tied to the suction line. While liquid-suction heat exchangers are common in refrigerant piping, here the exchanger is really necessary to be certain the returning suction line contains no liquid. Coolers commonly operate at light loads and a properly sized double suction riser is fairly standard.

Hot Gas Piping: Figure 10-14 is another schematic to illustrate a point in piping connection detailing. Hot gas loop piping is shown in three fairly common arrangements of multiple and single compressors. The loop prevents gas, which may condense when the compressor is stopped, from draining back into the compressor. It also prevents oil from draining from one compressor into another, nonoperating compressor.

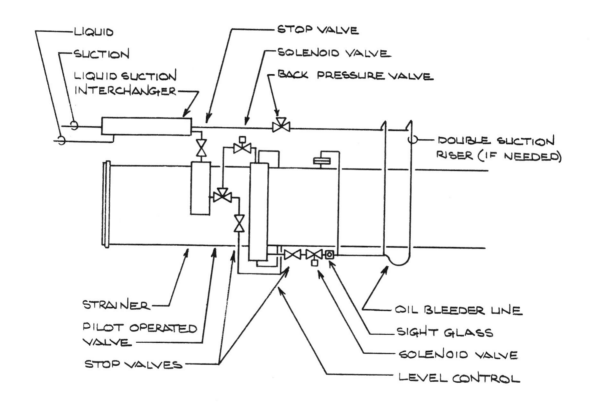

FLOODED COOLER
— NOT TO SCALE —

FIGURE 10-13

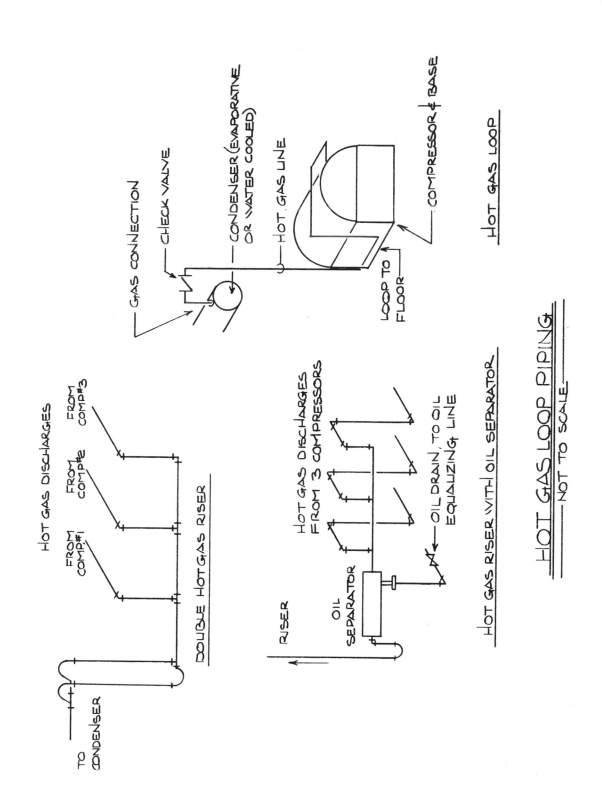

GAS CONNECTION

CHECK VALVE

CONDENSER (EVAPORATIVE OR WATER COOLED)

HOT GAS LINE

LOOP TO FLOOR

COMPRESSOR & BASE

HOT GAS LOOP

HOT GAS DISCHARGES

FROM COMP#1   FROM COMP#2   FROM COMP#3

TO CONDENSER

DOUBLE HOT GAS RISER

HOT GAS DISCHARGES FROM 3 COMPRESSORS

RISER

OIL SEPARATOR

OIL DRAIN, TO OIL EQUALIZING LINE

HOT GAS RISER WITH OIL SEPARATOR

HOT GAS LOOP PIPING
— NOT TO SCALE —

FIGURE 10-14

Liquid Piping: Figure 10-15 illustrates a balancing requirement on the now condensed hot gas which is leaving the condensers as liquid refrigerant. The drop into a common header that is 12 in or more below the condenser outlet resolves the balance problem. The loop precludes siphoning into the evaporator during shutdown. This loop can be, and usually is, replaced by a liquid line solenoid valve although both the loop and the solenoid valve are seen on some jobs.

Back Pressure: Figure 10-16 shows the installation of a back pressure valve. The selection and use of this valve is an engineering decision based on a desire to control suction pressure regardless of compressor suction variation, when lower suction pressure is required in some part of the system, or when there is a danger of evaporator freeze up. While it is almost self-evident, the detail makes the point that the valve goes on the evaporator's suction line, not on the liquid line.

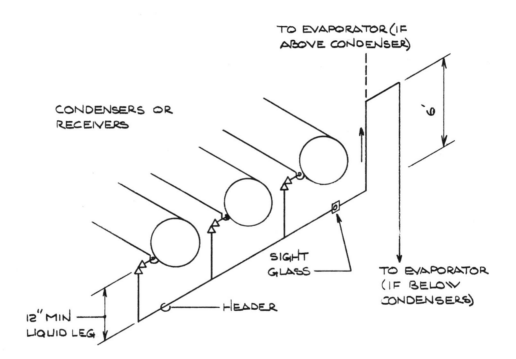

LIQUID PIPING TO INSURE CONDENSATE
FLOW FROM INTERCONNECTED CONDENSERS
NOT TO SCALE

FIGURE 10-15

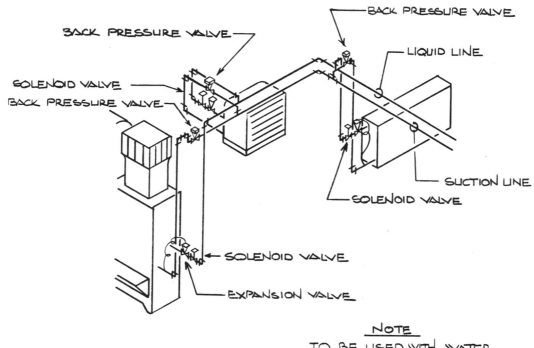

BACK PRESSURE VALVE

BACK PRESSURE VALVE

LIQUID LINE

SOLENOID VALVE

BACK PRESSURE VALVE

SUCTION LINE

SOLENOID VALVE

SOLENOID VALVE

EXPANSION VALVE

NOTE
TO BE USED WITH WATER
COOLING, AIR CONDITIONING
AND REFRIGERATION SERVICE

BACK PRESSURE VALVE APPLICATION
— NOT TO SCALE —

FIGURE 10-16

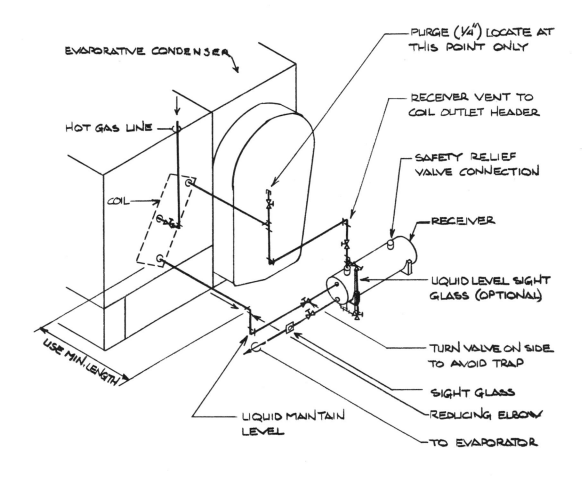

EVAPORATIVE CONDENSER

HOT GAS LINE

COIL

USE MIN. LENGTH

PURGE (1/4") LOCATE AT THIS POINT ONLY

RECEIVER VENT TO COIL OUTLET HEADER

SAFETY RELIEF VALVE CONNECTION

RECEIVER

LIQUID LEVEL SIGHT GLASS (OPTIONAL)

TURN VALVE ON SIDE TO AVOID TRAP

SIGHT GLASS

REDUCING ELBOW

TO EVAPORATOR

LIQUID MAINTAIN LEVEL

HOT GAS AND LIQUID PIPING
SINGLE COIL UNIT WITH RECEIVER VENT
NOT TO SCALE

FIGURE 10-17

Evaporative Condenser: Figures 10-17 through 10-19 are all related to the single subject of detailing gas and liquid piping connections at a remote evaporative condenser. The details are each fairly clear and obvious in their presentation. However, there are certain points that should be noted. While there are similar systems in operation that do not have a receiver, there are system problems that can be readily avoided if a receiver is utilized. There are systems that have no receiver vent. However, the equalizing line from the receiver to the condenser relieves gas pressure in the receiver and prevents liquid from being trapped in the condenser. Note that in the multiple unit installation detail there are individual vents for each unit. Figure 10-19 is essentially the same as Fig. 10-17 except we are getting extra mileage out of the condenser by running the liquid back through to subcool it. Actually there is more than this simple statement involved in the engineering of a subcooled circuit, but it does involve a second passage through the same condenser.

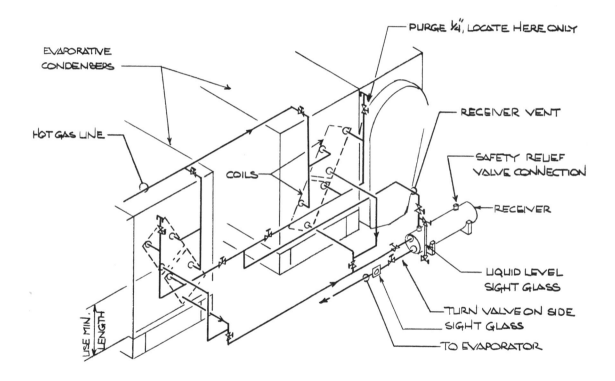

HOT GAS AND LIQUID PIPING
MULTIPLE DOUBLE COIL UNITS
——— NOT TO SCALE ———

FIGURE 10-18

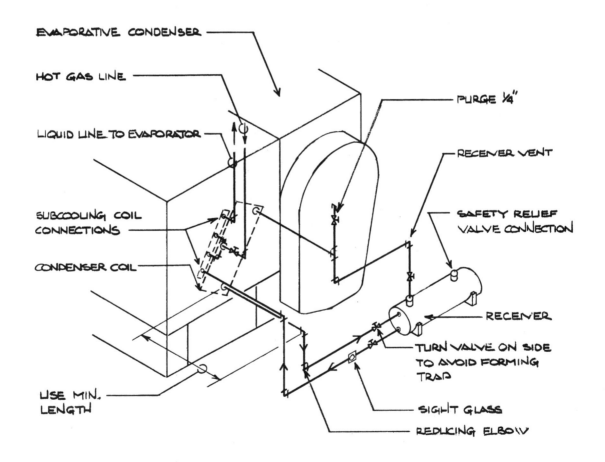

EVAPORATIVE CONDENSER

HOT GAS LINE

LIQUID LINE TO EVAPORATOR

SUBCOOLING COIL CONNECTIONS

CONDENSER COIL

USE MIN. LENGTH

PURGE ¼"

RECEIVER VENT

SAFETY RELIEF VALVE CONNECTION

RECEIVER

TURN VALVE ON SIDE TO AVOID FORMING TRAP

SIGHT GLASS

REDUCING ELBOW

SUBCOOLING COIL PIPING
NOT TO SCALE

FIGURE 10-19

Composite System: Figure 10-20 is a refrigerant piping detail of a small residential or commercial air conditioning system. Very commonly this is shown on the plans as an exterior air cooled condenser and an interior air handling unit with two lines labelled perhaps only as liquid and suction, and there is no detailing at all as to piping specialities. While this sort of plan has been omitted many times and somehow the system got more or less properly installed, the credit certainly did not go to the designer. This detail shows all the specialties required and, schematically, is a relatively foolproof piping arrangement. This detail, or a modified version of it, should be part of your standard detailing.

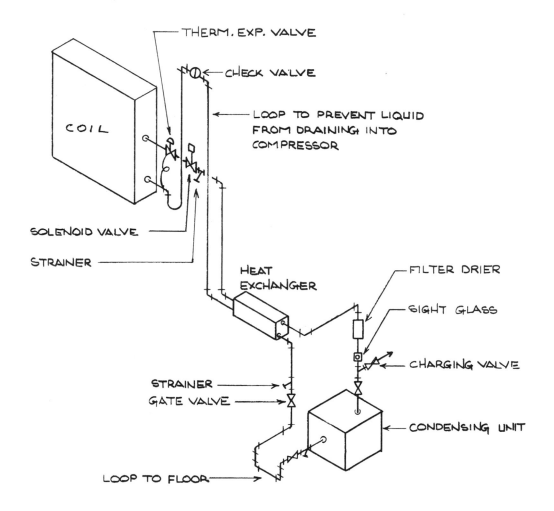

REFRIGERANT PIPING DETAIL
NOT TO SCALE

FIGURE 10-20

Heat Pump: Figure 10-21 illustrates schematically a typical heat pump cycle. If you have specified a packaged heat pump, there is no reason to use this detail. The piping is part of the package. As is true of most schematic diagrams, not all the related specialties are depicted. This detail is basically a flow diagram so that if you are separately locating some of the parts, you can follow the schematic as to refrigerant flow and further detail the connections at the coil, receiver, heat exchanger, and the like.

Hangers, Supports, and Special Connections

The detailing of hangers, supports, and special connections can be literally limitless. Nearly every engineer or designer has a special set of details that were taken from situations in which an installation problem was resolved. Only a few of these special situation details are depicted in the details that follow. Primarily the ones shown are common design detail situations that continually occur.

Anchors: Figure 10-22 illustrates a typical situation in which an expansion joint occurs beneath a typical steel framing situation. Usually the joint, if it is a slip-type or bellows-type joint, is located in the middle of the pipe run. Depending on a number of factors, one or more guides surround the pipe on both sides of the joint. This requires the pipe to be anchored at both ends. If this pipe was installed in a corridor ceiling, it might be possible to place an I beam across the corridor, anchored to two columns on opposite sides. The pipe could be locked to this beam.

Supports: Figure 10-23 assumes that this is not possible, and instead angles are welded to the joists and, through a bracket connection, the pipe is welded to the angles. This provides the necessary anchor. It is used on both ends opposite the expansion joint. If a U-

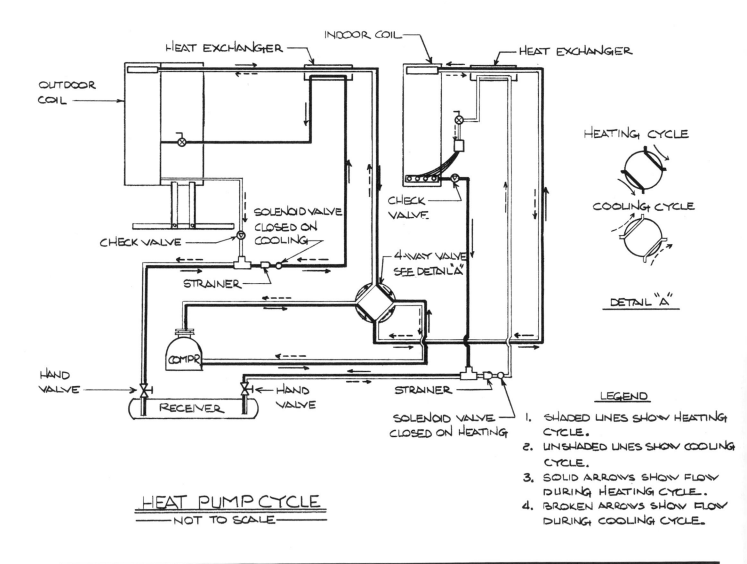

FIGURE 10-21

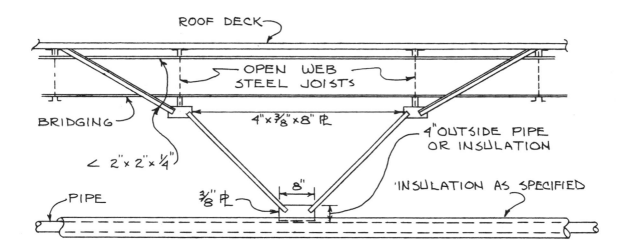

EXPANSION JOINT ANCHOR

NO SCALE

FIGURE 10-22

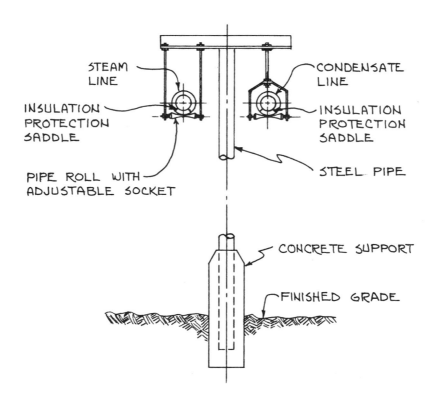

TYPICAL ABOVE GROUND PIPE SUPPORT

NO SCALE

FIGURE 10-23

shaped joint is used, you can also use this detail with a pipe system to anchor the center of the U-shaped expansion joint.

Figure 10-23 is a solution to steam and condensate piping or literally any piping that runs outside and above ground. For steam and condensate piping especially there is a lot of movement that is caused by temperature variation, and therefore the pipe is supported on a roller-type hanger. The main support is usually a 3-in or 4-in round column encased in concrete. This usually supports a 4-in × ⅝-in angle to which the properly sized pipe hangers can be attached.

Figure 10-24 illustrates one method of installing a hanger that is to be bolted into existing concrete or masonry. Frequently the material at the surface of the brick or concrete may appear questionable. The detail shows a 2-in long oversized sleeve set in a drilled hole and an expansion bolt inserted through the sleeve and deeper than usual into the masonry or concrete.

Wall Anchors: Figure 10-25 resolves another common installation error. When a pipe is supported from a vertical wall, it should not be clamped directly against the wall. Instead it should be supported free of the wall, as illustrated. The sizes of the bolts and angles can be varied depending on the size of the pipe and the loads involved.

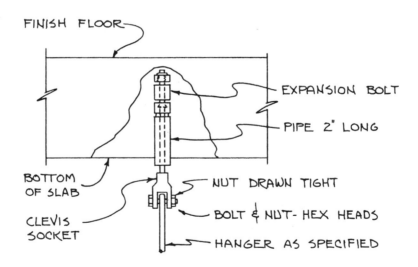

SUPPORT - FOR USE IN
OLD BRICKWORK AND CONCRETE
NO SCALE

FIGURE 10-24

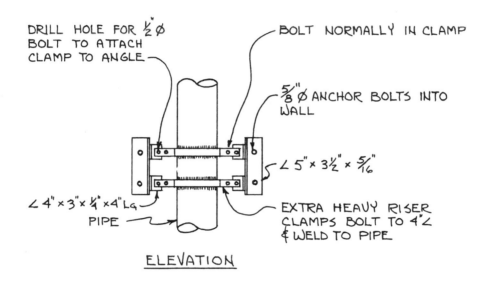

DRILL HOLE FOR $\frac{1}{2}$" Ø BOLT TO ATTACH CLAMP TO ANGLE

BOLT NORMALLY IN CLAMP

$\frac{5}{8}$" Ø ANCHOR BOLTS INTO WALL

∠ 5" × 3$\frac{1}{2}$" × $\frac{5}{16}$"

∠ 4" × 3" × $\frac{1}{4}$" × 4" LG

PIPE

EXTRA HEAVY RISER CLAMPS BOLT TO 4" ∠ & WELD TO PIPE

ELEVATION

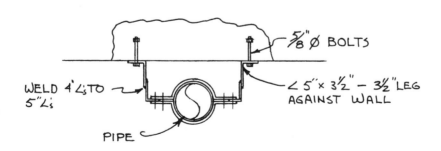

$\frac{5}{8}$" Ø BOLTS

WELD 4" ∠'s TO 5" ∠'s

∠ 5" × 3$\frac{1}{2}$" — 3$\frac{1}{2}$" LEG AGAINST WALL

PIPE

PLAN

ANCHOR TO VERTICAL WALL

NO SCALE

FIGURE 10-25

Roof Pipe Supports: One of the common problems of rooftop air cooled condensing units and the associated refrigerant piping is covered in Fig. 10-26. The pipe must run across the roof for some distance and then drop through the roof to the air handling unit at the ceiling of the floor below. Commonly one sees these pipes blocked up from the roof and clamped to the condenser frame. In Fig. 10-26 a simple U-shaped frame is made of 2-in × ¼-in bar stock. This is fastened and sealed to the roof deck. Simple U clamps hold the refrigerant lines.

Hangers: Figure 10-27 is a composite detail that provides certain valuable information. The breeching hanger detail is similar to the hanger detail for any heavy ductwork. The two enlarged partial details may be utilized when you feel the need, or are required, to more clearly depict the common clevis or roll hanger. Finally the standard horizontal spacing of the pipe

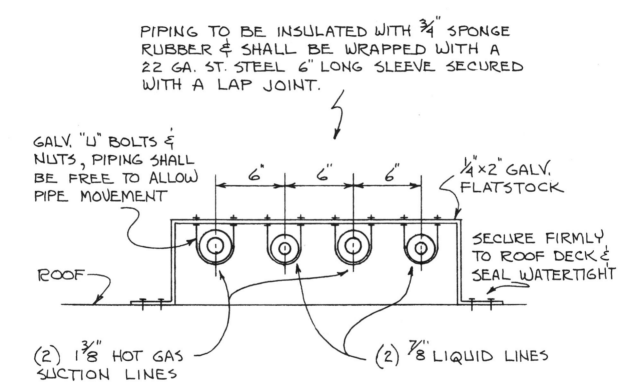

PIPING TO BE INSULATED WITH ¾" SPONGE RUBBER & SHALL BE WRAPPED WITH A 22 GA. ST. STEEL 6" LONG SLEEVE SECURED WITH A LAP JOINT.

GALV. "U" BOLTS & NUTS, PIPING SHALL BE FREE TO ALLOW PIPE MOVEMENT

¼"×2" GALV. FLATSTOCK

SECURE FIRMLY TO ROOF DECK & SEAL WATERTIGHT

ROOF

6"   6"   6"

(2) 1⅜" HOT GAS SUCTION LINES

(2) ⅞" LIQUID LINES

SUPPORTING PIPES OFF OF ROOF

NO SCALE

FIGURE 10-26

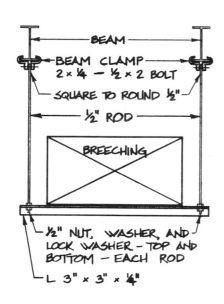

BEAM

BEAM CLAMP
2 × ¼ — ½ × 2 BOLT

SQUARE TO ROUND ½"

½" ROD

BREECHING

½" NUT, WASHER, AND
LOCK WASHER — TOP AND
BOTTOM — EACH ROD

L 3" × 3" × ¼"

BREECHING HANGER DETAIL
—— NOT TO SCALE ——

MAXIMUM HORIZONTAL SPACING OF PIPE HANGERS

PIPE SIZE	SPAN
1"	7'
1¼	8
1½	9
2	10
2½	11
3	12
3½	13
4	14
5	16
6	17
8	19
10	22
12	23
14	25
16	27
18	28
20	30
24	32

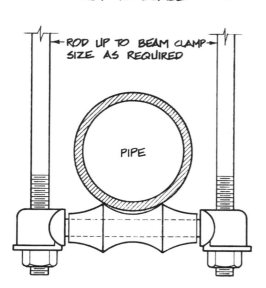

ROD UP TO BEAM CLAMP
SIZE AS REQUIRED

PIPE

STEAM PIPE ROLL HANGER WITH ADJUSTABLE SOCKETS
—— NOT TO SCALE ——

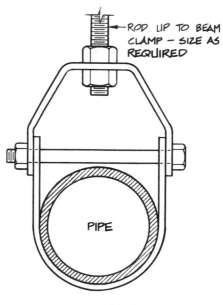

ROD UP TO BEAM
CLAMP — SIZE AS
REQUIRED

PIPE

CLEVIS HANGER
—— NOT TO SCALE ——

FIGURE 10-27

hangers is noted. A table such as this should appear in your plans or in your specification.

Condensate Piping: Figure 10-28 illustrates the use of a lift fitting or fittings on a vacuum return steam system. This does not work on a normal pumped or wet return condensate system. The key to the system is the special lift fitting. While some engineering calculations are involved, the system does work and will resolve problems of maintaining levels on vacuum condensate return systems.

Figure 10-29 is a series of self-explanatory details to be used in steam piping. These should be part of the

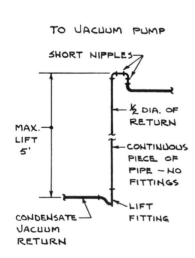

ONE-STEP CONDENSATE LIFT

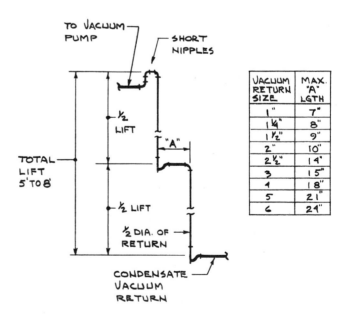

TWO-STEP CONDENSATE LIFT

VACUUM RETURN SIZE	MAX. "A" LGTH
1"	7"
1¼"	8"
1½"	9"
2"	10"
2½"	14"
3	15"
4	18"
5	21"
6	24"

FIGURE 10-28

SPECIAL PIPE DETAILS

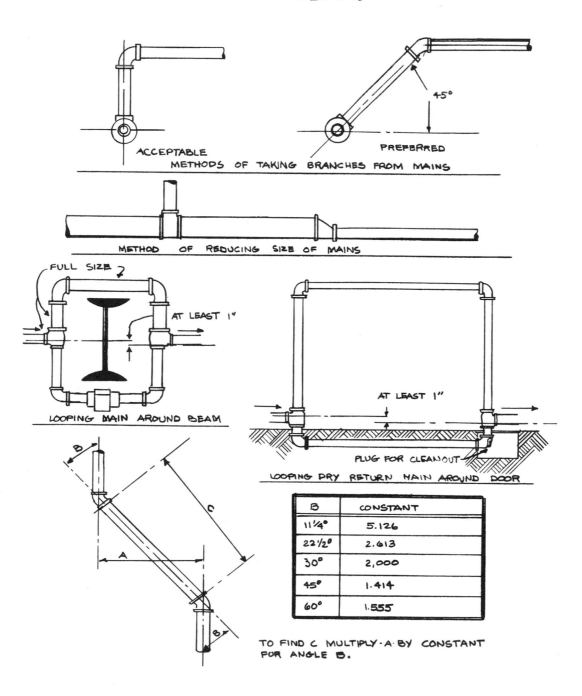

ACCEPTABLE   PREFERRED
45°

METHODS OF TAKING BRANCHES FROM MAINS

METHOD OF REDUCING SIZE OF MAINS

FULL SIZE
AT LEAST 1"

LOOPING MAIN AROUND BEAM

AT LEAST 1"

PLUG FOR CLEANOUT

LOOPING DRY RETURN MAIN AROUND DOOR

B	CONSTANT
11¼°	5.126
22½°	2.613
30°	2,000
45°	1.414
60°	1.555

TO FIND C MULTIPLY A BY CONSTANT FOR ANGLE B.

FIGURE 10-29

standard detailing for any piping system. Figure 10-30 contains other riser details, all of which are noted as to application for steam piping.

Compressed Air Traps: Compressed air piping, as illustrated in Fig. 10-31 also has certain problems that are frequently overlooked. In the compressed air system there are always water removal problems and, on occasion, combination oil and water problems. Dripping risers and the use of separators and receivers are part of the installation requirements. The various self-explanatory details on Fig. 10-31 should help you resolve your problems.

Exterior Underground Piping

In some ways the installation of exterior underground piping is one of the easier tasks for the designer and the detailer although the seemingly simple unobstructed layout may well be a source of considerable engineering investigation and calculation. Since our book is aimed at the detailer's and not the engineer's

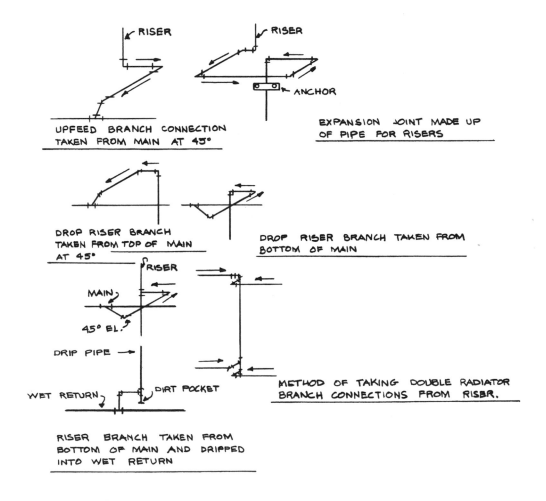

RISER DETAILS

FIGURE 10-30

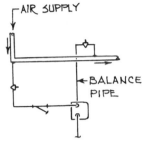

FLOAT TRAP DRAINING
FOOT OF VERTICAL LINE

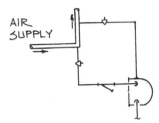

FLOAT TRAP DRAINING
HEEL OF RISE

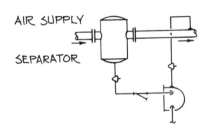

FLOAT TRAP
DRAINING SEPARATOR

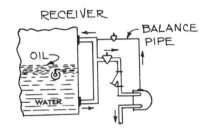

FLOAT TRAP DRAINING
OIL-CONTAMINATED RECEIVER

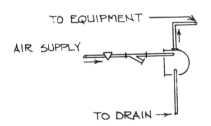

FLOAT TRAP DRAINING SMALL
LINE TO COMPRESSED AIR TOOL

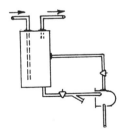

FLOAT TRAP
DRAINING RECEIVER

COMPRESSED AIR TRAPS
NO SCALE

FIGURE 10-31

problems, we begin in Fig. 10-32 with a simple detail of an underground insulated pipe passing through a wall. The pipe is assumed to be properly insulated for underground installation. The wall has a properly oversized hole through it to facilitate the passage of the pipe. The detailer's job is to illustrate how the installation will avoid creating a leak in the wall. In Fig. 10-32 there is a sleeve, either cast or properly caulked, in the wall that is slightly larger than the insulated pipe. The detailer must illustrate how this passage is to be sealed. Here it is done with a simple through-bolted mechanical seal which is specified to be properly caulked.

Wall Entry: Figure 10-33 is another solution to the sealing problem just mentioned. It is different in a number of respects. First, the *pipe* is in a metal conduit. Usually this section of the conduit is relatively short and is designed to be fixed in place in the wall forms *before* the concrete wall is poured. The concrete will *not* bond to the conduit. Over time water will migrate along the exterior surface of the pipe. The plate depicted is the seal. Water will also occur as condensed vapor inside the conduit, which also has to be sealed and drained as depicted. Otherwise, in time you would have the conduit acting as a water main and creating all sorts of problems.

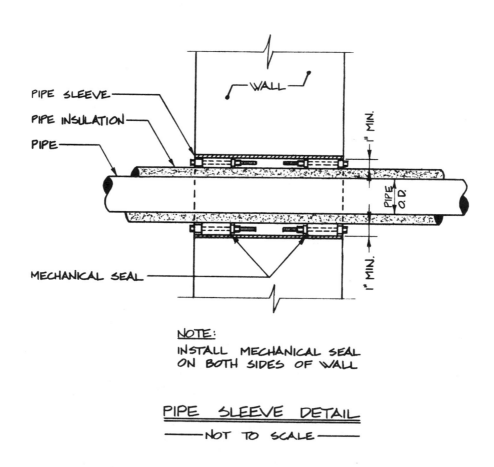

NOTE:
INSTALL MECHANICAL SEAL
ON BOTH SIDES OF WALL

PIPE SLEEVE DETAIL
—— NOT TO SCALE ——

FIGURE 10-32

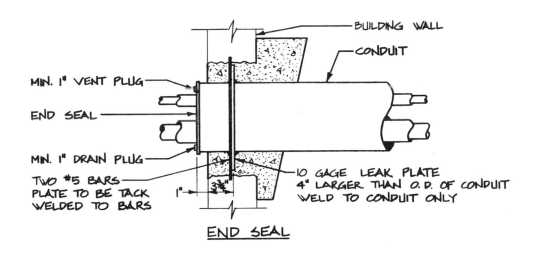

MIN. 1" VENT PLUG

END SEAL

MIN. 1" DRAIN PLUG

TWO #5 BARS
PLATE TO BE TACK
WELDED TO BARS

BUILDING WALL

CONDUIT

10 GAGE LEAK PLATE
4" LARGER THAN O.D. OF CONDUIT
WELD TO CONDUIT ONLY

1"   3¾"

END SEAL

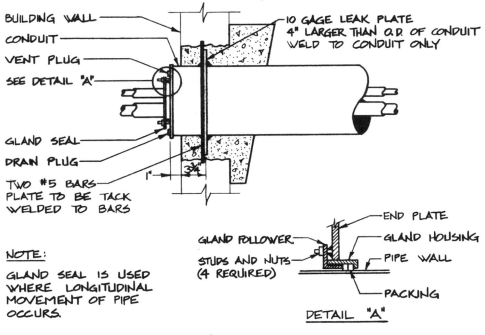

BUILDING WALL

CONDUIT

VENT PLUG

SEE DETAIL "A"

GLAND SEAL

DRAIN PLUG

TWO #5 BARS
PLATE TO BE TACK
WELDED TO BARS

10 GAGE LEAK PLATE
4" LARGER THAN O.D. OF CONDUIT
WELD TO CONDUIT ONLY

1"   3¾"

NOTE:

GLAND SEAL IS USED
WHERE LONGITUDINAL
MOVEMENT OF PIPE
OCCURS.

GLAND FOLLOWER

STUDS AND NUTS
(4 REQUIRED)

END PLATE

GLAND HOUSING

PIPE WALL

PACKING

DETAIL "A"

GLAND SEAL

DETAILS OF BUILDING WALL ENTRY
NOT TO SCALE

FIGURE 10-33

Trench Detail: Figure 10-34 seemingly depicts very little. All that is shown is a pipe in a trench. It shows very little, but it hints at a lot. As detailer your job is to show this detail and have your supervisor tell you exactly what to indicate as covering fill and exactly how deep it should be. This is subject to many variations. We chose not to show any of the various possible types of cover. We did not want to create the impression that any particular type or depth of cover was correct. This must be resolved by project and site design conditions.

Underground Conduit: Figures 10-35 and 10-36 illustrate two of the common conditions of any underground expansion joint. The first of the two figures, Fig. 10-35 is a somewhat enlarged view of the overall expansion loop shown in Fig. 10-36. The two details

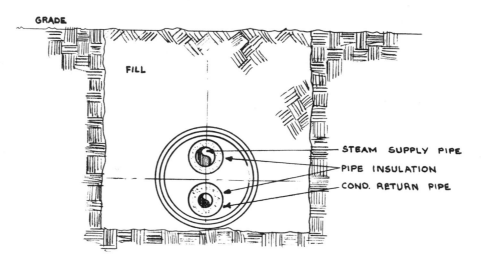

CROSS SECTION THRU PIPE TRENCH

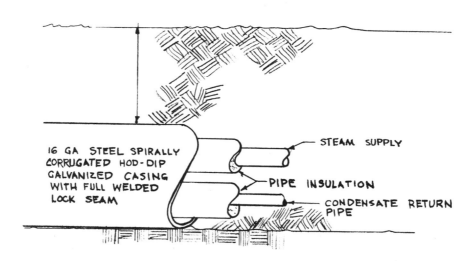

LONGITUDINAL SECTION THRU PIPE TRENCH

DETAILS OF PREFABRICATED UNDERGROUND CONDUIT

NO SCALE

FIGURE 10-34

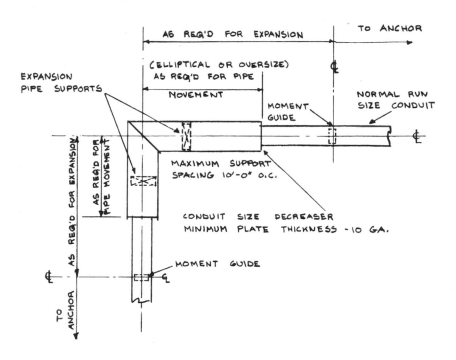

DETAIL OF PREFABRICATED UNDERGROUND CONDUIT
EXPANSION ELBOW

NO SCALE

FIGURE 10-35

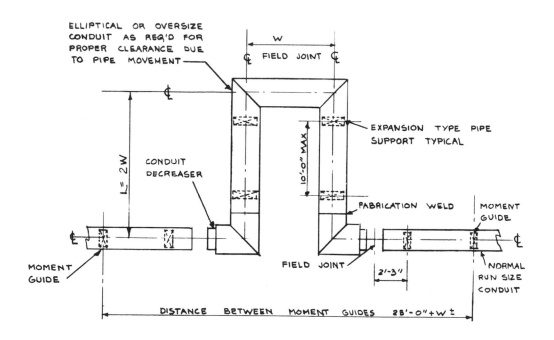

DETAIL OF PREFABRICATED UNDERGROUND CONDUIT
FABRICATED EXPANSION LOOP

NO SCALE

FIGURE 10-36

vary slightly in the quantity of information that describes the elbow. Figure 10-35, which depicts the elbow, could be combined with Fig. 10-36, which depicts the overall loop. The values of the length (L = 2W) and width (W) of the joint are fairly common standards. Proper calculations could alter these values to some degree. The distance between guides is also subject to some variation. Expansion pipe supports and guides should be carefully checked as to type and arrangement with the selected prefabricated conduit manufacturer. In theory the conduit, piping, supports, guides, and insulation could all be purchased separately and the total system could be custom designed. Or an underground conduit may already exist. If this is planned, than a considerable amount of engineering is required and details that are special to the design and

not depicted here would be required for hangers or supports, guides, insulation, and the like.

Figures 10-37 and 10-38 illustrate field joints and pipe anchors for an underground prefabricated pipe and conduit system. This is part of the type of detailing you would also need if your system, as noted in the previous paragraph, was a total customized installation. Whether your system is prefabricated or customized, the sections of straight pipe come in lengths dictated both by the standard practice of pipe manufacturers and by shipping, handling, delivery, and site restrictions. Figure 10-37 details carefully and clearly how the interior pipes are joined and how the pipe insulation and the enclosing conduit is joined. Note that the pipe joint must be tested before the insulation and conduit closure is applied. Be certain you do this

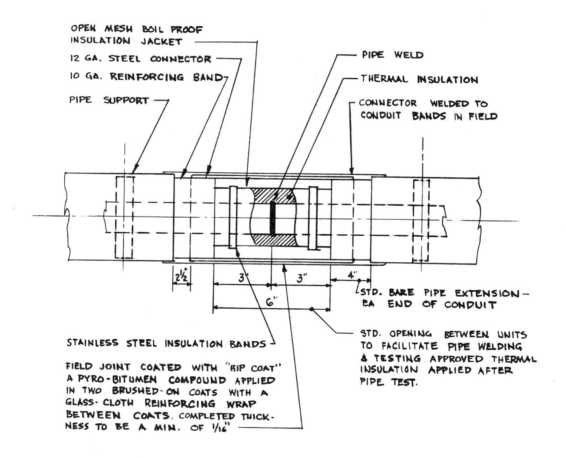

OPEN MESH BOIL PROOF INSULATION JACKET

12 GA. STEEL CONNECTOR

10 GA. REINFORCING BAND

PIPE SUPPORT

PIPE WELD

THERMAL INSULATION

CONNECTOR WELDED TO CONDUIT BANDS IN FIELD

2½"   3"   3"   4"

6"

STD. BARE PIPE EXTENSION — EA END OF CONDUIT

STAINLESS STEEL INSULATION BANDS

FIELD JOINT COATED WITH "RIP COAT" A PYRO-BITUMEN COMPOUND APPLIED IN TWO BRUSHED-ON COATS WITH A GLASS-CLOTH REINFORCING WRAP BETWEEN COATS. COMPLETED THICKNESS TO BE A MIN. OF 1/16"

STD. OPENING BETWEEN UNITS TO FACILITATE PIPE WELDING & TESTING. APPROVED THERMAL INSULATION APPLIED AFTER PIPE TEST.

DETAIL OF FIELD JOINT FOR UNDERGROUND CONDUIT

NO SCALE

FIGURE 10-37

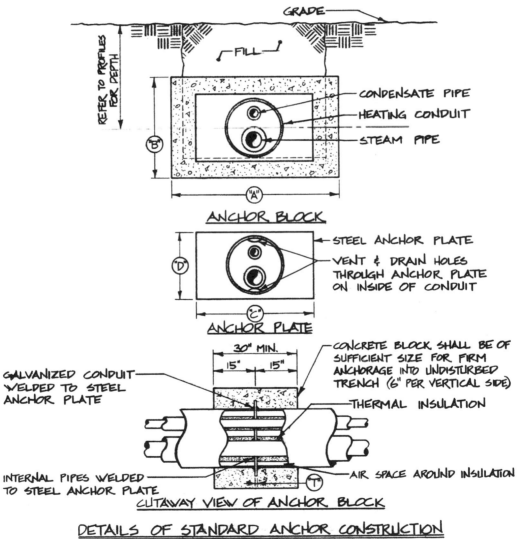

TYPICAL DIMENSIONS										
STEAM PIPE	COND. PIPE	INSUL. THKNS. STEAM	INSUL. THKNS. COND.	CONDUIT	TRENCH WIDTH	"A"	"B"	"C"	"D"	"T"
8"	5"	2½"	1"	24"	36"	63"	43½"	33"	25½"	¾"
6"	4"	2½"	1"	21"	33"	60"	40½"	30"	22½"	½"
6"	2½"	2½"	1"	18"	30"	57"	37½"	27"	19½"	½"
2"	2"	1½"	1"	12"	24"	51"	31½"	21"	13½"	½"

GRADE

FILL

REFER TO PROFILES FOR DEPTH

CONDENSATE PIPE
HEATING CONDUIT
STEAM PIPE

"B"

"A"

ANCHOR BLOCK

STEEL ANCHOR PLATE
VENT & DRAIN HOLES THROUGH ANCHOR PLATE ON INSIDE OF CONDUIT

"D"

"C"

ANCHOR PLATE

30" MIN.
15" 15"

GALVANIZED CONDUIT WELDED TO STEEL ANCHOR PLATE

CONCRETE BLOCK SHALL BE OF SUFFICIENT SIZE FOR FIRM ANCHORAGE INTO UNDISTURBED TRENCH (6" PER VERTICAL SIDE)
THERMAL INSULATION

INTERNAL PIPES WELDED TO STEEL ANCHOR PLATE

AIR SPACE AROUND INSULATION

"T"

CUTAWAY VIEW OF ANCHOR BLOCK

DETAILS OF STANDARD ANCHOR CONSTRUCTION
——— NOT TO SCALE ———

FIGURE 10-38

because you will have a very difficult time finding small leaks in the pipe once the closure is completed.

In Fig. 10-38 we not only depict the method of anchoring a pipe which is a required part of the pipe and expansion joint overall system, but we also give sizes of the anchor block itself. The concrete is standard 3000-lb. concrete. Note that the pipes and the conduit cover are all individually welded to the blocking, or locking, steel anchor plate. This facet of the pipe anchor is frequently poorly (or not at all) specified or depicted. There are no nonparticipants in the expansion joint or anchoring system. They all must move together and stop together. Anchor plates, as noted in the detail, are usually ½-in to ¾-in thick. This should be checked when pipes are very large or there is unusual pressure. The table is not complete, but similar tables can be found in conduit manufacturers' catalogs and, for unusual conditions, should be calculated both for anchor and for plate by a properly qualified structural engineer.

Seismic Piping

The answer the average engineer, designer, detailer, or draftsperson gives to questions about piping under possible earthquake conditions, is usually approximately, "Not really a problem in this area." Based on available data, that remark is a reasonable answer. Except for the California area none of the rest of the country really has had an earthquake problem that adversely affected normal pipe connections.

Thus Figs. 10-39 through 10-41, which are a set of three detail sheets all related to each other, have really

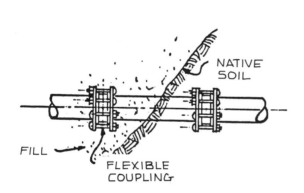

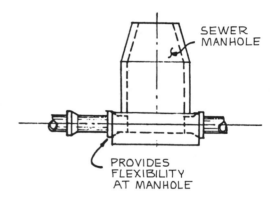

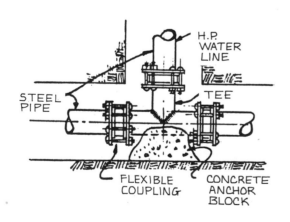

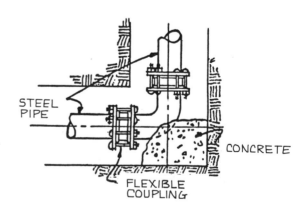

TYPICAL SEISMIC PIPING CONNECTIONS

1 OF 3

FIGURE 10-39

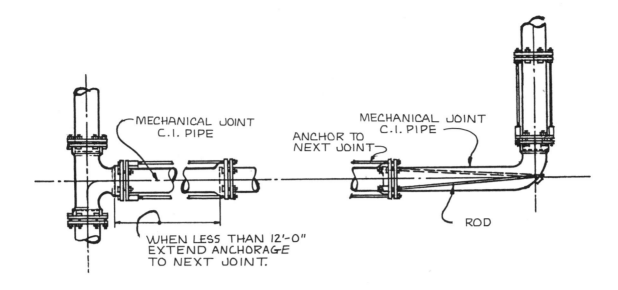

TEE   90° BEND

TYPICAL SEISMIC PIPING CONNECTIONS
2 OF 3

FIGURE 10-40

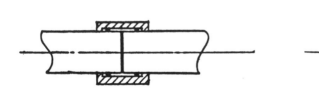

ASBESTOS-CEMENT
COUPLING

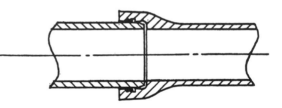

BELL & SPIGOT JOINT
WITH GASKET CONNECTION

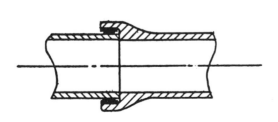

VCP CONNECTION

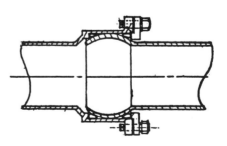

BALL JOINT

TYPICAL SEISMIC PIPING CONNECTIONS
3 OF 3

FIGURE 10-41

little or nor application generally. As such their presentation could easily and logically have been omitted. But on occasion all engineers face situations in which, because of surface pressures and subsoil peculiarities, there is the possibility of more than normal pipe movement.

The three figures show a set of details which are by no means complete and certainly are *not* intended as details that could be copied without thought or engineering investigation or calculation. Their purpose is to illustrate the type of seismic design detail that may be used when, for example, an underground sewer or water line is subject to unusual movement.

Figure 10-39 shows three typical water pipe connections that commonly occur. These are connections across different soils and show tees and elbows that require an anchor point. In each case the solution is the same. All show the proper flexible connections. This figure also illustrates a way of achieving flexibility at the manhole by using the well-known bell and spigot connection on both sides of the manhole. It looks a little odd, but it works.

Figure 10-40 is another way of getting flexibility and some strength by using extended rods between joints.

Finally Fig. 10-41 illustrates some typical piping connections between lengths of steel, asbestos-cement, plastic, and cast iron piping.

Ductwork

In presenting our details on various types of duct connections we are assuming that the following fixed information will be in your basic specifications. Commonly the standard ductwork specification contains specific descriptions of the types of material permitted in low, medium, and high velocity ductwork. These include tables showing gauges of copper, aluminum, and galvanized iron that are related to specific dimensions of round, oval, and rectangular ductwork; specific descriptions of the types of connections and supports required; specifications for acoustical linings and thermal insulation; and materials, connections, and supports for specialized exhaust systems such as kitchen hoods, chemical fumes, wood chips, and industrial abrasives.

The detailing of special or unusual duct connections is literally endless. Many, possibly the majority, of your projects will require special details that could not be covered in a book such as this. Even books devoted

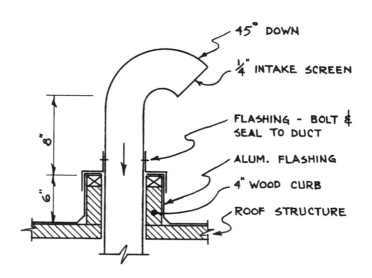

TYPICAL FRESH AIR ROOF INTAKE

NO SCALE

FIGURE 10-42

only to ductwork do not begin to cover all the possibilities. The following details and discussion are based on details that illustrate items in a standard specification that are normally part of the standard detailing on any project in which some ductwork is part of the design.

Outside Air Intake: Figure 10-42 is a typical rooftop fresh air intake. The height of the curb and the height to the bottom edge of the intake are based on average values. Depending on snow accumulation and blowing roof dust and dirt, these may vary or be varied to suit your local conditions. Most often the duct is braced by a wooden roof curb, which provides an easy place to attach the all important flashing, as illustrated. Commonly this flashing and cap flashing are aluminum although many engineers and clients request copper. Your specification should spell out the metal gauge you desire. Your specification should also describe

what we have carefully not shown: the work of the roofing contractor in bonding the flashing to the roof.

Duct Supports: Figure 10-43 shows a typical floor-level support for a vertical duct riser. This simple riveted or bolted angle support is normally described in your standard duct installation specification paragraphs. The object of the detail is to forestall any arguments. This is what your words mean, and this is what you expect and intend to have installed.

Hangers: Figure 10-44 also resolves a problem interpretation. Your specifications describe a good set of direct hangers, their spacing, and connection. You do not want to have the type A hanger installed in large ducts, and you want to make clear where type A, the least expensive hanger, will be permitted and, most importantly, where it will *not* be permitted. Finally, you already have a hanger spacing schedule in the

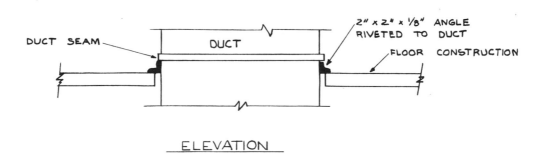

ELEVATION

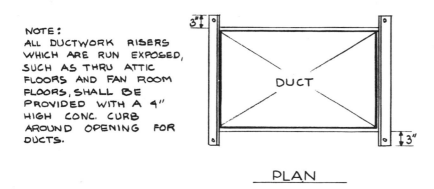

NOTE:
ALL DUCTWORK RISERS WHICH ARE RUN EXPOSED, SUCH AS THRU ATTIC FLOORS AND FAN ROOM FLOORS, SHALL BE PROVIDED WITH A 4" HIGH CONC. CURB AROUND OPENING FOR DUCTS.

PLAN

TYPICAL SUPPORT FOR VERTICAL DUCTS

FIGURE 10-43

specification. Figure 10-44 ends the excuse "I don't carry the spec book around when I'm working."

Take-Offs: Figure 10-45 illustrates rectangular branch direct take-off details. All of us have seen some truly poor connections that fortunately worked out satisfactorily. You may have your own pet way that varies from the four versions we have illustrated; if they work, we have no argument with them. All we are

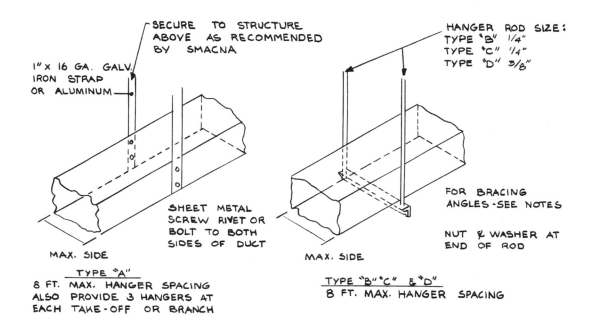

DUCT SCHEDULE	
DUCT DIMENSIONS INCHES	TYPE HANGER
UP THRU 12	A
13 18	A
19 30	A/B
31 42	B
43 54	B
55 60	B
61 84	C
85 96	C
OVER 96	D

NOTES:

1. FOR SEVERAL DUCTS ON ONE HANGER TYPE "B"-"C" OR "D" MAY BE USED. SIZE OF HANGER WILL BE SELECTED ON THE SUM OF DUCT WIDTHS EQUAL TO MAX. WIDTH OF DUCT SCHEDULE.

2. SCHEDULE FOR ANGLES FOR BRACING: TYPE "B" 1½" X 1½" X ⅛" ANGLE. MAX. SPACING 8'-0" CENTERS; TYPE "C" 1½" X 1½" X 3/16" ANGLE MAX SPACING 8'-0" CENTERS; TYPE "D" 2" X 2" X ¼ MAX SPACING 4'-0" CENTERS.

DUCT HANGERS

FIGURE 10-44

RECTANGULAR DUCT TAKE-OFF. SEE FIGURE A,B,C,D FOR ORDER OF
PREFERENCE. PROVIDE A SPLITTER DAMPER FOR EACH TAKEOFF.

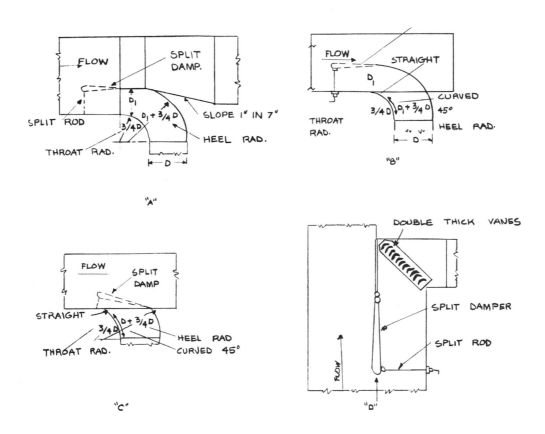

RECTANGULAR DUCT TAKE-OFF DETAILS

FIGURE 10-45

saying is that the four versions we show work. You can use some, part, or all in your own version, but do show something.

Flexible Connections: Figure 10-46 shows a typical flexible connection in about as elaborate a detail as you will normally ever need. Many engineered plans do not show this detail at all. The plans will contain a note that calls for 6-in canvas, or for a 6-in flexible connection. Probably if your specification says, for example, the flexible connection is part of the air handling unit of the manufacturer's standard factory-supplied package, you will get a decent result. If not, and you do not include this detail, only the low bidder knows what you will get.

Transitions: Figure 10-47 illustrates duct size transition that is related to the connection to a branch outlet. We have illustrated safe lengths for a change of shape from square to round in branch return connections and in tight-to-main-duct ceiling diffuser connections. These are not the only ways to handle these items and in the following detail we show another version of the tight-to-main-duct ceiling diffuser.

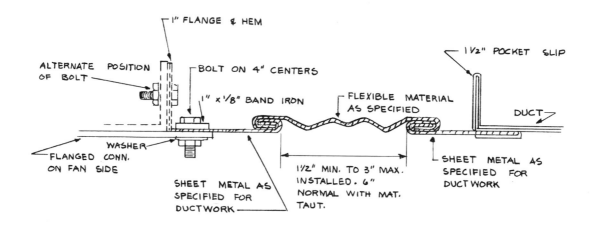

RECTANGULAR FLEXIBLE CONNECTION

FIGURE 10-46

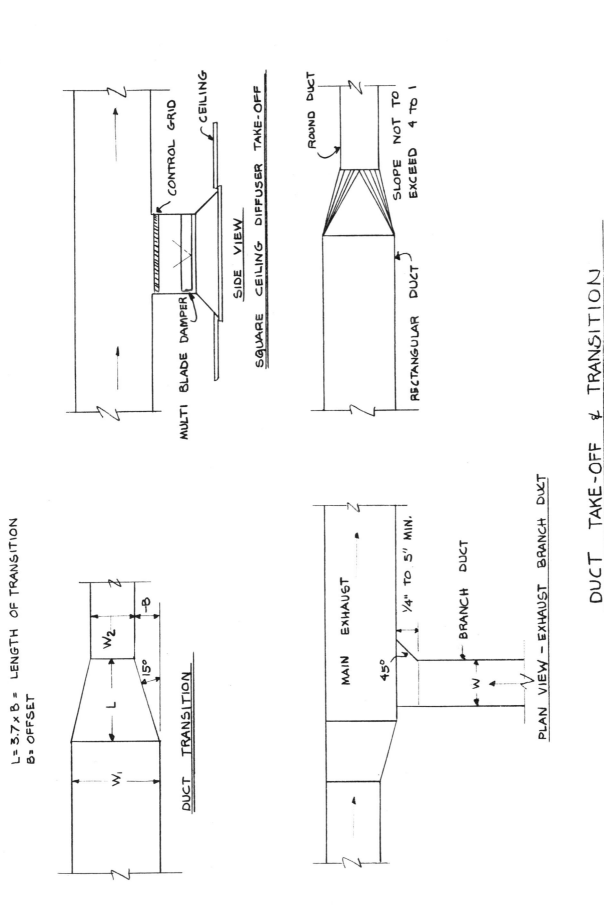

L = 3.7 × B = LENGTH OF TRANSITION

B = OFFSET

CONTROL GRID

CEILING

MULTI BLADE DAMPER

SIDE VIEW

SQUARE CEILING DIFFUSER TAKE-OFF

ROUND DUCT

SLOPE NOT TO
EXCEED 4 TO 1

RECTANGULAR DUCT

W₂

B

15°

L

W₁

DUCT TRANSITION

MAIN EXHAUST

¼" TO 5" MIN.

45°

BRANCH DUCT

W

PLAN VIEW — EXHAUST BRANCH DUCT

DUCT TAKE-OFF & TRANSITION

FIGURE 10-47

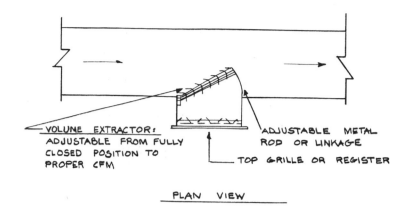

VOLUME EXTRACTOR:
ADJUSTABLE FROM FULLY
CLOSED POSITION TO
PROPER CFM

ADJUSTABLE METAL
ROD OR LINKAGE

TOP GRILLE OR REGISTER

PLAN VIEW

SUPPLY GRILLE OR REGISTER TAKE-OFF

FIGURE 10-48

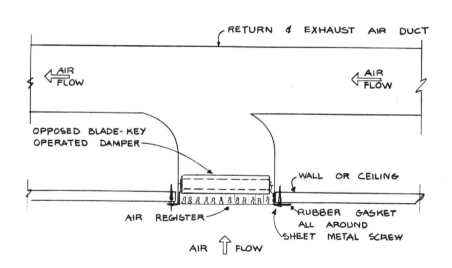

RETURN & EXHAUST AIR DUCT

AIR FLOW

AIR FLOW

OPPOSED BLADE-KEY
OPERATED DAMPER

WALL OR CEILING

AIR REGISTER

RUBBER GASKET
ALL AROUND

SHEET METAL SCREW

AIR ⇧ FLOW

RETURN OR EXHAUST REGISTER INSTALLATION

FIGURE 10-49

Figure 10-48 is a single detail showing another way to solve the close-coupled supply grille or register. This situation is always a problem, especially when the grille is the first outlet in a long supply duct and the air does not want to stop and make a quick 90° turn. The volume extractor is a proprietary item and does cause problems on occasion, but it is one solution that you should investigate further.

When you have the room, Fig. 10-49 is almost an ideal book solution to the control of return air from a branch duct. With the volume damper behind the grille of the air register, you can really balance the return duct system. We strongly recommend the use of this detail whenever possible.

Multiple Supports: We previously presented the proper sort of detail for duct hangers when the ducts are hung individually. Figure 10-50 is a version of the B C D hanger mentioned in Fig. 10-44. This detail is an anticipatory detail. What you are saying is that you know the space for all your ducts is tight, and with

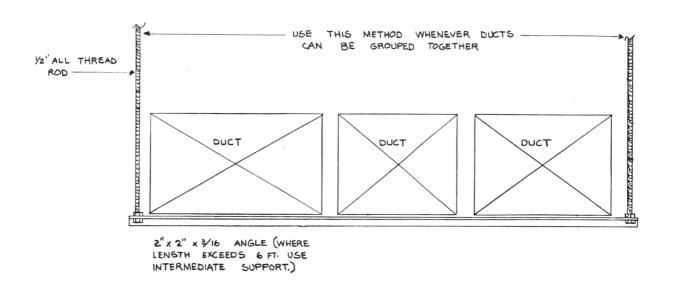

NOTE:
DUCTS TO BE SUPPORTED AT
NOT LESS THAN 10' O.C.

½" ALL THREAD ROD

USE THIS METHOD WHENEVER DUCTS CAN BE GROUPED TOGETHER

DUCT   DUCT   DUCT

2" x 2" x 3/16 ANGLE (WHERE LENGTH EXCEEDS 6 FT. USE INTERMEDIATE SUPPORT.)

MULTIPLE DUCTS ON TRAPEZE HANGERS

FIGURE 10-50

individual hangers the collection of ducts might not fit. This is an acceptable solution to the problem of squeezing the ducts together.

Air Control: In our previous duct connections and duct take-offs we have shown situations in which splitter dampers are used. Figure 10-51 shows a splitter-type damper regulator that implies the virtue of your specification. Not only will it control air, but it

will not vibrate. Nothing is more embarrassing or frustrating than to have a job, which is ready to be turned over to the client, that has a rattling damper.

Mixing Boxes: Figure 10-52 depicts a dual-duct high velocity mixing box. Your floor plans already show flexible connections so why bother with this sort of detail? Flexible connection is not necessarily a definitive phrase. This detail shows exactly what you mean

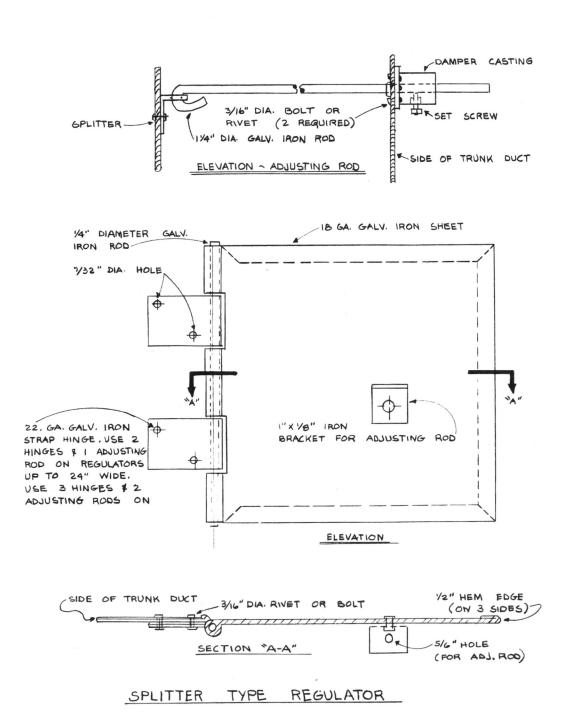

SPLITTER TYPE REGULATOR

FIGURE 10-51

COLD (OR HOT) AIR FLOW

HIGH PRESSURE ROUND
FLEXIBLE CONNECTIONS
HOT & COLD DUCT
CONNECTIONS SHALL
NORMALLY BE AS SHOWN-
HOWEVER BOX SHALL HAVE
INTERCHANGEABLE CONNECTIONS
SO THAT A CHANGE CAN BE
MADE IF NECESSARY TO SUIT
DUCT CONNECTIONS

ALL FLEXIBLE DUCTS
SHALL BE OF THE PRE-
INSULATED TYPE. INSULATION
SHALL BE AT LEAST 1" THICK
& 3/4 LB. DENSITY GLASS
FIBER BLANKET COMPLETE
WITH EXTERIOR VINYL VAPOR
BARRIER.

FOUR MOUNTING HOLES

CEILING DIFFUSER

NECK

SINGLE MOTOR MIXING BOX

COLD (OR HOT) AIR FLOW

PLAN VIEW

1/2" Ø ALL THREADED RODS
ATTACH TO STRUCTURE IN
AN APPROVED MANNER

LOCKING NUTS

NECK

SET BOX DEAD LEVEL

CEILING DIFFUSER

CEILING LINE

PROVIDE ACCESS TO
EACH BOX AND
CENTER AS
RECOMMENDED BY MFR.

SIDE ELEVATION

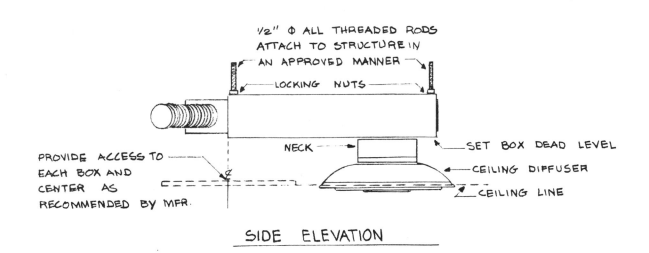

DUAL DUCT HIGH VELOCITY MIXING BOX

FIGURE 10-52

by a flexible connection. It also shows how you want the mixing box supported. And finally it shows something frequently forgotten. It shows an access door to get at the mixing box connections. All you need is one box problem that is above a plastered ceiling or in a splined tile ceiling and you will be convinced of the need for this detail.

Fire Dampers: Figure 10-53 illustrates two typical types of fire dampers in two typical situations. Again this is the sort of illustration which seems to be

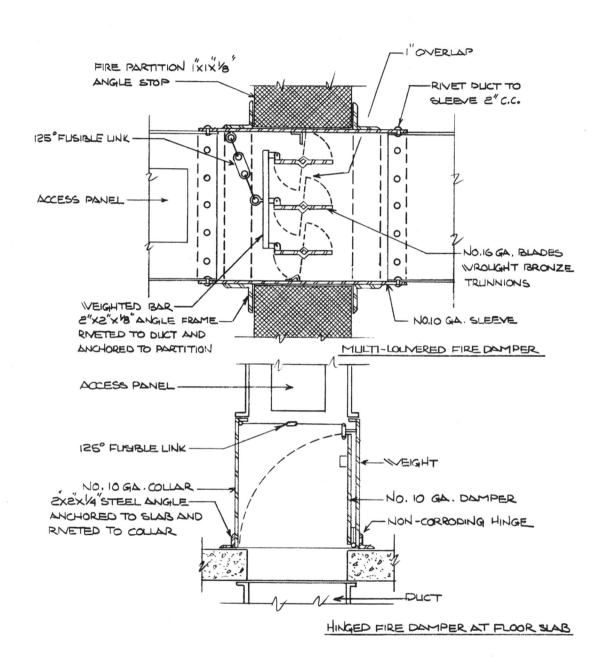

FIGURE 10-53

redundant. A fire damper is a fire damper, and when called for on the plans or required by the code inspector, it will get installed. What kind, what types, and what type of access is quite another matter. The code inspector or your construction manager may demand and with luck get at no extra charge what is really needed. However, the whole construction installation process will be far more readily expedited if you show clearly what is expected. Either of these dampers (and there are other proprietary-types available) can be made to work in a horizontal wall or vertical floor penetration situation. You may, or may not, require both types. Do not show both types unless you are prepared to accept either one in all situations or unless you note where each type will be permitted.

Specialized Exhaust Systems

Many of the various systems designed by the practicing consultant require exhaust systems that do more than merely remove a certain portion of the supply air stream and replace it with an equal amount of outdoor air. The last few words of the above sentence, "replace it with an equal amount of outdoor air," are vital to every system. Even though, as detailer, you may not be responsible for system engineering and design, you can perform a vital checking function for your firm if you observe, and bring to your superior's attention, the fact that your exhaust details are not complemented by equivalent air intake louvers or details.

The imbalance of outdoor air supply and space exhaust is a very common problem. There may be reasons why the design creates this problem. The fact remains that no structure is airtight. You can place a negative or positive air pressure in any given area, but overall there must be an air balance.

The most common exhaust system is generally described as dilution ventilation. In an ordinary office building, for example, the exhaust of the total structure may be resolved via the toilet exhaust system. The toilets get their makeup air from the corridors. The corridors get their air from the offices. And the offices get the required makeup air from the fresh air intake to the supply system.

Again as detailer you can note the position of the exhaust and supply connections. The object of all exhaust systems is to remove air by pulling it away from people and arranging it so that no one is in the direct path of the exhaust air stream.

The details we are presenting here are generally schematic and illustrate the specific or general solution in an overview approach. The actual sizes and precise shapes must be designed and dimensioned to fit your specific requirements. A large amount of design data is contained in the four volumes of the *ASHRAE Handbook*. Specialized exhaust systems that are generally used with some type of exhaust filter and collector have velocities in the ductwork of 2000 to 5000 fpm. In those systems the crucial term is "capture velocity," which is the speed the exhausted air must be be traveling to cntrain (capture) the material or fume to be exhausted. One of the best books devoted to this specific subject is *Industrial Ventilation* published by the Committee on Industrial Ventilation, P. O. Box 16153, Lansing, Michigan 48901.

General Smoke Exhaust: Figure 10-54 is a schematic detail that usually accompanies plans, specifications, and the sequence of operation instructions. There are many sensors for temperature, smoke, and pressure that are tied into the supply, return, and exhaust fan system that must be shown and explained. The detail shown here depicts, using circled plus and minus signs, the areas pressurized and evacuated when a fire occurs on a given floor. The stairwell usually, or the freight elevator area at times, is pressurized to provide a way out for occupants. The toilet-smoke exhaust fan pulls out smoke and prevents it from being recirculat-

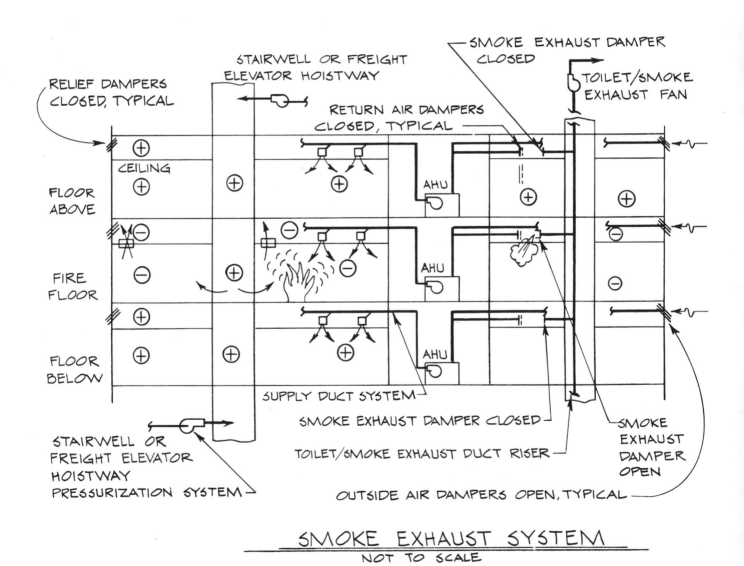

SMOKE EXHAUST SYSTEM
NOT TO SCALE

FIGURE 10-54

ed in the affected area. Your system may differ, as there are other possible solutions. Very definitely some solution should be depicted on your detail sheet.

Kitchen Smoke Exhaust: Figure 10-55 is a typical commerical kitchen hood exhaust system detail. Your first reaction to the detail may be that the supply air system is short-circuited into the head, which is what we want it to be, more or less. We want to have the air movement occur around the people who are working around the cooking under the hood. The air should make a circling 360° loop, passing over the occupants, entraining the hood fumes, and discharging everything

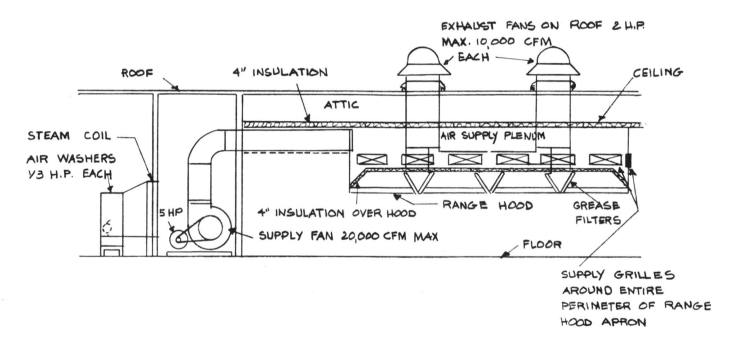

SUPPLY — EXHAUST ARRANGEMENT FOR COMMERCIAL KITCHEN HOODS

FIGURE 10-55

out through the roof. The exhaust ducts depicted are 18-gauge welded black iron inside a stainless steel hood enclosure.

Exhaust Hood: Figures 10-56 and 10-57 are fume exhausts for an open tank. Figure 10-56 illustrates that the fumes are captured on the two sides of the tank and pulled down into a collector at one end of the tank and then expelled out via the exhaust system. Figure 10-57

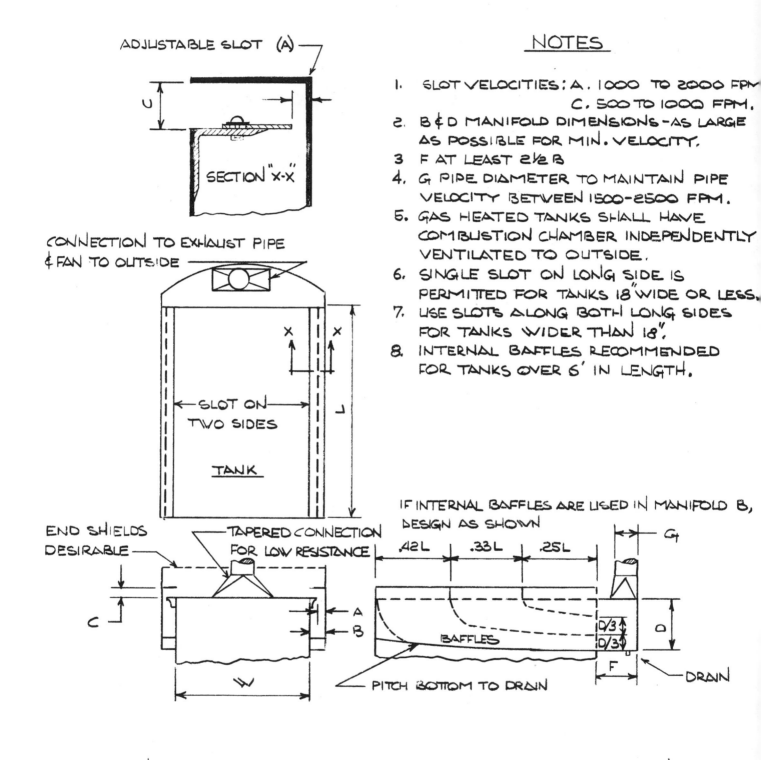

NOTES

1. SLOT VELOCITIES: A. 1000 TO 2000 FPM
   C. 500 TO 1000 FPM.
2. B & D MANIFOLD DIMENSIONS – AS LARGE AS POSSIBLE FOR MIN. VELOCITY.
3. F AT LEAST 2½ B
4. G PIPE DIAMETER TO MAINTAIN PIPE VELOCITY BETWEEN 1500-2500 FPM.
5. GAS HEATED TANKS SHALL HAVE COMBUSTION CHAMBER INDEPENDENTLY VENTILATED TO OUTSIDE.
6. SINGLE SLOT ON LONG SIDE IS PERMITTED FOR TANKS 18" WIDE OR LESS.
7. USE SLOTS ALONG BOTH LONG SIDES FOR TANKS WIDER THAN 18".
8. INTERNAL BAFFLES RECOMMENDED FOR TANKS OVER 6' IN LENGTH.

EXHAUST HOOD FOR INDUSTRIAL VENTILATION

LATERAL TYPE FOR OPEN SURFACE TANKS

NOT TO SCALE

FIGURE 10-56

depicts the exhaust duct intake at one of the long sides of the tank or table; from there it goes out through the exhaust system. These two seemingly simple details are very difficult to design in a fashion that will work.

Figure 10-56 shows a typical example of industrial fume tank ventilation, and Fig. 10-57, with its rear collector, is typical of both industrial and laboratory fume hoods.

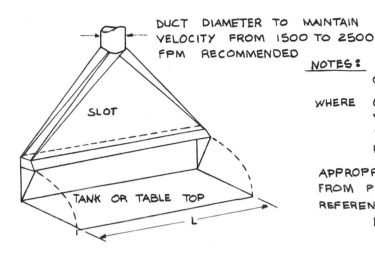

DUCT DIAMETER TO MAINTAIN VELOCITY FROM 1500 TO 2500 FPM RECOMMENDED

SLOT

TANK OR TABLE TOP

L

NOTE:

MAXIMUM WIDTH W = 4 FEET
MINIMUM RATIO, LENGTH TO WIDTH L/W, 2 TO 1
MAXIMUM LENGTH OF TRANSITION PIECE, 4 FT
(USE MORE THAN ONE FOR LARGER HOODS

NOTES:

$$Q = W \times L \times K$$

WHERE   Q = AIR QUANTITY REQUIRED, CFM
W = TANK OR TABLE WIDTH, FEET
L = TANK OR TABLE LENGTH, FEET
K = VENTILATION RATE, CFM PER SQ. FT.

APPROPRIATE VALUES OF K CAN BE DETERMINED FROM PROCEDURES OUTLINED IN THE FOLLOWING REFERENCES:

1. INDUSTRIAL CODE RULES 12 AND 18 FOR OPEN SURFACE TANK OPERATIONS; DIV. OF INDUSTRIAL HYGIENE, NEW YORK STATE DEPARTMENT OF LABOR.

2. A.S.A. STANDARD Z9.1, "VENTILATION AND OPERATION OF OPEN SURFACE TANKS."

3. "INDUSTRIAL VENTILATION"; AMERICAN CONFERENCE OF GOVERNMENTAL INDUSTRIAL HYGIENISTS.

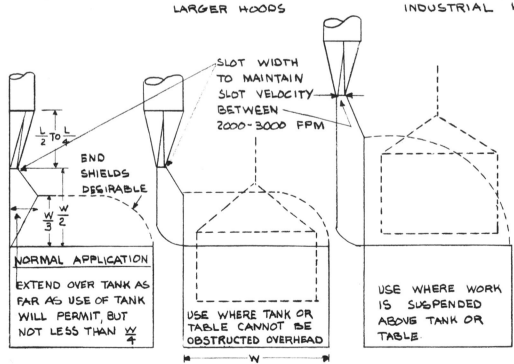

SLOT WIDTH TO MAINTAIN SLOT VELOCITY BETWEEN 2000-3000 FPM

$\frac{L}{2}$ TO $\frac{L}{4}$

END SHIELDS DESIRABLE

$\frac{W}{3}$ $\frac{W}{2}$

NORMAL APPLICATION

EXTEND OVER TANK AS FAR AS USE OF TANK WILL PERMIT, BUT NOT LESS THAN $\frac{W}{4}$

USE WHERE TANK OR TABLE CANNOT BE OBSTRUCTED OVERHEAD

W

USE WHERE WORK IS SUSPENDED ABOVE TANK OR TABLE.

REAR HOOD FOR OPEN SURFACE TANKS

NO SCALE

FIGURE 10-57

While we have said frequently that this is not a book on engineering or design, we feel the detail itself in both of the previous cases is the major part of the design. On each of the two details we have lettered a series of notes. These notes cover both recommendations and design criteria. They are *not* to be copied verbatim. Instead they are guides to what should be incorporated in notes on your detail, based on your design and specific size recommendations.

Hoods of this nature should be carefully evaluated as to type of fume and the quantity, density, and volume discharged to the air per minute so that the proper volume of discharged air may be specified. As noted, Fig. 10-57 refers to the book, *Industrial Ventila-*

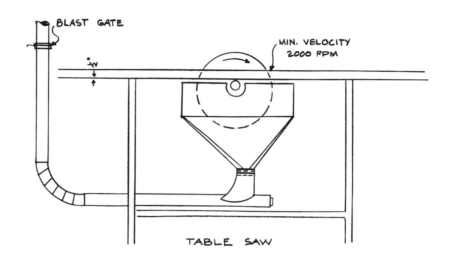

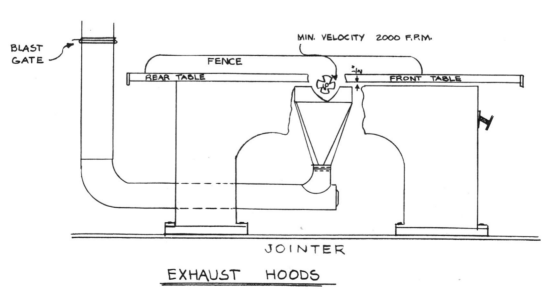

EXHAUST HOODS

RECOMMENDED DESIGNS FOR VARIOUS EQUIPMENT

FIGURE 10-58

tion. The Conference of Governmental Industrial Hygienists regularly evaluates and tries to improve designs of this type.

Figures 10-58 through 10-60 depict commonly used collectors in industrial grinding and cutting operations. They are certainly not the only ways to provide the proper collection of exhausted grinding or cutting particles. Some forms of grinding operations, grinding graphite for example, create a product to be exhausted that is very similar to blowing dust.

In all of these details there is a common denominator. The collector encloses half or more of the grinding

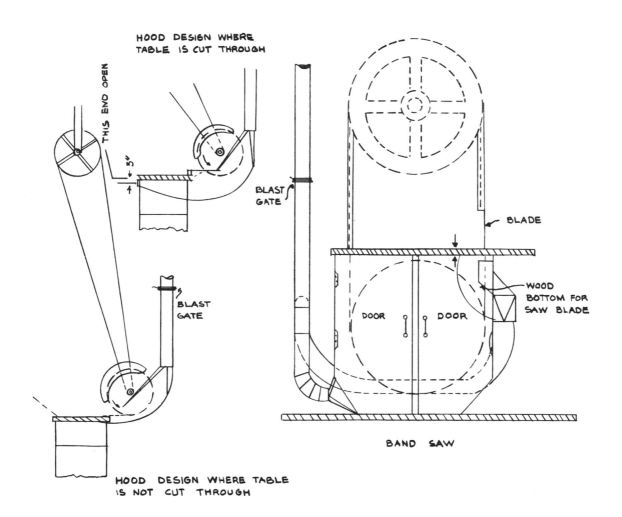

EXHAUST HOODS

RECOMMENDED DESIGNS FOR VARIOUS EQUIPMENT

FIGURE 10-59

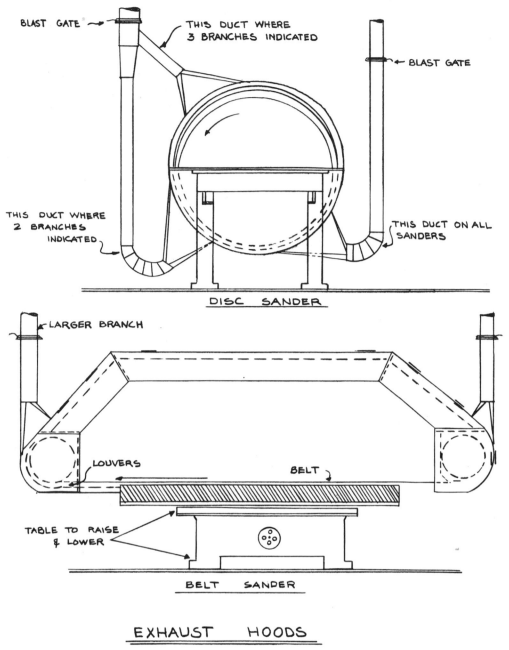

BLAST GATE

THIS DUCT WHERE
3 BRANCHES INDICATED

BLAST GATE

THIS DUCT WHERE
2 BRANCHES
INDICATED

THIS DUCT ON ALL
SANDERS

DISC SANDER

LARGER BRANCH

LOUVERS

BELT

TABLE TO RAISE
& LOWER

BELT SANDER

EXHAUST HOODS

RECOMMENDED DESIGNS FOR VARIOUS EQUIPMENT

FIGURE 10-60

wheel. This is not always possible. Sometimes the collector cannot, for work performance reasons, enclose any of the wheel. This invariably increases the degree of design and engineering difficulty. Ideally the particles should be directed toward the collector and should discharge as near to the collector as possible since there are two velocities to resolve. The direction and velocity of the particle as it leaves the surface on which it was located and the direction and velocity of the air toward the opening of the collector is very important. Your detail must demonstrate that these two directions result in the mixture of air and particles entering the collector and not, even in part, bypassing the collector and mixing with the room air.

ELEVEN

Equipment Using HVAC and Refrigeration

Presented as details in this chapter are the various items of equipment that are most commonly thought of when one refers to mechanical details. In our view mechanical details should not only include all the types of radiation, unitary, and field-assembled equipment but also the piping arrangements that precede or are directly related to the particular installation.

In this chapter we begin with the detailing that is usually expected. As such we include radiation, unit ventilators, unit heaters, and standard heating and ventilating units. We continue with radiant ceiling coils and then, in chilled and hot water piping details, show coil connections.

The total number of details is fifteen, which may not seem large or sufficient in variety. Actually that is not the case. Many of the possible different details that could be presented are merely rearranged versions of the details we have depicted. In addition other details you might expect are probably in Chap. 10. This is especially true of refrigerant piping. In Chap. 4 we presented some detailing at the compressor and between the separated compressor and condenser. Technically, if we strictly followed our outline, the air supply unit and its connections would have been shown in this chapter. But, to reiterate what we said previously, the close arrangement of the compressor, condenser, and related evaporator seemed to dictate that it all be shown in Chap. 10.

Wall Fin Radiation

Piping to most radiation commonly does not require much of a detail. Many firms, because of long standing design procedures, will show typical piping of radiators, cabinet unit heaters, and the like primarily to illustrate valves, traps, control valves, and other spe-

cialties. As a consequence, we are depicting some of these details in areas in which we feel the detailing is a little out of the ordinary.

In Fig. 11-1 we illustrate the piping to a two-tier section of finned radiation. Of and by itself the piping connections to wall fin radiation are not a very demanding detail task. Whether the radiation is wall fin, convector, or a cast iron radiator, the detail shows the installation of a manual or automatic control valve on the supply side and a trap, if steam is involved, on the return side.

Figure 11-1 is typical but includes an added twist. For the sake of economy and piping simplicity both the supply and return hot water heating mains are also in the cover. This could be an economical installation whether the building was one-story or multi-stories. It very definitely is economical in a long one-story classroom or office wing.

There are several items to note and to carefully engineer before designing or detailing. The hottest water is in the supply main. Thus, for best convection action, put the supply main at the bottom. The reverse is true for the return main. The system, even though it has an automatic control valve, can be much more readily fine-tuned if a balancing valve or fitting is installed, as we depict.

The sizing and selection of finned radiation and its cover can get tricky. Be certain that you have ample space around the largest supply and/or return pipe combination so that you do not unduly restrict air flow over the principal source of heat output, the finned coil.

The output of the coil is going to be somewhat reduced because the supply pipe blocks part of the flow. In addition your overall design must account for the uncontrolled heat that is given off by the supply and return piping. Never try to insulate the supply and return mains because you will end up blocking all air flow and producing a no-heat situation.

Our detail is based on a stock size 4½-in fin radiation section. The supply and return are not sized since they vary depending on what area the detail purports to represent. In our case the largest size of either the supply or return is 2 in.

Unit Ventilators

As we noted in our previous discussion, the piping of a single piece of radiation, regardless of whether it is located in a cabinet or not, is not a very trying task.

This is equally true if you are piping up a single-unit ventilator. Thus we are not depicting that sort of detail. However, in Fig. 11-2 we are depicting another detail which we feel is a little out of the ordinary.

Figure 11-2 is an illustration of a unit ventilator installation that has supplemental fin radiation and its piping within the finned radiation enclosure. Again, this works well in a hot water system but not in a steam system.

The number of units per loop that can be supplied in this fashion takes some design and engineering calculation, as well as some thinking about the size of the extended fin radiation. Usually this is the sort of system that is supplied from an overhead corridor or

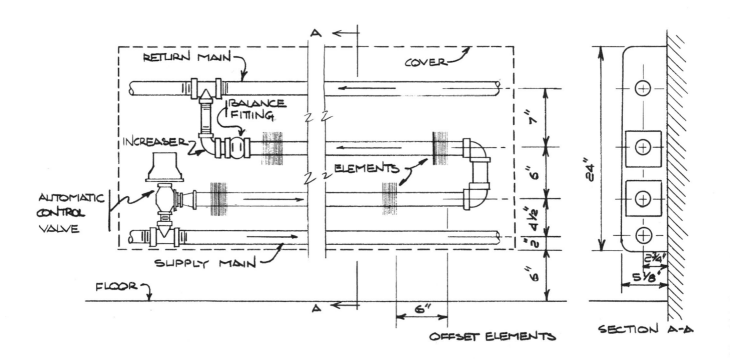

PIPING DETAIL OF FINNED RADIATION
— NOT TO SCALE —

FIGURE 11-1

room system with the riser in the room partition wall that branches out to feed systems on both sides. One or possibly two systems are supplied per loop. The extended fin radiation is incorporated as part of the supply main.

Some expansion provision may be required and the system shows small expansion loops which could very easily be replaced by expansion joints. In the usual arrangement there is the possibility of air binding, and an air vent at the system high point has been included.

Ventilator, Convector, and Cabinet Heater

As we have previously noted, frequently a steam-supplied unit ventilator, convector, or cabinet heater may have a slightly different connection for maintenance or security reasons.

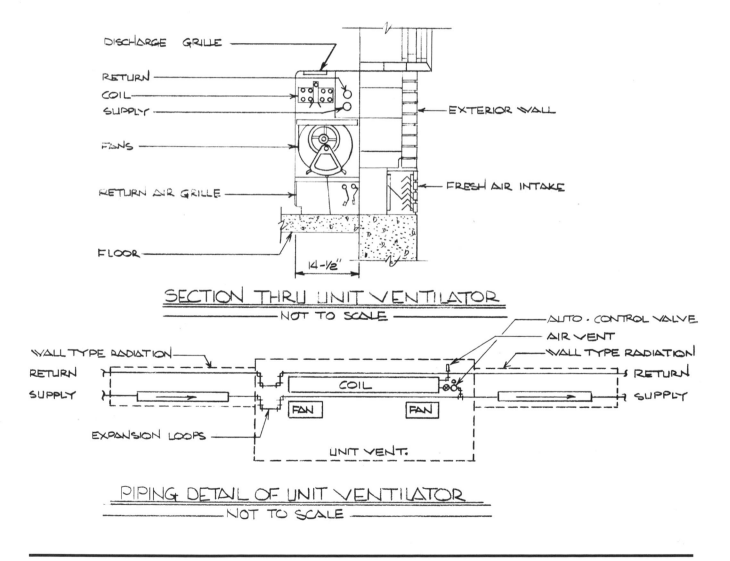

SECTION THRU UNIT VENTILATOR
— NOT TO SCALE —

PIPING DETAIL OF UNIT VENTILATOR
— NOT TO SCALE —

FIGURE 11-2

In Fig. 11-3 we illustrate a steam-supplied heating element that could be in any of the three above noted situations. The trap is located where you would normally expect the trap to be located. The trap on occasion has to be serviced, but suppose a heating element installation is in a prison, a mental institution, or somewhere else in which the people can be a problem.

If you are knowledgeable, you can take apart the trap, but you might be scalded by the steam, and the billowing steam would attract instant attention. It is also likely that the control valve would be damaged or disarranged if it was placed in its usual location within or adjacent to the heating element. Therefore the vital control valve is in the pipe tunnel and available only to maintenance personnel.

This detail not only illustrates a security piping detail, but it also indicates the typical steam piping to radiation. And if we removed the trap and substituted a balancing fitting and an air vent on the piping high point at the element, we would have a perfectly usable hot water system detail.

Radiant Ceiling Coils

Commonly radiant ceiling coils are thought of as coils whose task is to provide the solar source of

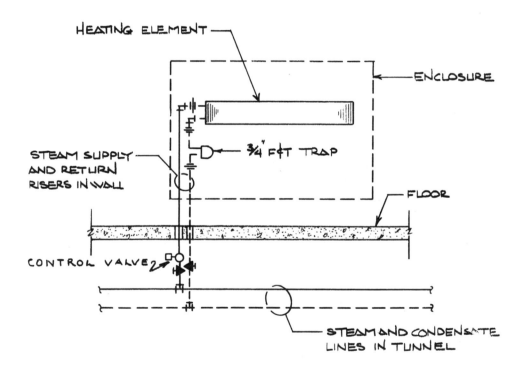

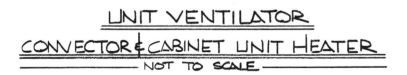

NOT TO SCALE

FIGURE 11-3

heating for a given area. In many projects, especially many institutional-type projects such as hospitals and other health care centers, the use of radiant coils is to offset the basic envelope heat loss or heat gain. In effect the base load is resolved by the coil installation, and the varying load of people and ventilation air is covered by a separate system.

Figure 11-4 is one way to resolve the control situation in which there are sources of both hot and chilled water to one coil. The primary problem to be resolved is to be certain that hot and chilled water do not get mixed up. The three-way valves are set up to go directly from full open to full bypass when they switch from cooling to heating and vice versa, and they operate simultaneously.

Usually at a predetermined outdoor temperature the coil controls sequence from full open, at which as in Fig. 11-4 they supply and return chilled water, to a full bypass, at which they supply hot water. This control sequence must be part of the overall comfort control

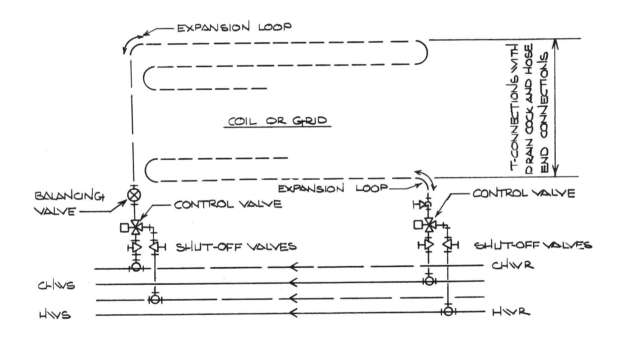

PIPING DIAGRAM FOR RADIANT CEILING
NOT TO SCALE

FIGURE 11-4

system. The modulation of room temperature, the fine tuning, is effected by varying the temperature of the separate discharge air supply.

Unit Heaters

As we previously stated, we are trying not to show equipment piping connections that are fairly obvious and well-known. In Fig. 11-5 we show steam connections to both horizontal and vertical (projection) unit heaters in a composite detail in which we also illustrate some other pertinent points of interest. If you happen to require both types of detailing, we suggest you show these two details with more separation. We had to crowd them a little to fit the book's space requirements.

Note first that both steam and condensate enter the top of their respective mains at a 90° angle. To work on either the unit heater or the trap merely close the supply and return gate valves. Using strainers with valved drains is always a good practice, as is including unions for ease of disassembly.

Frequently you see traps installed without the 12-in cooling leg that also assures that condensate, and not steam, is entering the trap. Gate valves are normally full pipe size and traps are usually sized at three times the condensate discharge capacity of the heater coil. If

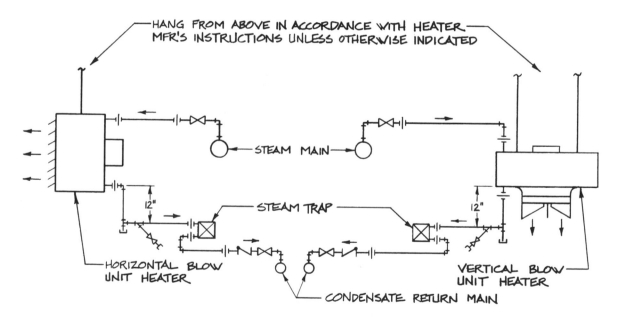

HANG FROM ABOVE IN ACCORDANCE WITH HEATER MFR'S INSTRUCTIONS UNLESS OTHERWISE INDICATED

STEAM MAIN

STEAM TRAP

HORIZONTAL BLOW UNIT HEATER

VERTICAL BLOW UNIT HEATER

CONDENSATE RETURN MAIN

ELEVATION

HORIZONTAL & VERTICAL BLOW STEAM UNIT HEATER PIPING
——— NOT TO SCALE ———

FIGURE 11-5

you do not use specific hanger details, our note will resolve most hanger problems.

Air Handling Unit Coils

Steam Piping to Heating Coil: Commonly the steam piping installation can very easily be improperly detailed and, unless it is carefully checked, present a seemingly complete, satisfactory arrangement. In the series of details that follows this discussion we present a number of different installation detail techniques.

Figure 11-6 is a simple, straightforward horizontal blow fan coil installation that is in a limited vertical space location and has a motorized control valve. Notice that both the steam and condensate lines pitch up or rise vertically where possible to facilitate the flow of steam and condensate. Even in small units there should be a valved bypass around the control valve. And the controlling manual bypass valve, parallel to the control valve, should be a globe, not a gate, valve. Purists would have us put both a gate and a globe valve in this position, but in small situations the globe will suffice. It can be adjusted to throttle steam flow; a gate really cannot.

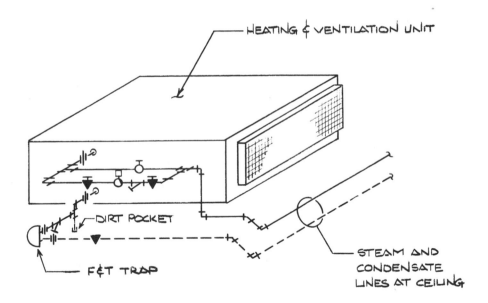

HEATING AND VENTILATION UNIT
NOT TO SCALE

FIGURE 11-6

The installation is obviously in a tight space, and we really cannot fit in the 6-in to 12-in drop from the condensate outlet to the trap. But we do the best we can. We have at least a short cooling leg and a dirt pocket. We also show the trap with an indication of the height drop through the trap.

Figure 11-7 is a composite detail showing a collection of common piping arrangements. These include

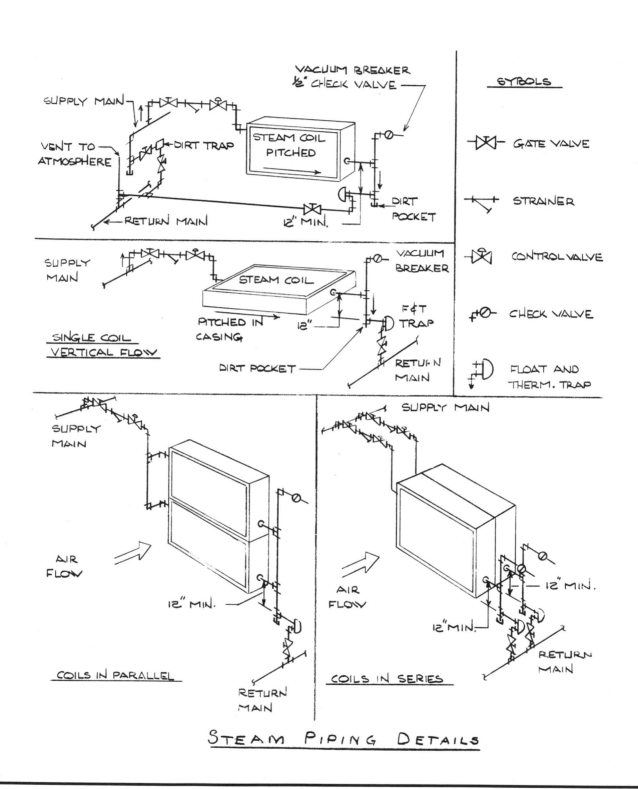

STEAM PIPING DETAILS

FIGURE 11-7

single coils with both normal and vertical flow, coils in parallel, and coils in series. All of these arrangements are fairly common except perhaps for the coils in series. Generally the coils in series would only be used when an especially high temperature at discharge was required and a space restriction precluded getting the flow through a single standard coil. In essence the design breaks no new ground when compared to our previous fan coil illustration except for one especially important item, the check valve.

These coils are assumed to be in a fairly large unit heater or air handling unit. A case could easily be made that the one control valve should have been divided into two control valves, one rated at one-third capacity and one rated at two-thirds and that we should have shown bypass valves around the control valves. Both statements are true. We could have and perhaps should have. But we felt we made our point in the previous detail and, because of space limitations, simply omitted these two items.

Very commonly, in cost-saving installations, one control valve with no bypass is installed. This is especially common when the coils are not too large and the control manufacturer feels the situation can be handled with one valve. But regardless of how many valves and bypasses you install, there exists the constant possibility of pulling a vacuum on larger coil installations. This occurs when the control valve is of the modulating-type and is nearly closed. The small amount of steam expanding in the large coil can create a vacuum in the coil that is holding the condensate. If the coil is a 100-percent fresh air installation and the temperature is 20°F or lower, you will freeze the coil, even a nonfreeze coil. The check valve acts as a vacuum breaker. It lets atmospheric air into the coil and permits the condensate to flow by gravity to the trap.

Steam Coil Piping Schematic: Figure 11-8 is a completely detailed air handling unit coil installation. Some governmental agencies insist on this type of

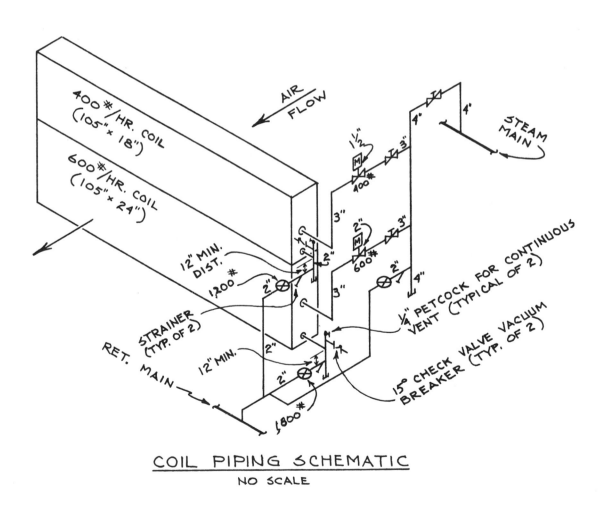

COIL PIPING SCHEMATIC
NO SCALE

FIGURE 11-8

detailing. The overall dimensions and capacity of each coil is shown. The number of rows is not noted since the performance is stated. The control valves are both sized and performance-rated. The length of the trap drip leg is carefully noted. The coil supply main is dripped and its trap is shown. This drip trap could have been smaller but a full line-size trap was requested. Each coil trap is rated at three times the coil capacity. Not only do we have vacuum breakers on each coil, but we also have a ¼-in continuous vent petcock to be certain that the coils are never air bound. There are no bypass lines around the coils as these are 100-percent fresh air coils, and if a control fails, the fresh air damper is automatically closed and the fan for the air handling unit is stopped. Nothing is perfect. The coil could still possibly freeze if the fresh air damper stuck open or the fan motor failed to stop. But it would be highly unlikely since the condensate system would clear the coil and, in the situation in which these coils were installed, the plant operators would very, very quickly be aware that they were in a freezing draft.

Our point in this detail, as in all others, is that you always will have to slightly modify these details to precisely fit your installation requirements.

Coil Water Piping Connections: While there is no such thing as a truly simple piping detail that requires little or no thought or engineering ability, the nature of a hot or chilled water system is such that it will tolerate a lot of poor connection arrangements.

In Fig. 11-9 we show the typical arrangement of piping connections for either a hot or chilled water coil. With a few possible eliminations this is the piping arrangement for any coil, anywhere. Note that the two coils could both be for hot water or chilled water. The two halves of the figure depicted are identical.

Also note that the typical three-way motorized valve controller is on the return, not the supply, line. The valve works more accurately in this position. Coils, especially big vertical coils need draining at times for a variety of reasons. Balancing valves in all hot-water-using equipment should on the return, not the supply. And for larger installations you need thermometers to verify what is occuring within the coil.

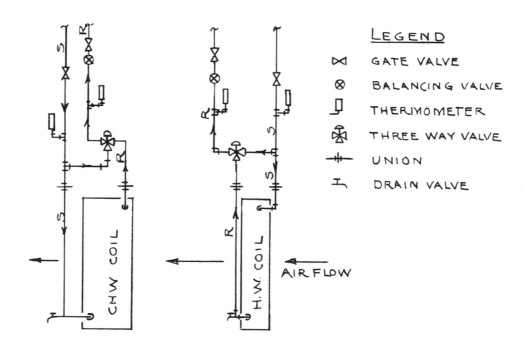

LEGEND

⋈ GATE VALVE
⊗ BALANCING VALVE
▯ THERMOMETER
⧓ THREE WAY VALVE
⊣⊢ UNION
⊥ DRAIN VALVE

TYPICAL HEATING/COOLING COIL
PIPING CONNECTIONS
NO SCALE

FIGURE 11-9

In certain cases, especially in certain processes, the coil temperature is to be maintained and when not in use the coil is to be shut off. In the lower half of Fig. 11-10 we depict such an arrangement. There are also occasions in which for special reasons the supply and return lines may be required to assure given, fixed quantities at the full open position of the three-way valve. This can be done with an arrangement of globe valves as we have depicted in the upper half of Fig. 11-10.

Chilled Water Piping Systems

When chilled water is delivered from a central refrigeration plant to a series of separated structures or

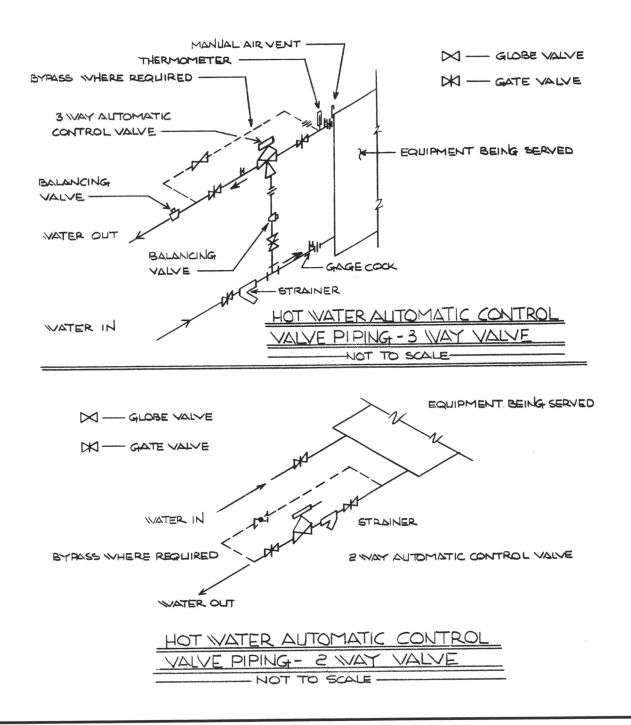

FIGURE 11-10

specialized users, there are a number of ways this source can be utilized. The control source has its own set of primary distribution pumps. Commonly each individual using the source has its own source pumps, called secondary pumps. Generally the task of the secondary pumps is to provide the proper capacity and head to be certain that the proper flow is maintained. This subject is very complex and quite properly belongs in a design textbook. There are a number of engineering books on this subject; you can also get information from pump manufacturers and the *ASHRAE Systems Handbook*.

As this is not a book on design, we will not go into the complex engineering aspects. But in the next three details we present overall schematic installation details. Our first detail, Fig. 11-11 illustrates a constant

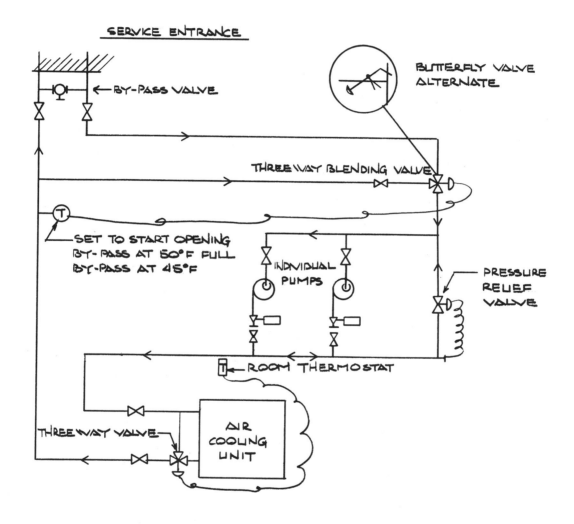

SERVICE ENTRANCE

BUTTERFLY VALVE ALTERNATE

BY-PASS VALVE

THREE WAY BLENDING VALVE

SET TO START OPENING BY-PASS AT 50°F FULL BY-PASS AT 45°F

INDIVIDUAL PUMPS

PRESSURE RELIEF VALVE

ROOM THERMOSTAT

THREE WAY VALVE

AIR COOLING UNIT

CHILLED WATER SYSTEM
CONSTANT VOLUME VARIABLE TEMPERATURE
NOT TO SCALE

FIGURE 11-11

volume, variable temperature installation. Besides the standard three-way valve, mixed water controller on the using equipment, the overall building supply system has a three-way controller on the suction side of the building's pumps. As noted in our detail, this can be a standard three-way modulating valve or can be a butterfly valve, depending on supply temperature consistency. Balance is important in the using system.

This is achieved by a pressure relief valve and by a manually set globe valve at the chilled water service entrance.

Figure 11-12 is another possible version of a using system in which the temperature of the chilled water service is guaranteed but because of a collection of user varying loads, the volume varies. Here there is a different problem that is resolved in a different fashion.

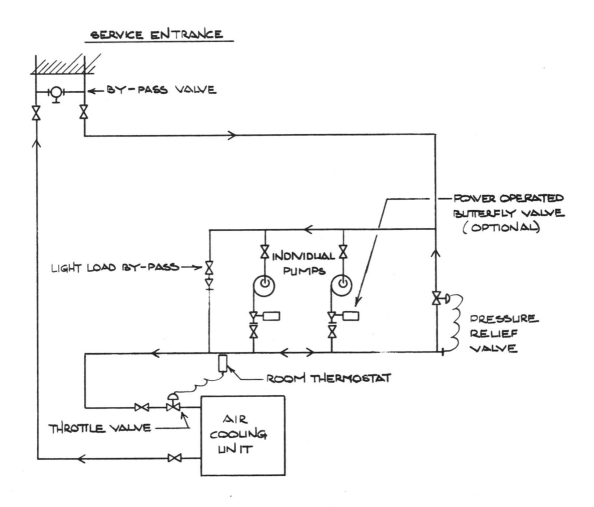

CHILLED WATER SYSTEM
CONSTANT TEMPERATURE VARIABLE VOLUME
—NOT TO SCALE—

FIGURE 11-12

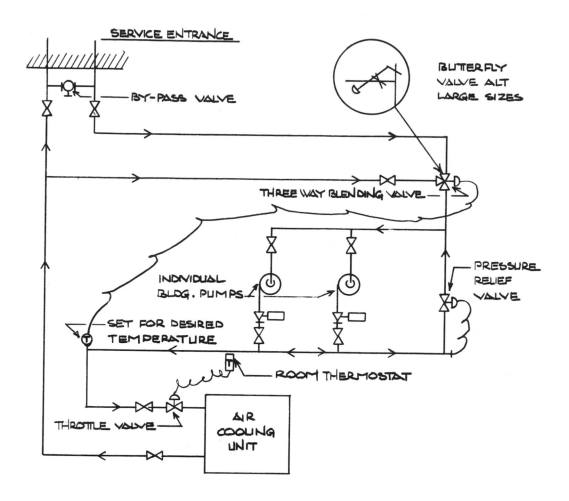

SERVICE ENTRANCE

BY-PASS VALVE

BUTTERFLY VALVE ALT LARGE SIZES

THREE WAY BLENDING VALVE

INDIVIDUAL BLDG. PUMPS

PRESSURE RELIEF VALVE

SET FOR DESIRED TEMPERATURE

ROOM THERMOSTAT

THROTTLE VALVE

AIR COOLING UNIT

CHILLED WATER SYSTEM
SELECTIVE TEMPERATURE VARIABLE VOLUME
— NOT TO SCALE —

FIGURE 11-13

Again we have a problem because of the throttle valve and the system source. Generally most of the problem, if not all of it, is created by the user. We added one other feature. In times of light load on the using side, we can stop our pumps and let the service pressure solve our distribution requirements.

Figure 11-13 is another version of using the chilled water service. It is a further refinement of Fig. 11-12 or a modified version of Fig. 11-11. The key control item is the throttling valve. It creates certain balance requirements. We cannot for a design reason depend on

light load bypass. Thus, in our design we utilize a three-way blending valve which in slight water temperature variations tends to keep the throttling valve operating in a narrower range.

These three details are basically system schematics. They seem, at first glance, to be very similar and not all that complex. They are quite the contrary. The engineering behind the selection of any of the systems is not simple and each should only be selected after a thorough review of all aspects of your requirements.

TWELVE

Energy Conservation Systems

Since the OPEC-induced oil crisis of the early 1970s, energy conservation has been the most talked about subject in the HVAC industry. Every segment of manufactured equipment used as part of an assembled HVAC system has been redesigned and updated for the purpose of minimizing energy usage. All of this is a function of design. It does not create new details, merely new equipment in the same application. Given this fact, it is reasonable to assume that the firm employing the detailer carefully investigates each item of equipment specified, as well as the basic design calculations, to be certain that the proposed solution uses the least amount of energy that is practical. While building construction and orientation also play a large role in energy conservation, this is not a subject for a book on mechanical detail application.

Actually a number of energy conservation ideas have been in use for a very long time. Nowhere is this more evident than in the design of an electrical power generating plant. For more than 50 years utility companies and their design engineers have concentrated their efforts on extracting the last possible Btu out of each unit of fuel consumed. Items such as flue gas economizers, which are currently regarded as a new energy conservation idea, have been used in electric utility plant design for a long long time.

One of the details we are presenting in this chapter should not only help solve a detail problem but also stimulate your imagination. Industry of all types typically discharges reasonably safe hot liquids and gases to sewers and to the atmosphere. Our heat reclaimer detail is but one idea that could be readily modified to solve other problems. Even our standard air-to-air toilet exhaust heat exchanger has many applications besides toilet exhaust systems.

In many parts of the country in which process air

conditioning is required all the time for 12 months of the year, engineers and designers are rediscovering the value of the cooling tower. Given a sufficiently low wet bulb temperature, the tower can easily provide 3000 or more hours of full load cooling without any use of the refrigeration compressor. The detail as we depict it can readily be modified and used very successfully with an air cooled or a water cooled chiller. This detail is, we repeat, not a truly new idea.

For the clean room the process discharge air is frequently below that of the normally discharged air conditioning system. The same run-around coil application that we depict in our details can be used between the discharge air and the fresh air intake with surprisingly large energy consumption savings.

The point we are making is that our few details have many applications. The engineer's job is to think through the design and clearly understand conditions of temperature and humidity at each step. Between the run-around coil and the air-to-air heat exchanger there are a great many application possibilities.

Heat Reclaimer

For the detailer there are only a few details that are basic to all systems. These details really involve the extraction of heat being lost from a liquid stream or an air stream. In the air stream application there are standard manufactured devices that use no moving parts to transfer heat, or, in the case of the heat wheel, involve a standard manufactured moving-part device.

To illustrate some of the basic types of details that would require an appearance on a detail sheet we are presenting a typical waste water heat recovery system, a run-around coil detail, an air-to-air heat exchanger and a cooling tower source of free cooling.

Figure 12-1 is a typical heat reclaimer system for hot

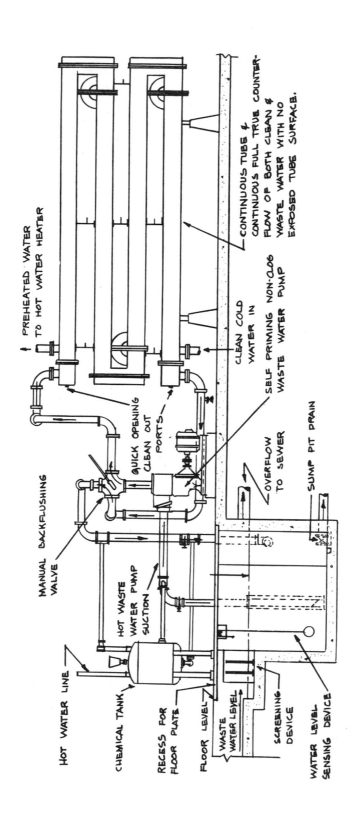

HEAT RECLAIMER SYSTEM

FOR TEXTILE MILLS, LAUNDRIES, OR WHEREVER HOT
WATER IS DUMPED "DOWN THE SEWER".

FIGURE 12-1

waste water in laundries, textile mills, and similar processes. Waste water that is still hot enough to be usable as a tempering medium is collected in a sump prior to its discharge to the sewer and utilized in a heat exchanger coil to temper (heat) the incoming cold water feed to the water-using process. Flows may vary and the system pump insures a steady amount of waste water to the heat exchanger. The overflow of the sump insures that sudden surges in waste water output will not be held up simply because the sudden large volume increase cannot be effectively handled by the exchanger.

Heat Recovery Loop

Figure 12-2 is a typical run-around coil detail that is very commonly used in energy recovery systems in which the purpose is to extract heat from air that must be exhausted. Normally about 40 to 60 percent of the heat being wasted can be recovered. This seemingly simple detail has two points that should be carefully noted. The difference in fluid temperatures in this closed system creates small expansion and contraction problems and an expansion tank is required. Most importantly the temperature of the incoming supply air can create a coil temperature so low that the coil in the exhaust air stream begins to ice up. Normally this begins at some 35°F. This is when the three-way bypass valve comes into play and the glycol is not circulated through the outside air coil. Obviously if the outside air temperature is low enough, the system will go into full bypass and the circulating pump should be stopped.

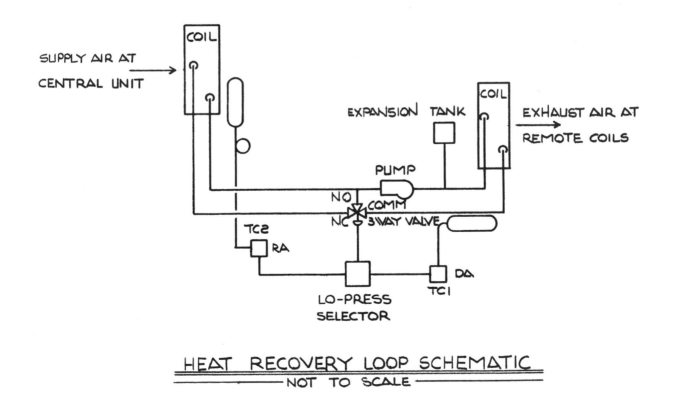

HEAT RECOVERY LOOP SCHEMATIC
NOT TO SCALE

FIGURE 12-2

Air-to-Air Heat Recovery

Figure 12-3 is the classic design example of an air-to-air heat exchanger. This system is used when there are no contaminents involved and you are simply exhaust-ing a portion of the return air stream. This is commonly done in many current office system designs. However, if the proper arrangement of makeup air, outdoor air, and return air streams is not carefully detailed, in

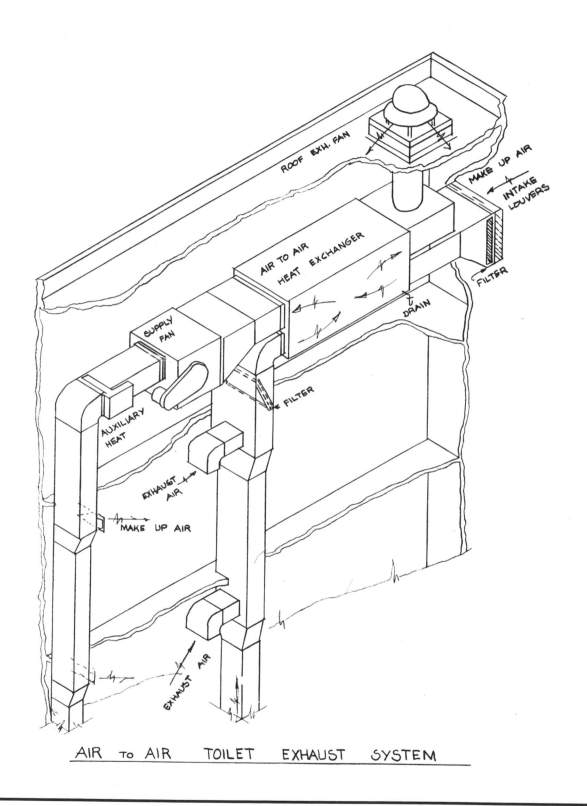

AIR TO AIR TOILET EXHAUST SYSTEM

FIGURE 12-3

many cases the proper heat interchange effect is not achieved. Figure 12-3 is a careful detail of just how the two air streams, intake and exhaust, should be arranged for maximum results.

Figures 12-1, 12-2, and 12-3 do not illustrate all the possibilities of HVAC energy recovery, but they do illustrate the basic methods of interchange that would be involved if you were saving energy through an economizer in a boiler stack, by recapturing energy from a water or air cooled refrigeration condenser, or with any of the many other possibilities that require

further design investigation before you complete your final system plans.

Tower Source Cooling

Figure 12-4 is really a 30-year-old concept of something for almost nothing that is a current energy conservation concept. The idea is deceptively simple. In the spring and fall, when the air conditioning chiller is needed and the air temperature results in a tower water temperature equal to that generated by the chiller, why not use the tower water directly? Obvi-

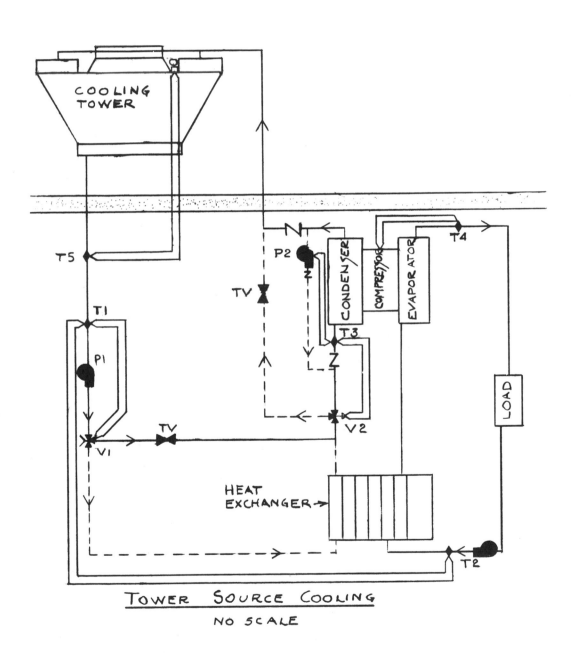

FIGURE 12-4

ously you can; 45 to 50°F water is usable regardless of the source. However, the tower water usually contains all sorts of contaminents. Therefore, do not use it directly; use it indirectly through a heat exchanger.

Generally the tower water heat exchanger operates at a 4°F spread between tower water and heat exchanged, circulating chilled water. At about 61°F air temperature and on 8° drop through the tower you get tower water at 53°F, and with a 4° rise through the exchanger you get 57°F chilled water—without the chiller. It seems very simple and it basically is. However, there is a balancing act that needs to be addressed, and the pressure drop through the heat exchanger is a problem to be resolved since the pressure loss through the exchanger should not exceed that through the chiller's condenser circuit. If it does, the tower will get less than design flow and tower efficiency will be a problem.

In Fig. 12-4 we illustrate one solution. As the temperature drops, the load sensed by T1 approaches that at T2. When T1 goes below T2, valve V1 lets the flow go through the heat exchanger. Meanwhile throttling valve TV ensures that pump P1 sees a constant head and flow. This is a concept used in lots of secondary pumping circuits of hot or chilled water primary-secondary pumping circuits. To solve low flow chiller-condenser problems, T3 modulates V2 to a less than full bypass position and T3 also separates pump P2 to maintain condenser water at an acceptable level.

At some point T4 senses a temperature lower than the system load requires and stops the compressor and the interlocked pump P2. At this point you have the tower providing "free" cooling. Nothing is truly free since you still have to pay for the tower pump and fan electrical energy, but that energy costs a lot less than the energy required to operate the compressor.

We have not detailed the premise of free cooling using 100 percent outside air because that is really a standard control function in your control design of fresh and return air dampers. And, obviously, you cannot use free tower cooling when you have air cooled condensers and no tower. But this free tower cooling concept which, as we noted before is a 30-year-old idea, merits serious attention when applicable.

THIRTEEN

Control Systems

Control systems are one of three basic types: electric, pneumatic, or direct digital. You may have thought we have omitted a system but we did not. In the standard application of the computer to the electric or pneumatic system the computer is either the master recording device or the master programming device. It is not the controller or the control system. This is an area that is very commonly misunderstood.

In simple straightforward applications of control systems for residences and small commercial buildings the electric thermostat connects to a packaged system with built-in electrical controls. However, in the past three decades larger, custom engineered systems have been pneumatically rather than electrically operated. These systems are made up of field-assembled components that are installed by the control manufacturer's personnel. Their operational source of energy is either the building electrical system or an air compressor that delivers a controlled source of compressed air.

Operation is smoother, quicker, and more positive with the pneumatic devices. Slight changes in electric power needed to gradually or more precisely change valve, damper, and other control device settings always turned out to be more trouble free in a pneumatic installation than in an electric one. We are not trying to belittle electrical controls but merely to state fairly well accepted observations. The control device either had an air supply or it did not, and it clearly worked or did not work for that specific reason. There were and are situations in which the electrical device is obviously the correct selection and our details begin with these situations.

All control systems have as their basis of design a series of control segments or elements designed to perform a specific task. There already exists an excel-

lent source of these design elements in the current *ASHRAE Systems Handbook*, which is one of the four books published by the American Society of Heating, Refrigeration and Air Conditioning Engineers. Merely repeating these details or a version of them would not assist either designer or detailer.

Far too frequently the design engineer receives very good temperature control diagrammatic solutions to his or her design problems from the excellent services provided by the representative of the temperature control manufacturer. Sometimes these are simply given to the drafting group to copy as is or to enlarge, reduce, clarify, or whatever with little or no thought.

We are not going to get involved in the ethics or logic of this common situation, and since this is a book on details and not design, we are not going to involve ourselves in the various steps of the design process. For all except very simple situations, there is no standard design. This left us with a dilemma over what to present as details in this book. Our solution was to present one solution to a typical control situation that included a representative collection of the design elements and that was taken from an actual installed project.

Replacing time clocks and other manual or semiautomatic devices with computer supplied control sequences is the most rapidly growing area of control application. The computer becomes the supervisory agent, monitoring the automatic control system, and is able to reset the set points of the controller *which is noncomputerized*. This arrangement has certain limitations. The basic control loop is still under the command of the existing controller and is still only as good or as responsive as the existing controller. Frequently the linkage between the computer and controller is mechanical or electromechanical with built in inaccuracies. The computer's sensor and the controller's sensor may differ, which leads to confusion. Rather than depict this sort of supervisory system in our details, we elected to depict the latest state-of-the-art system.

In essence the usual control system has a sensor that sends a signal to a controller to maintain, for example, the position of a three-way chilled water valve. This valve controls the flow of chilled water to the coil and thus creates the proper coil temperature so that the sensor, which is sensing a moving air stream, is satisfied that the temperature it senses is as desired—at the moment. This must happen continually and fast enough to maintain the desired set point. The entire case for digital control is based on the need for speed and accuracy.

Common to all nondigital control systems is the problem of "hunting." Nondigital devices cannot act sufficiently quickly or precisely to secure exact condi-tions. By the time the three-way valve adjusts to the new requirement, a further new requirement is requested by the sensor. Consequently there is a steady overshooting and undershooting of the proper value.

The same type of controls, powered by direct digital operators, use rapid-fire, varying on-and-off signals to quickly reach the desired set point. In effect a change in position is accomplished by a rapid-fire set of stop and go signals that are directly generated by the computer through its built-in operational software.

All digital computers work with binary (on-off) information. The process of using binary information is called pulse-width modulation (PWM). The computer's binary outputs are directly connected to the modulation device. PWM uses open-close pulses of varying time duration to position controlled devices very precisely. Wide pulses are used for major corrections. As less correction is required, the PWM becomes progressively shorter. Our details illustrate various direct digital applications.

Standard Control Systems

Residential Control Systems: Figure 13-1 illustrates the simple control of any hot water heating system. the basic boiler-burner controls maintain the boiler water temperature. The temperature differential alone will cause the system to circulate. For many years this was the gravity hot water system method of heating. Add a space thermostat and a circulator and you have the common residential circulating hot water heating system. The *ASHRAE Systems Handbook* lists four control principles for energy conservation. They are: operate equipment only when needed; use sequence heating and cooling, which is the continuous supply of heating and cooling effect; provide the heating and cooling needed, which will be further discussed; and supply heating and cooling from the most efficient source.

Since we assume that the boiler shown in the detail is being run as efficiently as possible, our only concern in this case is to provide the heating needed. We do this by adding an outdoor control, normally called an outdoor anticipator. Our typical design calls for the boiler water to be 180°F when the outdoor temperature is 0°F. If we kept the boiler water at this temperature at all times, the system would be constantly clicking on and off at an outside temperature of about 55°F. Usually the arrangement is to reduce, one for one, the boiler water temperature as the outdoor temperature rises. At 0°F the boiler water temperature is 180°F and at 55°F the boiler water is 180 minus 55, or 125°F. The controller may also be arranged to shut off the burner at 115°F, or when the outdoor rises to 65°F (180 minus 65). Since the space radiation will give off less heat at a lower temperature, the circulating pump will run longer and the system operation will not only be smoother,

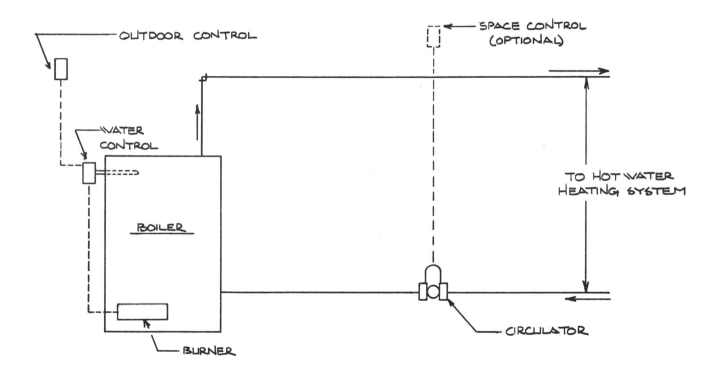

CONTROL OF HOT WATER HEATING WITHOUT DOMESTIC WATER SERVICE
NOT TO SCALE

FIGURE 13-1

but it will also waste less fuel because of temperature override avoidance.

Duct Heater Control: Figure 13-2 is a typical electric duct heater control that accomplishes some of the same things that were accomplished in Fig. 13-1. The temperature sensor in the duct calls for heat to be added to the air. When appreciable amounts of heat are required, this is usually accomplished by a series of heaters with step controllers so that we can produce more nearly only the heat we need. But electric heaters are very fast acting and before the sensor got the message the space could be overheating. Therefore we put a high limit control in the discharge air. This reset is similar to our outdoor control. It is set by varying space conditions. At times we really only need the air slightly warmed. The reset could perform that function in conjunction with other controllers, or it could assume a fixed, constant discharge air temperature. Either way it is avoiding override and fuel wastage.

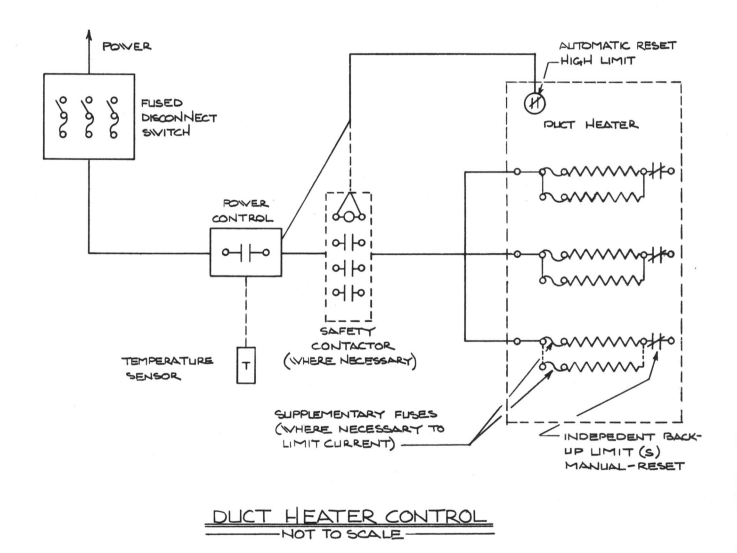

DUCT HEATER CONTROL
—NOT TO SCALE—

FIGURE 13-2

Control Schematic: Figure 13-3 is a typical heating and cooling system with an energy recovery system. The system is a one-pass system. That is, there is no recirculated air. This is a typical hospital operating room or industrial clean room application. Obviously if you continually heat and cool air and continuously exhaust the treated air, you are going to use a lot of fuel. Before 1970 fuel was so inexpensive that no one was overly concerned about it. Since the 1970 energy shortage, the run-around coil installation is normally the first element to be designed. In theory, if the incoming air was 0°F and the leaving air was 70°F, the run-around coil could make the entering and leaving temperatures both be 35°F, and in summer if the entering air was 95°F and the leaving air 75°F, the run-around coil could make them both equal at 85°F.

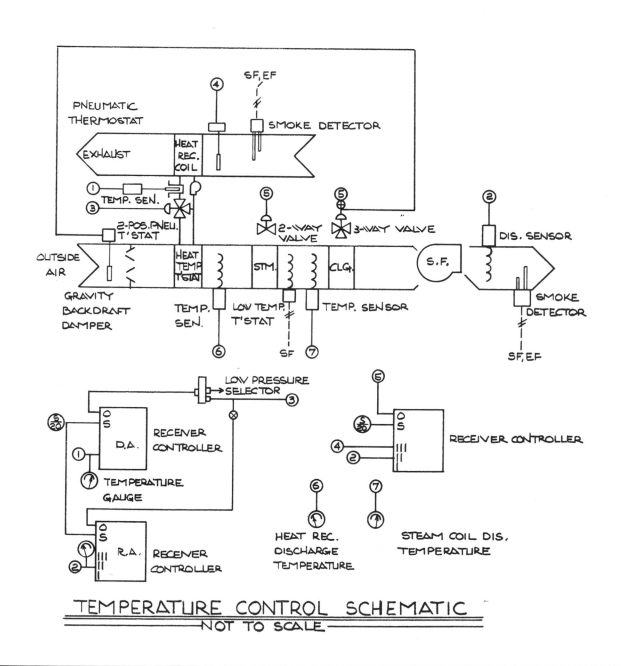

TEMPERATURE CONTROL SCHEMATIC
NOT TO SCALE

FIGURE 13-3

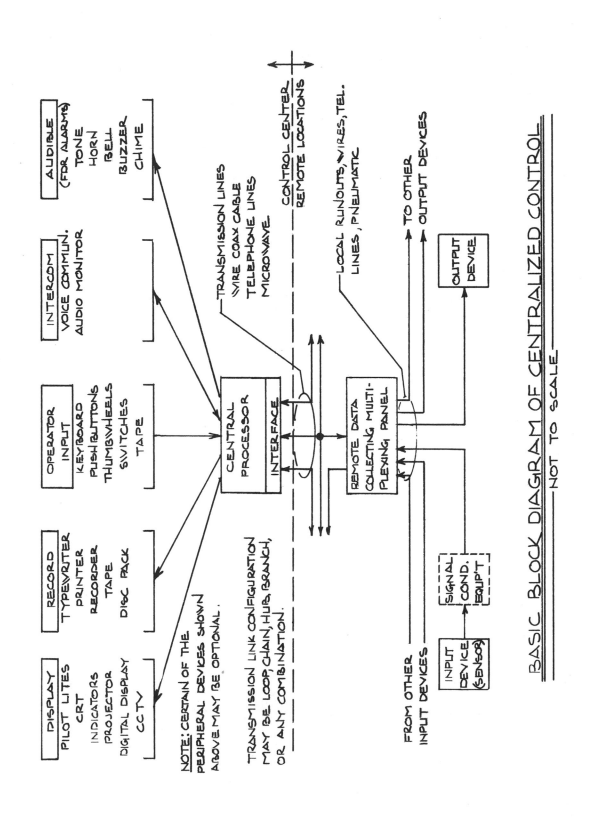

BASIC BLOCK DIAGRAM OF CENTRALIZED CONTROL

NOT TO SCALE

FIGURE 13-4

The actual run-around design produces conditions near these numbers and will save about 60 percent of the energy that was previously being wasted on a true calculated basis. Our next device is the steam coil controlled by a two-way, not a two-position, valve. The two-way valve is a modulating valve controlled by a sensor in the discharge air. If something goes wrong, another sensor (low temperature) stops the fan when the air gets too cold for proper system use. The cooling coil is a chilled water coil controlled by another sensor to maintain proper summer air temperature. Finally in both supply and exhaust ducts we have smoke detectors. Obviously we do not want to circulate smoke. If smoke is detected, the fan is stopped.

These devices are tied together through receiver controllers to relate air pressure to proper air-pressure and electrical-operating signals. This is essentially a standard HVAC control diagram. If the heating medium was hot water instead of steam, the hot water controller would be a three-way valve, similar to that on the chilled water controller.

Computer Control Systems

With the rapidly, almost explosively, growing computerized system of control there has come an entirely new language and concept of control and also considerable misunderstanding of it by people who are, and are not, connected with the overall building operational system industry. Computerized systems are becoming a whole new service industry for building operation and management.

Let us take a very simple case that is illustrated by Fig. 13-4. This is a block diagram of the centralized computer control system. We disagree for reasons that follow with the title of this detail, but we have used the current, commonly accepted title so that the following discussion will be clear.

In our detail we start, at the bottom, with signals, electric or pneumatic converted to electric, being received at a master multiplexing panel. These signals are then sent to a control processor. The implication is that the processor (microprocessor) can also be receiving signals from other points as well. The processor has taken the place of what used to be a master controller. The processor may display information received on its CRT (cathode-ray tube) screen, record received data on its related printer, send data through either programmed or human operator input, have voice transmission capability, and sound alarms.

All of the above could be done, is being done, and has been done many times without a computer. What the computer, properly programmed, can do is to make the necessary decision efficiently, immediately, and without emotion, day or night, 365 days a year. After the computer makes its preprogrammed decision or one is made by an operator who overrides the automatic decision, the control signal goes back through the microprocessor and the output device to the system.

In the vast majority of cases input and output signals are put to work through standard pneumatic or electric control devices. Our centralized computer control system does not control anything. It *manages* the system. If the controls it manages are inaccurate, the control system is inaccurate. The analogy is the same as for quality control of an industrial plant production line. The president and his administrators can promise anything, but if the workers do not deliver, the promise is worthless. Quality control begins with the workers—or in our case, the controllers.

You can interface a computer to a controller. It is both costly and may be inaccurate. The computer's sensor and the controller's sensor may not agree, creating confusion and lack of confidence in one system or the other.

Direct Digital Control: Figure 13-5 is the first of five details on direct digital control. The system is again a block diagram. In abbreviated form we have not only shown the central control system but also how the central system can be tied back through telephone lines to remote stations, service centers, and manufacturers' diagnostic centers. In the detail shown we are controlling air handling units by using local control computers which operate direct digital control devices.

Air Handling Systems: Figure 13-6 contains a lot of things that are self-evident if you read our previous discussion on a typical HVAC control system. Instead of a one-pass system this system has both return and fresh air and is a typical variable volume system. More importantly it shows a local computer control panel which, with its direct digital devices, is the heart of the new and far more accurate control system.

Basically the computer's signal is digital, on and off. Since most controls require modulation, which is adjustment to changing conditions, the computer supervisory system requires a complicated interface between the on-off digital signal and the varying (analog) operating signal requirement. Recent technological advances, called pulse-width modulation, have resolved that problem. A modulating device is now operated directly by the on-off digital action. By varying the time of the on and off phases, the computer can directly control changing functions, slowly or rapidly. The speed of the computer is such that as the control approaches a new set point, the very short duration of the on and off pulses can very precisely control the final setting. It is somewhat similar to a truck driver "fanning" his brakes to come to a quick, controlled stop, only it is far more accurate. The driver presses and quickly releases his air brake pedal to avoid the skidding and uncontrolled sliding that would result from a steady pressure. The computer does the same thing and avoids overshooting or undershooting a set point. This creates limits or spreads around control points. Direct digital, since it still usually operates through an air pressure activated control device that is not mechanically perfect, is not absolutely perfect. But control within one-half of one degree is not uncommon with direct digital.

There are terms used in this detail and subsequent details that require explanation. PID is proportional integral derivative. This is a type of device that reacts to temperature change by the difference between set point and actual temperature (proportional), by how long the difference has existed (integral), and by how fast the temperature is changing (derivative). The local computer can resolve all three problems simultaneous-

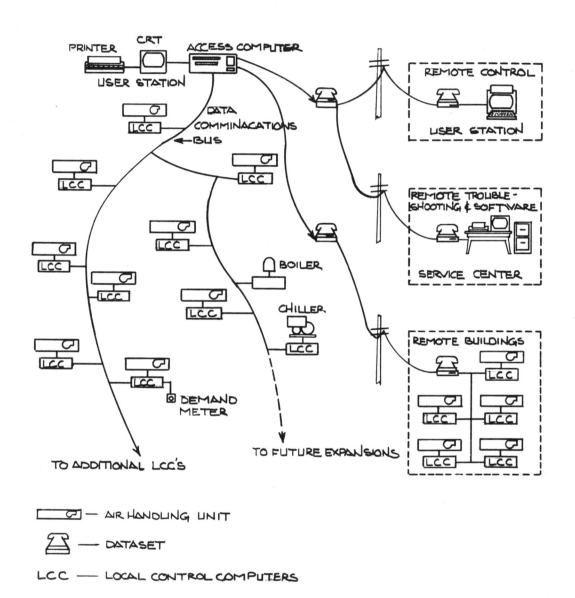

A TYPICAL DDS SYSTEM
— NOT TO SCALE —

FIGURE 13-5

ly. PWM is pulse-width modulation. Having digested the three conditions, the computer, through its timing of start-stop signal pulses, controls the width (time) of each pulse to precisely reset the control to the desired condition.

As far as the control diagram is concerned, the controls themselves are special to accept the pulsed signals. But they are the same in function and location as for a conventional system. The sequence of operation comes from the access computer software. The software has been designed to operate at one or many different sequences to produce the results you specify. If you want to change the results for any mode of operation—occupied, unoccupied, set back, or opti-

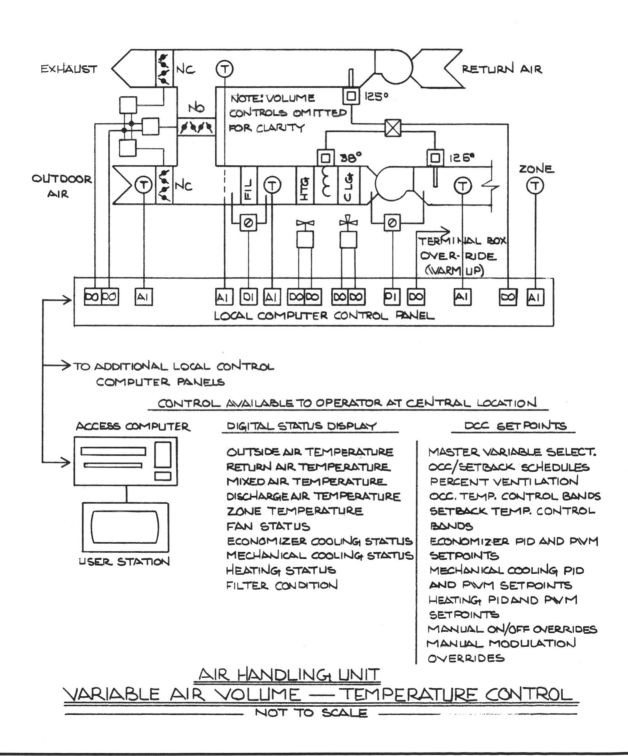

AIR HANDLING UNIT
VARIABLE AIR VOLUME — TEMPERATURE CONTROL
NOT TO SCALE

FIGURE 13-6

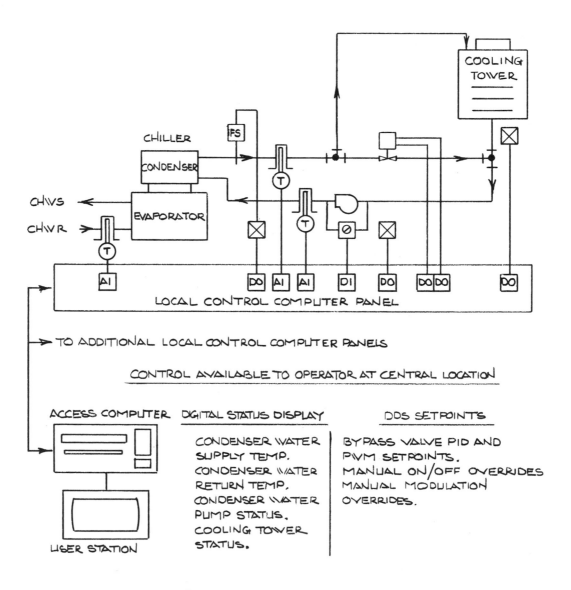

CHILLED WATER PLANT
COOLING TOWER AND PUMP CONTROL
— NOT TO SCALE —

FIGURE 13-7

mal start—you can easily adjust set points to do this.

Water Systems: Figure 13-7 is an illustration of a direct digital control operating a chilled water plant, cooling tower, and pumps. Again the difference between it and a conventional system is in the controls directly operated by the computer.

Figure 13-8 illustrates controls for the boiler and pump in a hot water plant. Figure 13-9 illustrates controls for the heat exchanger and pump in a hot water plant.

You might say, "Now I see how the detail for a direct digital control system differs from a conventional control system. It seems somehow simple. What's the big deal! The same thing can be accomplished in many cases with a conventional system and a programmable seven-day clock." True! The direct digital control system is not doing something different. It is doing the same thing much more accurately. Its advantages are speed and accuracy of response. These two simple facts create very large energy savings and,

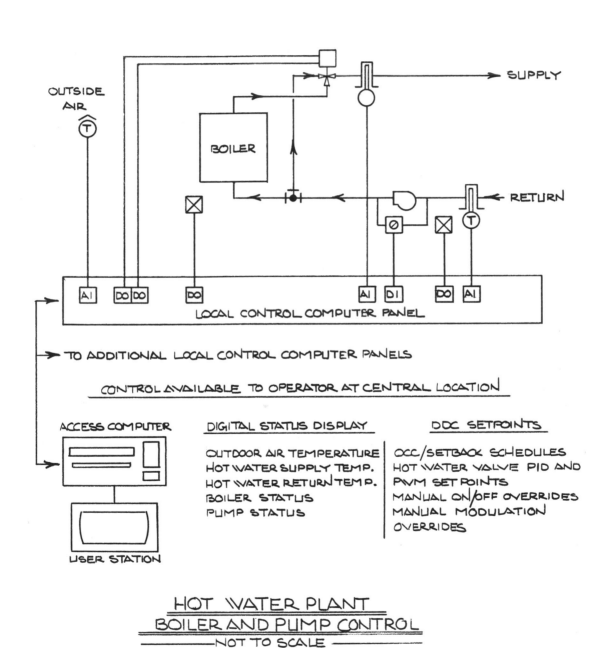

HOT WATER PLANT
BOILER AND PUMP CONTROL
— NOT TO SCALE —

FIGURE 13-8

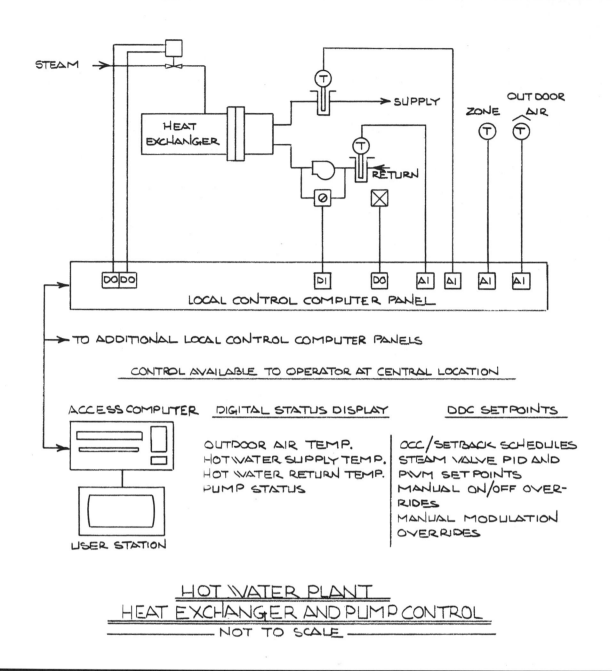

STEAM

HEAT EXCHANGER

SUPPLY

RETURN

ZONE

OUTDOOR AIR

DO DO

DI

DO

AI AI AI AI

LOCAL CONTROL COMPUTER PANEL

TO ADDITIONAL LOCAL CONTROL COMPUTER PANELS

CONTROL AVAILABLE TO OPERATOR AT CENTRAL LOCATION

ACCESS COMPUTER

USER STATION

DIGITAL STATUS DISPLAY

OUTDOOR AIR TEMP.
HOT WATER SUPPLY TEMP.
HOT WATER RETURN TEMP.
PUMP STATUS

DDC SETPOINTS

OCC/SETBACK SCHEDULES
STEAM VALVE PID AND
PWM SET POINTS
MANUAL ON/OFF OVER-
RIDES
MANUAL MODULATION
OVERRIDES

HOT WATER PLANT
HEAT EXCHANGER AND PUMP CONTROL
NOT TO SCALE

FIGURE 13-9

when precise temperature and humidity control are critical, provide that the operating result *is* the designed condition, not approximately the designed condition.

The details as depicted illustrate systems controlled by a control computer. Newer systems now being marketed have been further refined by the inclusion of stand-alone programmable local computer devices which can inexpensively control a given local control requirement and can also be overriden and controlled, at some future date, by a centralized computer system.

FOURTEEN

Emergency Power

There have always been requirements for an alternative source of electrical power for people, processes, and structures. Many of today's electronic installations specifically require, at the time of installation, an uninterruptible power source. In the narrow context the emergency generator is not an uninterruptible power source since even the fastest acting set of generator controls still require a short delay to get the generator operative. Only a battery powered system that will perform satisfactorily, or some form of inverter and/or alternative operable power source can be used to provide true uninterruptible power. But the generator is a power source that can operate motors for all types of systems for long periods of time, which the battery system can only do to a very minor degree and for a limited time. Thus both systems have their special applications.

Battery and inverter powered systems are usually covered in an electrical specification. Regardless of where they are covered, no mechanical detail is involved. The electrical detail is primarily a simple schematic showing the alternating current source, the transfer switch, and the battery powered source.

In essential elements the emergency generator system consists of an internal combustion engine, a generator, a starting battery, an instrument panel, and a control system. This is all part of the generator package and needs no detailing. But the source of the generator's fuel, the type of fuel, the removal of the products of combustion, and the dissipation of the generator's heat very definitely do require detailing.

There are three types of fuels that are used by generators: gas, gasoline, and diesel. If the source is natural or propane gas, the detailing is mostly provided by the manufacturer of the generator, and when propane gas is used, by the connections at the gas storage tanks that are a standard part of the propane supplier's equipment. But when the fuel is gasoline or diesel, we need to detail the tank and piping, which is what this chapter covers.

Finally there is the problem of room ventilation. This is fundamentally a design problem. If the generator is air cooled, the air outlet is usually the full size of the engine's radiator or larger. As in any other movement of air, the intake, normally not fan assisted, is also as large as or larger than the discharge. The designer's problem is to size these intake and exhaust louvers. The sizing of the intake and the fan assisted exhaust louvers for the water cooled generator is equally important. There is still a lot of heat to be dissipated even with a water cooled radiator installation.

Emergency Generator Systems

As in all mechanical systems there will always be special items to detail or parts the designer feels must be detailed. In the seven details that follow the major common features of the typical installation of emergency generators are depicted. For those special items not depicted it is suggested that the designer refer to the catalogs of the manufacturer whose equipment is being specified. Some of the special connections are special to a particular product and are not necessary or even desirable on a competing product.

Figures 14-1 and 14-2 are two basic overall details of the emergency generator installation. Figure 14-1 is a typical gasoline engine driven emergency generator system with an air cooled radiator. A diesel engine and diesel fuel could also be used. The fuel tank is assumed to be adjacent to the generator and on, or above, the level of the generator. A primer, or day tank, is noted but not shown. The piping for these not-depicted tanks is covered in Fig. 14-2.

To keep the generator radiator cool (it is similar to your automobile radiator), air must be introduced into the room and the air from the radiator must be ducted to the outside to avoid overheating the room. As noted, the air intake may have a thermostatically controlled louver damper instead of a simple, self-closing louver.

As in your auto, the generator has an exhaust pipe. Since it is rigidly supported, a flexible connection is required, as well as a drip and drain leg with drain plug for exhaust moisture removal. Also as in your auto, a muffler (not depicted) is required. Finally, again as in your auto, the exhaust pipe is hot when the engine is running and the thimble pipe sleeve through the wall must have insulation between the muffler pipe and sleeve (thimble). Those tiny holes in the circle around the muffler pipe represent sleeve insulation.

Figure 14-2 is a companion to Fig. 14-1. It illustrates other points of the generator installation. It also indi-

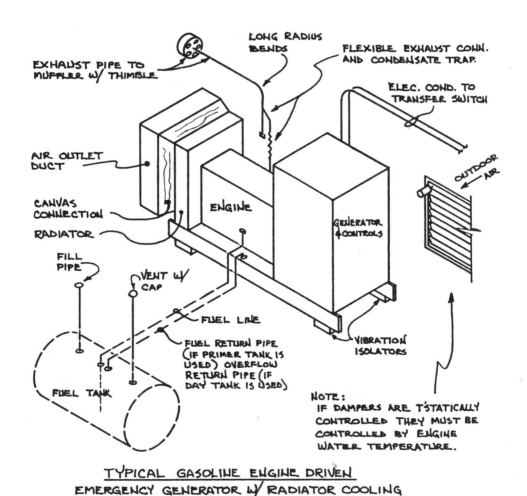

TYPICAL GASOLINE ENGINE DRIVEN
EMERGENCY GENERATOR W/ RADIATOR COOLING

FIGURE 14-1

cates an alternative way of cooling the radiator using a city water supply. As you can see, if we tried to put all of this detailing on one small drawing, it would be difficult to follow. Even if you use city water, stream water, or well water for cooling, the room will overheat if the engine operates for more than ½ hr. *Room ventilation is still required.*

This detail shows the piping to the day tank that is commonly used in fuel systems as an assured engine priming source. It shows the muffler, commonly called the silencer, bracket mounted on the wall. The engine starts with battery power and a battery charger is noted. This charger assures that the battery is always charged.

The generator is electrically tied through the noted electrical feed line to the power source. When power fails, the electrically held-open switch is mechanically closed and the engine starts. The power generated by the engine driven generator is conducted to the emergency panel and through the panel branch circuits it is connected to systems requiring this alternative source. Within the limitations of the total generator output these connected users can be anything–motors, pumps, elevators, lighting, and the like. These users

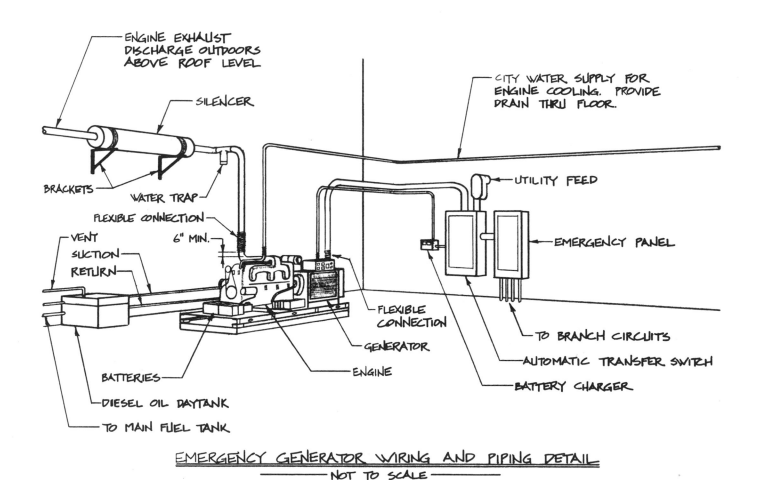

EMERGENCY GENERATOR WIRING AND PIPING DETAIL
— NOT TO SCALE —

FIGURE 14-2

have the same alternating current as before; only it is from a different source.

Gasoline Primer: Figure 14-3 illustrates the resolution of a problem created by long periods of idleness and relates to the statement made in connection with Fig. 14-2 about primers. During long shutdowns gasoline evaporates from the carburetor bowl. The primer tank feeds fuel by gravity to the bowl and ensures an available supply of fuel to start the engine. The solenoid valve opens when the engine starts and also prevents gasoline from draining into an idle engine. Not depicted is a restrictive bushing in the return line which pressurizes the tank. The return line to the tank also serves as a vent line. It will not gravity feed if there are any dips in the line which trap fuel and block the free flow of air through the line.

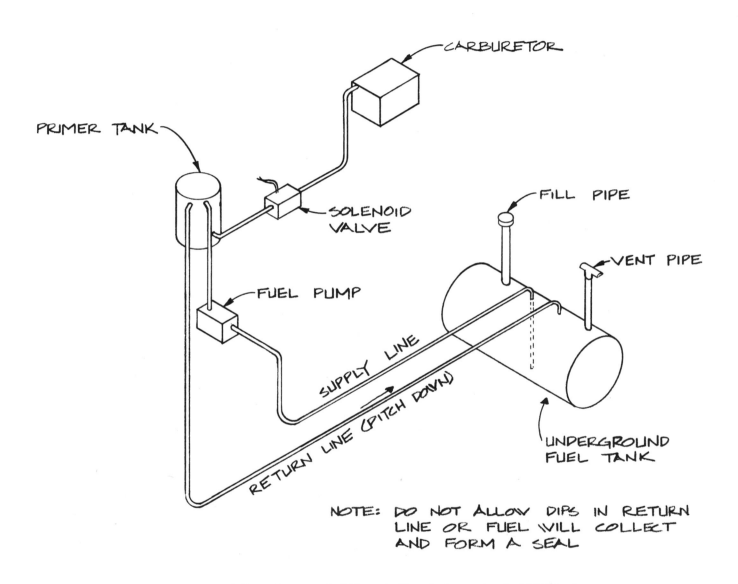

GASOLINE PRIMER TANK SYSTEM

NO SCALE

FIGURE 14-3

Diesel Systems: Figure 14-4 is a typical diesel fuel system with the supply tank below the emergency generator. It is for a specific unit, a Detroit diesel. Other diesel systems are very similar. The difference usually is whether the return goes directly to the day tank or can be bypassed directly to the fuel tank as is done in our illustration. In all cases the day tank is assumed to be near the engine generator and the lift required is within the capability of the engine's fuel pump. The fuel pump lifts fuel from the storage tank to the day tank and is controlled by a float switch in the day tank. The return, if it goes to the day tank, must be below minimum level of the tank or drain-back from the engine and filters may occur. The overflow solves

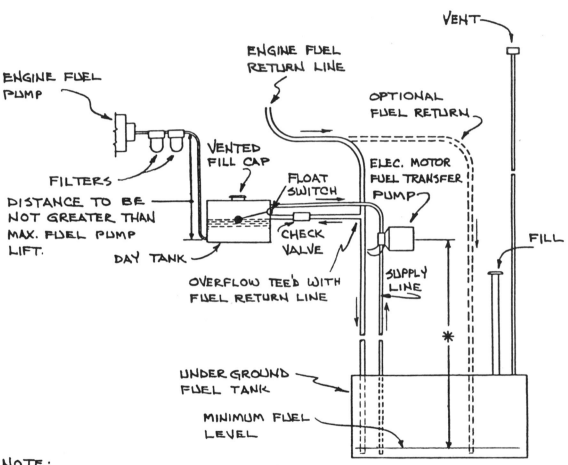

TYPICAL DETROIT DIESEL FUEL SYSTEM WITH
SUPPLY TANK BELOW GENERATOR SET

FIGURE 14-4

control failure and constant operation of transfer pump problems.

Radiator Details: Figure 14-5 illustrates the solution to the problem of a remotely located radiator. Generally there is no problem if the two separated items are less than 15 ft (dimension B in the detail) apart. This detail solves the common problem that occurs when the horizontal distance, B, is over 15 ft but the vertical distance, A, *does not* exceed 15 ft. Two items are required: first an auxiliary pump to create circulation, and second a surge tank as shown to resolve differences between engine pump and auxiliary pump operational conditions. Note that the makeup water now goes to the surge tank.

Figure 14-6 illustrates the solution to the other side of the problem, the high remote radiator installation.

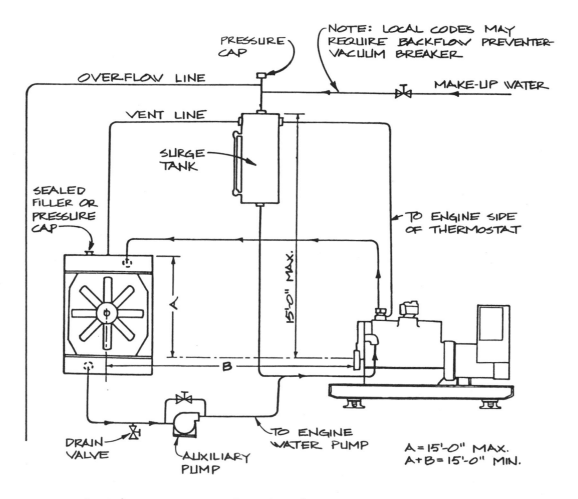

LONG REMOTE RADIATOR INSTALLATION
WITH SURGE TANK

NO SCALE

FIGURE 14-5

Here head pressure is a problem, and you need a two-section tank with a partial baffle separating the hot side and the cold side. Hot fluid flows to the hot side, is pumped from the hot side up to the radiator, and then flows by gravity from the radiator to the cold side and back to the engine's radiator.

Hot Well: Figure 14-7 illustrates not only the hot well described in Fig. 14-6 but also a method of standpipe water cooling. The hot well as depicted must be sized to contain the full water capacity of the system that is required to keep all inlets and outlets submerged. Since the heated water expands, almost 8 percent of the volume calculated must be added to this value. As can be seen in the previous high remote radiator detail, the radiator is above the hot well and

drains into the cold well after shutdown. Therefore the baffle as depicted must have an opening large enough to allow free flow up to the rate of the engine or the auxiliary pump, whichever has the largest rating. The hot well is vented to atmosphere.

The hot well must not exceed the 15-ft 0-in height limitation, and its supports must be strong enough to hold the normal weight of water plus approximately 65 percent of the cooling system capacity when the engine is operating. Mount the auxiliary pump below the tank to preclude air binding.

Figure 14-7 also shows a standpipe cooling arrangement which is highly recommended if city water is available. In the detail depicted the engine pump pulls water from the standpipe. City water pressure forces

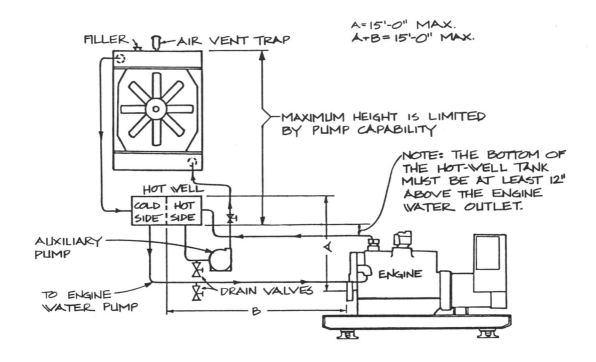

HIGH REMOTE RADIATOR INSTALLATION
NO SCALE

FIGURE 14-6

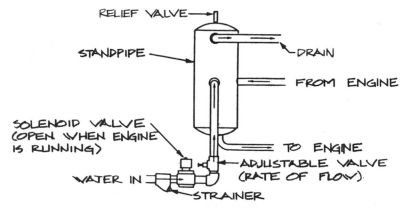

RELIEF VALVE

STANDPIPE

DRAIN

FROM ENGINE

SOLENOID VALVE
(OPEN WHEN ENGINE
IS RUNNING)

TO ENGINE

ADJUSTABLE VALVE
(RATE OF FLOW)

WATER IN

STRAINER

STANDPIPE COOLING SYSTEM
NO SCALE

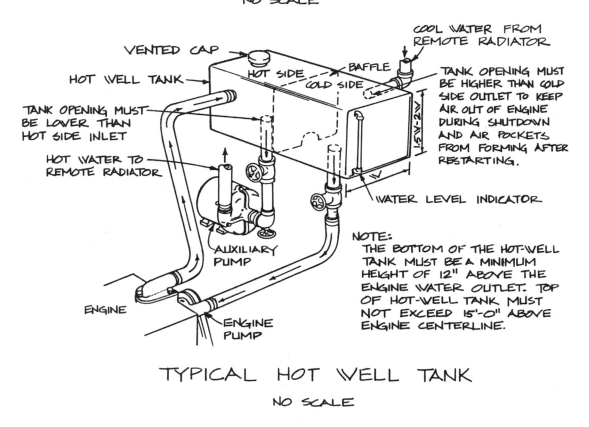

COOL WATER FROM
REMOTE RADIATOR

VENTED CAP

HOT WELL TANK

HOT SIDE

BAFFLE

COLD SIDE

TANK OPENING MUST
BE HIGHER THAN COLD
SIDE OUTLET TO KEEP
AIR OUT OF ENGINE
DURING SHUTDOWN
AND AIR POCKETS
FROM FORMING AFTER
RESTARTING.

TANK OPENING MUST
BE LOWER THAN
HOT SIDE INLET

HOT WATER TO
REMOTE RADIATOR

WATER LEVEL INDICATOR

AUXILIARY
PUMP

NOTE:
THE BOTTOM OF THE HOT-WELL
TANK MUST BE A MINIMUM
HEIGHT OF 12" ABOVE THE
ENGINE WATER OUTLET. TOP
OF HOT-WELL TANK MUST
NOT EXCEED 15'-0" ABOVE
ENGINE CENTERLINE.

ENGINE

ENGINE
PUMP

TYPICAL HOT WELL TANK
NO SCALE

FIGURE 14-7

water into the pipe and out an overflow line. Heated and cooled water continuously mix to provide a proper water temperature to the radiator. The relief valve is a vacuum relief valve which eliminates possible siphoning effects created by long lines to the sewer drain connection. The city water flow rate may be controlled by adding a solenoid shutoff on the incoming line to preclude flow when the engine is not operating. Assuming the city water flow rate is adequate, a manual globe valve at the entry to the standpipe can be hand adjusted for design flow conditions.

FIFTEEN

Special Situations

No collection of words could possibly equal an introduction composed of tape recordings of any group of engineers, designers, or detailers that begin invariably with the words "if only." Those of you who are knowledgeable and experienced in the mechanical field know what is coming next and what is about to be written here. The "if only" is followed by a description, frequently in unprintable language, of the situation that developed on a given project all because of some item that was thought to be the contractor's responsibility, the equipment supplier's responsibility, or just plain obviously did not need to be detailed. It was not and the arguments could be heard a block away. What is worse, the owner stuck the engineer with the bill for the added work.

It would be very rewarding if we could assure you that every peculiar problem that you may encounter is covered in this chapter. Unfortunately such a list of details seemingly has no end. We have tried to cover some of the obvious problems, but we have to admit that we do not begin to cover everything, everywhere, for every situation. The details we have picked are the ones that in our 30 years of experience have most frequently been brought up in conversations with our fellow consultants as particularly aggravating problem areas. At the very least we hope the details will not only assist you but will also act as a reminder to investigate other odd items on your plans that seem a little vague. This is where your investigation should always begin. This is where a good detailer who simply questions the clarity of some point of the design can very nearly be worth his or her weight in gold.

Specialized Piping Details

Condensate: Figure 15-1 is a typical situation in which condensate, for some process reason, must be wasted but is too hot for the normal fittings and materials of the waste piping system. The solution is to use a water-to-water heat exchanger. This simple system has become a hot idea in the energy conservation business and can be used where you have a use for the hot water.

Food Grinder: Figure 15-2 is a typical water supply connection to a specialized restaurant-type food grinder. The common household garbage disposal unit is similar in nature, and you probably never gave it a second thought when you attached the cold water supply. But these devices can be a source of water supply contamination, and the purpose of the vacuum breaker is to prevent this problem.

Washing Machine: Figure 15-3 is another typical connection to, in this case, a hospital washing machine. Again it has simple hot and cold piping to a mixing valve, but these machines usually have spring-loaded devices that really can slam shut. And they, too, are a source of water supply contamination. Thus our little detail shows shock absorbers (water hammer arrestors), as well as a vacuum breaker.

Shower: Figure 15-4 shows a multiple shower and eye wash detail that might resemble the work of a berserk Christmas tree decorator. Actually this shower and sink gadget collection serves a very serious industrial safety purpose. Workers using dangerous chemicals in laboratories and process applications can

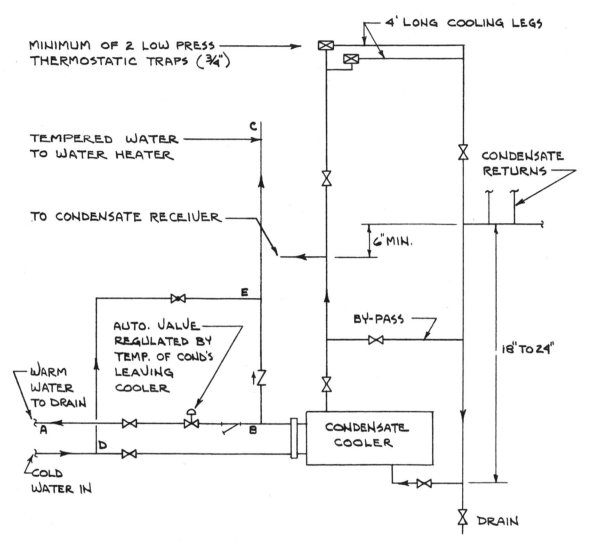

MINIMUM OF 2 LOW PRESS THERMOSTATIC TRAPS (¾")

4' LONG COOLING LEGS

TEMPERED WATER TO WATER HEATER

CONDENSATE RETURNS

TO CONDENSATE RECEIVER

6" MIN.

C

E

AUTO. VALVE REGULATED BY TEMP. OF COND'S LEAVING COOLER

BY-PASS

18" TO 24"

WARM WATER TO DRAIN

A

D

B

COLD WATER IN

CONDENSATE COOLER

DRAIN

NOTES:

1. DELETE SECTION A-B IF CONTROL OF LEAVING CONDENSATE TEMPERATURE IS NOT REQUIRED

2. DELETE SECTIONS B-C & D-E IF PRE HEATING OF HOT WATER IS NOT REQUIRED

TYPICAL STEAM CONDENSATE COOLER PIPING

FIGURE 15-1

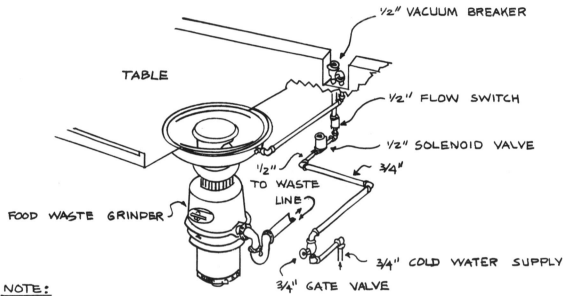

½" VACUUM BREAKER

TABLE

½" FLOW SWITCH

½" SOLENOID VALVE

¾"

½"
TO WASTE
LINE

FOOD WASTE GRINDER

¾" COLD WATER SUPPLY

¾" GATE VALVE

NOTE:
ALL FOOD WASTE GRINDERS MUST
HAVE WATER TO OPERATE. WATER
SUPPLY MAY COME FROM FAUCET
ABOVE OR FROM DISCHARGE OF
DISWASHING MACHINE PRE-WASH.

TYPICAL PIPING & CONTROLS FOR FOOD WASTE
GRINDER

FIGURE 15-2

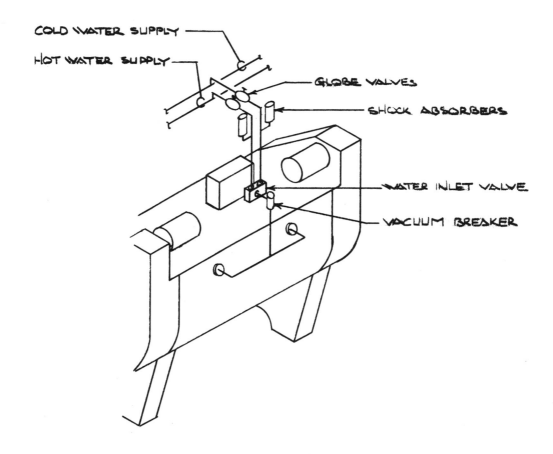

COLD WATER SUPPLY

HOT WATER SUPPLY

GLOBE VALVES

SHOCK ABSORBERS

WATER INLET VALVE

VACUUM BREAKER

WATER SUPPLY PIPING FOR HOSPITAL WASHING MACHINES
— NOT TO SCALE —

FIGURE 15-3

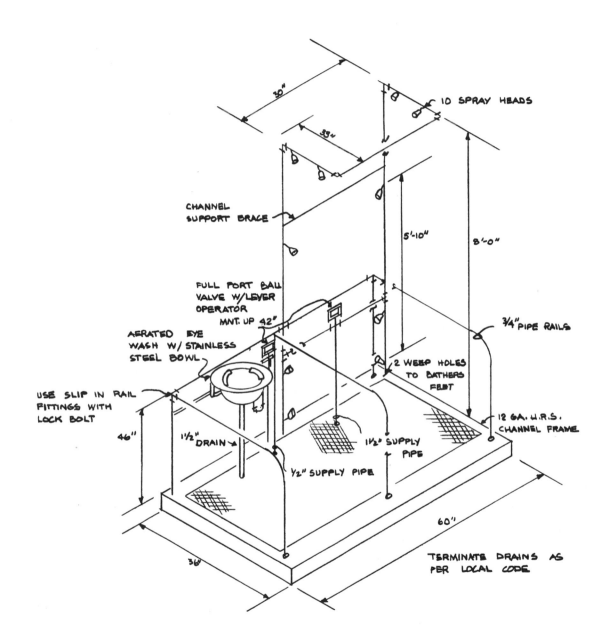

30"

10 SPRAY HEADS

33"

CHANNEL
SUPPORT BRACE

5'-10"

8'-0"

FULL PORT BALL
VALVE W/LEVER
OPERATOR
MNT. UP 42"

3/4" PIPE RAILS

AERATED EYE
WASH W/ STAINLESS
STEEL BOWL

2 WEEP HOLES
TO BATHERS
FEET

USE SLIP IN RAIL
FITTINGS WITH
LOCK BOLT

12 GA. H.R.S.
CHANNEL FRAME

46"

1 1/2" DRAIN

1 1/2" SUPPLY
PIPE

1/2" SUPPLY PIPE

60"

TERMINATE DRAINS AS
PER LOCAL CODE

36"

MULTIPLE SPRAY SHOWER & EYE WASH DETAIL

FIGURE 15-4

spill or get splashed by these chemicals. The chemical reaction can be very rapid and there is not time for them to run to a locker room, strip off their clothes, and find an available shower. Instead this device is located near the workers; they jump in the shower, clothes, shoes, and all, and wash off the chemicals, or they stick their faces in the eye sink and rinse them. Not shown is an acid waste drain that is underneath the platform, and connected to an acid waste drainage system.

Tank Heating: Figure 15-5 is a detail of the boiler piping connections to prevent system freeze ups in a fire protection water storage tank. Usually this sort of installation serves a dual purpose. As illustrated the system is a basic low temperature hot water system. In addition the boiler has a heat exchanger that is usually

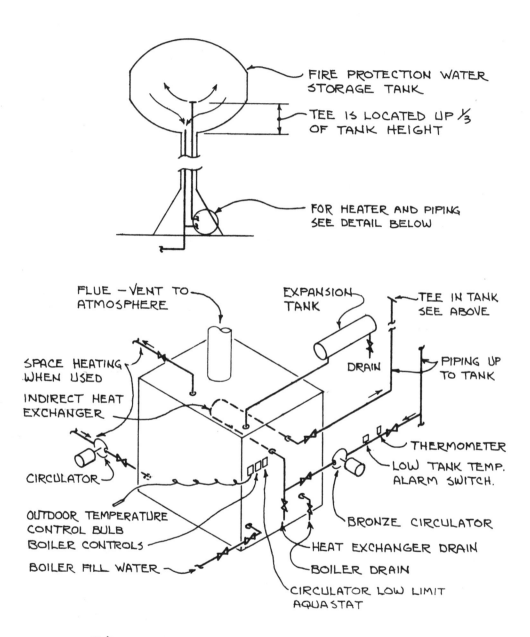

FIRE PROTECTION WATER STORAGE TANK

TEE IS LOCATED UP ⅓ OF TANK HEIGHT

FOR HEATER AND PIPING SEE DETAIL BELOW

FLUE — VENT TO ATMOSPHERE

EXPANSION TANK

TEE IN TANK SEE ABOVE

DRAIN

PIPING UP TO TANK

SPACE HEATING WHEN USED

INDIRECT HEAT EXCHANGER

THERMOMETER

LOW TANK TEMP. ALARM SWITCH.

CIRCULATOR

BRONZE CIRCULATOR

OUTDOOR TEMPERATURE CONTROL BULB BOILER CONTROLS

HEAT EXCHANGER DRAIN

BOILER DRAIN

BOILER FILL WATER

CIRCULATOR LOW LIMIT AQUASTAT

TYPICAL BOILER PIPING FOR HEATING OF FIRE PROTECTION WATER STORAGE TANK

NO SCALE

FIGURE 15-5

external; it is controlled by an outdoor bulb to maintain conditions above freezing in the hot water storage tank.

Figures 15-1 through 15-5 show the typical sort of detail that, when the need arises, you generally will have to invent on the spot.

Specialized Structural Details

"Specialized Structural Details" seems hardly an appropriate title to appear in a book on mechanical details. These details are, as we described in the preceding section, one-of-a-kind, seldom-used details. They come in an endless variety that no book could

cover. The following details are seven samples of what we mean by specialized structural details. We call them structural details because they relate to structural support or protection of mechanical equipment.

Air Handling: Figure 15-6 is a fairly typical HVAC air handling arrangement of a fan module, filter module, and coil module. The fan is a large fan that handles 40,000 cfm at 7-in of static pressure. To simplify vibration mounting the fan section is isolated by flexible connections and is spring-mounted for vibration elimination. One small and important problem remains. The free floating fan section at the high static head tends to rotate on start-up. To restrain this

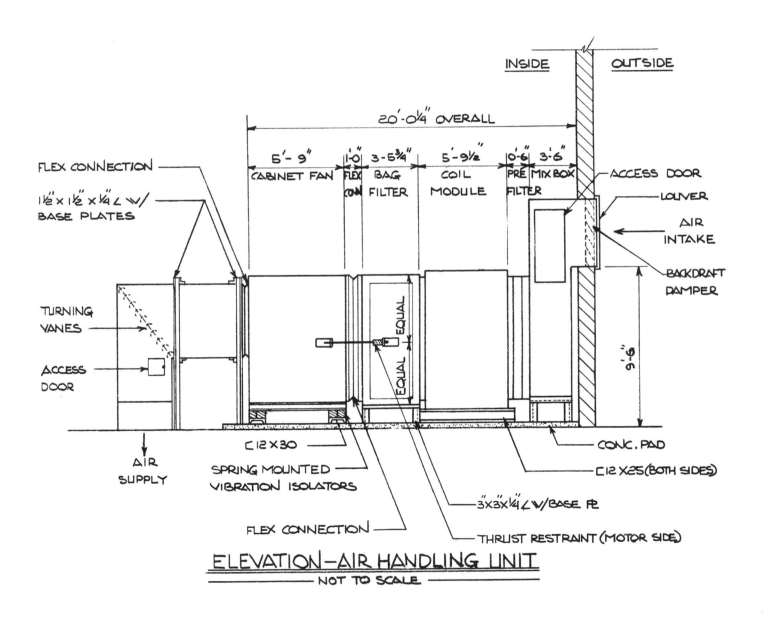

ELEVATION—AIR HANDLING UNIT
— NOT TO SCALE —

FIGURE 15-6

common start-up thrust effect we have a simple, very useful 3-in angle thrust restraint on the motor side of the fan housing.

Coil Supports: Figure 15-7 is a typical resolution of a coil installation problem in which the coil is remotely located somewhere in the supply and/or return ductwork. In the detail depicting the coil we show a cooling coil squeezed into a location between two

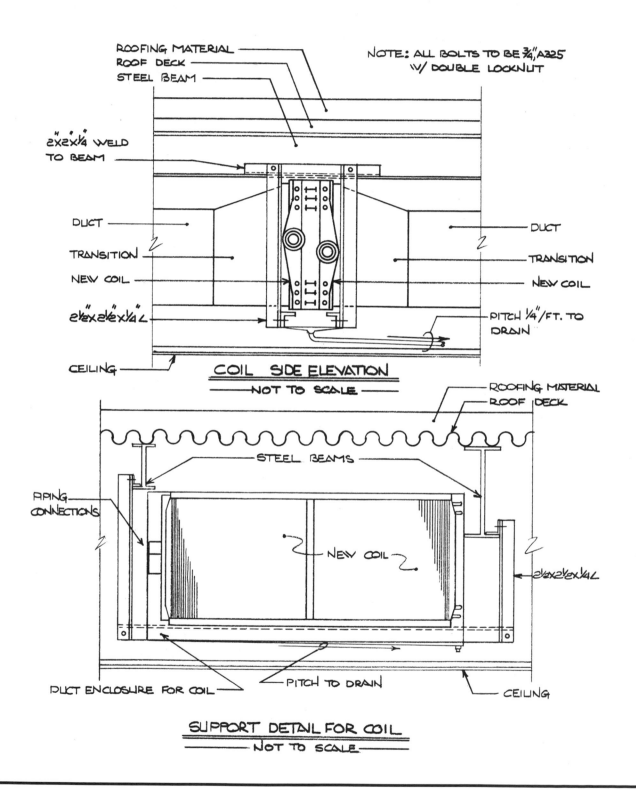

FIGURE 15-7

unequally sized beams. Note the detailing which depicts how the job-built angle-iron frame for the coils has to be fastened with angle clips to the beams. Note also and do not forget that a cooling coil needs, in this case, a field-built drain pan and a drain line. The ductwork to and from the coil must fasten to the angle-iron coil frame.

Grounding: Figures 15-8 and 15-9 illustrate a protection for an above ground fuel tank. We could have put these two details in our piping or fuel tank detail sections, but we used this location in the book to call this important item to your attention. Above ground fuel tanks should always be grounded and in the tank ground detail of Fig. 15-8 you will note that the grounding is not a casual affair. The ground wire is specified and the ground rod is described.

Figure 15-9 is a detail in large scale of the grounding connection to the tank. Far too frequently the total subject of tank grounding is described with a casual sentence in the tank specification. This can create a very serious problem of fires and explosions. Note the great lengths we have gone to in describing how the connection is made to both insulated and uninsulated tanks. We urge you to place this sort of detailing on your drawings whenever you are involved with above-ground fuel storage tanks.

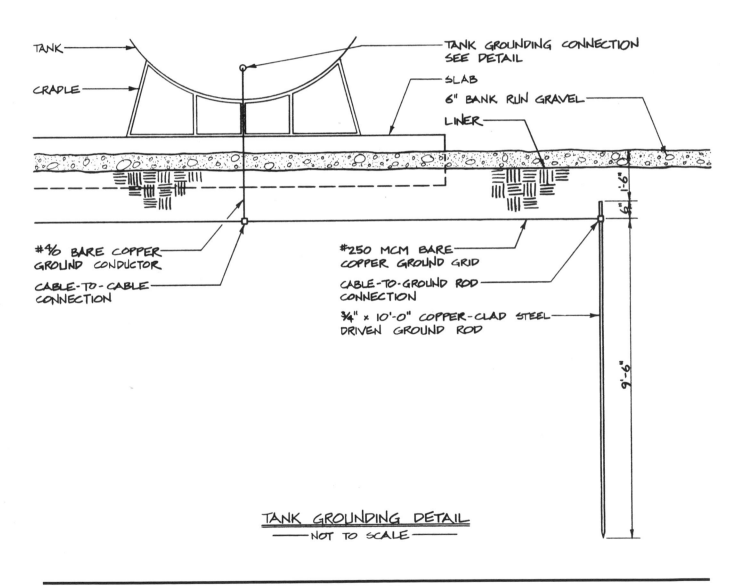

TANK GROUNDING DETAIL
— NOT TO SCALE —

FIGURE 15-8

Guard Post: Figure 15-10 is a very simple detail that the mechanical detailer usually assumes is covered by the structural detailer, and the structural detailer assumes it is covered by the architectural detailer. Frequently it is not covered by any of the above detailers. Mechanical equipment is frequently put outdoors by mechanical designers, and it is frequently in or near a parking lot. Somebody has to protect the equipment by a simple row of guard points, a concrete wall, or something similar. Preferring to be safe rather than sorry, we suggest that unless you are certain some protective enclosure is provided you show a line of guard posts around your outdoor mounted equipment and use Fig. 15-10 as a typical guard post detail.

Bases: Figure 15-11 is another typical example of the mechanical designer's structural problem. Perhaps this time there is no structural engineer on your mechanical rehabilitation project, and the air cooled, exterior mounted condenser needs a concrete base. We have presented a typical base to hold up most units

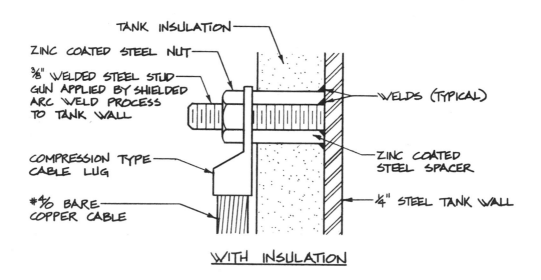

WITH INSULATION

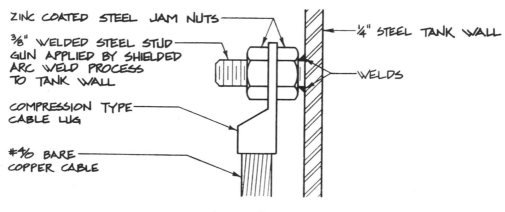

WITHOUT INSULATION

NOTE:
EPOXY COAT ENTIRE CONNECTION FOR CORROSION PREVENTION

TANK GROUNDING CONNECTIONS
—— NOT TO SCALE ——

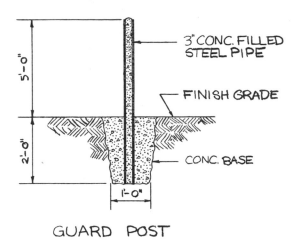

3" CONC. FILLED STEEL PIPE

FINISH GRADE

CONC. BASE

5'-0"

2'-0"

1'-0"

GUARD POST

TYPICAL PROTECTIVE VEHICLE BUMPER FOR EXTERIOR SURFACE MOUNTED EQUIPMENT.

FIGURE 15-10

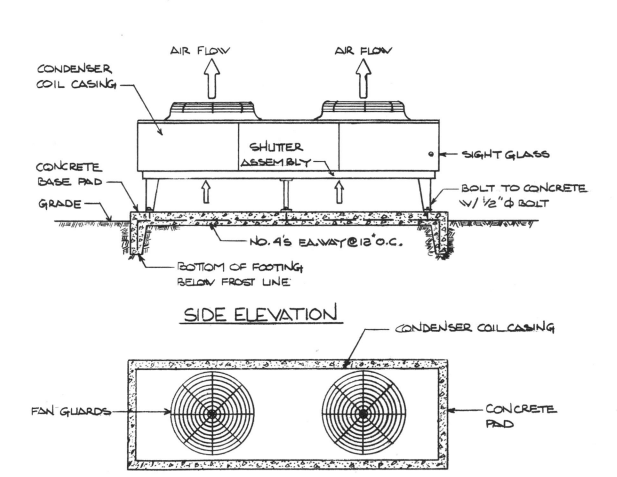

AIR FLOW

AIR FLOW

CONDENSER COIL CASING

SHUTTER ASSEMBLY

SIGHT GLASS

CONCRETE BASE PAD

GRADE

BOLT TO CONCRETE W/ ½" ⌀ BOLT

NO. 4's EA. WAY @ 12" O.C.

BOTTOM OF FOOTING BELOW FROST LINE.

SIDE ELEVATION

CONDENSER COIL CASING

FAN GUARDS

CONCRETE PAD

SIDE ELEVATION

AIR COOLED CONDENSER BASE

NOT TO SCALE

FIGURE 15-11

301

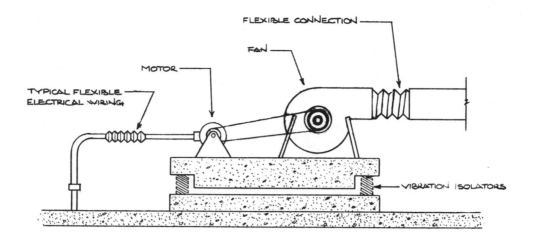

INERTIA BASE
NOT TO SCALE

FIGURE 15-12

in normal soil conditions. We suggest the use of this detail as is. However, if you have a very poor soil, we strongly urge you to secure the services of a qualified structural engineer to design a base suitable for the poor soil.

In Fig. 15-12 we show an inertia base for a high velocity exhaust fan. Inertia bases are common for many vibration transmission solutions. Commonly the engineer calls for one in the specifications and then does not give the slightest clue as to what is intended. The fan, pump, or compressor sits on a specially sized and designed concrete pad to both dampen and evenly absorb vibration forces. The pad sits on specially selected springs to absorb this force. The springs then

sit on their own pad which acts as the final resolution and precludes the vibration forces from being transmitted to the floor. We have shown a simple flexible connection, but the subject of flexible connection is also complex. Frequently flexible connections are required both horizontally and vertically to resolve all vibration problems in pipe or high velocity duct systems. This entire system requires serious engineering investigation. Our detail would be enhanced with further descriptive connections and notations. In effect our detail displays the basic bare facts. Never treat this subject lightly. Thoroughly investigate and carefully note all requirements.

SIXTEEN

Plan Specification Data

Two of the most important elements of detailing are the symbol list and the equipment schedules. While you may think it is obvious, no drawing using symbols to represent various valves, fittings, and other specialities has any clear-cut meaning without a list explaining what each symbol means.

The symbol list, or legend as it is sometimes called, is a necessary part of every set of drawings that is used for installation and construction purposes. In the figures that follow this discussion we present a condensed version of this type of information using reasonably standard symbols. Not every symbol for every type of device is indicated. However, we feel we have covered the symbols most commonly used.

To clarify the type and use of symbols we must always begin with a very simple, yet vital premise. Your symbol list must be the sole source of explanation of the symbols used on all the plans which the symbol list covers. However, do not carry this thought to its extreme, and foolish, end. If you have only one symbol list to cover every type of mechanical and electrical plan, your symbol list ends up as a composite list that covers symbols for every category of mechanical (plumbing, heating, air conditioning, compressed air, etc.) and electrical (power, lighting, signals, communications, etc.) equipment. Obviously the composite list, while possible, would not be very feasible or logical.

In the preparation of the contract the drawings that are used in both bidding and subsequent construction, equipment, and performance schedules play a very vital role. Commonly many large, privately financed construction projects are constructed with owner furnished equipment that is purchased in advance. The schedule forms an even more vital role when this occurs. The owner's advance equipment purchase order usually includes a limited, precise, and condensed description of the equipment to be purchased plus the schedule which lists the pertinent performance data. The construction plans will usually list the same schedule with the notation that the equipment described will be furnished at the job site by the owner for installation by the contractor.

What makes a good schedule and how many different kinds are needed? It is easy to say that a good schedule should be clear, complete, and correct. But in any schedule, and especially in any mechanical schedule, clarity of lettering is important. A 3 should not resemble a 5 because of sloppy work. Secondly the schedule must completely cover all the pertinent performance data. Without the proper overall size, pressure drop (coil fluid and air), system temperature differential, coil fluid and air leaving and entering temperatures, and number of rows, you do not have a complete coil specification. Location, material, and method of installation might also be added to the list. And, as should be obvious, all the data should be correct.

Literally the entire specification could be expressed as a series of schedules. Materials, methods, installation features, and just about anything you may want to specify could be in a schedule. Sometimes hangers, sheet metal construction, and similar standard items are probably best expressed in the specification and schedules are best used where they really make a contribution, such as in hanger spacing, types of valves for certain situations, and the like.

Most contractors and suppliers constantly request the consultant to specify quantities. For example, in an air distribution job that has a large variety of

JOB ____________________ **EQUIPMENT LIS**

BOILER

INSTALLED RADIATION ____________ X 25% = ____________

MANUFACTURER ____________

QUANTITY ____________

MODEL ____________

NET RATING ____________

GROSS RATING ____________

DIMENSIONS ____________ long X ____________ wide X ____________ high

CHIMNEY ________ X ________ X ________ high

DESCRIPTION—

BURNER

MANUFACTURER ____________

MODEL ____________

GPH ____________ of ____________

ELECTRICAL ____________

I.D. FAN

MANUFACTURER ____________

MODEL ____________

ELECTRICAL ____________

EXPANSION TANKS

MANUFACTURER ____________

QUANTITY ____________ SIZE ____________ DIMENSIONS ____________ long X ________

OIL TANK

SIZE ____________ gallons DIMENSIONS ____________ long X ____________ diameter

FIGURE 16-1

diffusers, grilles, and registers they want not only the size, type, etc., but also how many of each kind. It makes their job both simple and safe, and it can make the engineer's job a *nightmare*, especially in a last minute, late night effort to get the job out on time. We think the engineer, designer, and detailer should *never* specify quantities. His or her job is to design the project not to count its parts. Counting is the responsibility of the supplier and contractor.

Schedules

Design Schedules: As we previously said, there are all sorts of schedules that are both possible and useful. We think the engineer needs a schedule for his own office on each project. Responsibilities change, people change, and a busy office has more than one project under construction at the same time. Figures 16-1 through 16-3 are three samples of a long list of possible schedules. Figures 16-1 and 16-2 are one way of doing things. They are only a part of what you should develop for your own use. For example, a secretary with a complete list such as these two schedules partially portray could in opening shop drawing submissions quickly note on the submission whether the item listed is as specified or is something different. And the question, "What was specified?" is easily answered by such a schedule.

Another version is begun in Fig. 16-3. Design conditions are not always the same and even a simple listing of the equipment as we show on this figure could

readily be placed on a page or two and used for many projects. A large amount of time can be saved if this sort of list is available. We show space for only two items. Obviously you could trace or photocopy this schedule and extend the lines down the page to cover a whole list of items.

Performance Schedules: Figures 16-4 through 16-7 are examples of limited concise performance schedules. Figure 16-4 covers steam and hot water converters. Our only concern, regardless of whether it is steam-to-water or water-to-water, is that the performance of both fluids meets our stated requirements. In steam the trap capacity is commonly four times the steam supply capacity. Figure 16-5 covers a fairly common HVAC air system with an air handling unit (usually with built-in cooling and heating coil and filters; note that coils are not covered here) plus some special filters. Figure 16-6 covers an expansion tank and specialized air filters. Figure 16-7 covers circulating pumps, chilled water coils, and a steam humidifier. The chilled water coil schedule can, with an appropriate title change, cover preheat, precoil run-around, brine, reheat run-around, and reheat coils. Normally there is a pump in the run-around heat recovery coil module circuit, and you may have need for a humidifier. Usually the humidifier is steam. However, you may utilize an air washer system, which would require a schedule tailored to the air washer. We hope these four details put the premise of schedules in proper focus.

JOB ________________________________ EQUIPMENT LIST

CONVERTOR

MANUFACTURER ______________________

MODEL ______________________

WORKING PRESSURE ______________________

CAPACITY ______________________

DIMENSIONS __________ long X __________ diam.

HOT WATER STORAGE TANK & HEATER

MANUFACTURER ______________________

MODEL ______________________

CAPACITY ______________________ gallons

CAPACITY __________ gpm _____° — _____° water

DIMENSIONS __________ long X __________ diam.

PRESSURE RELIEF VALVES

MANUFACTURER ______________ MODEL ______________ QUANTITY ________

CAPACITY __________ btu/hr PIPE CONNECTIONS __________ inlet ________ outlet

FILL VALVE

MANUFACTURER ______________________ MODEL ______________________

AIR SCOOP

MANUFACTURER ______________________ MODEL ______________________

LOW WATER CUT-OFF

MANUFACTURER ______________________ MODEL ______________________

FIGURE 16-2

AIR CONDITIONING DESIGN DATA

DESIGN AREA	SUMMER				WINTER			
	OUTSIDE		INSIDE		OUTSIDE	INSIDE		
	DB	WB	DB	WB	DB	DB	%HUMIDITY	
ALL AREAS	92	74	73	61	-11	73	50	

MAJOR AIR CONDITIONING EQUIPMENT ITEMS

EQUIPMENT	LOCATION	PERFORMANCE MINIMUM CAPACITY	MOTOR		REMARKS
			NOM. HP	PHASE VOLT.	

FIGURE 16-3

Mechanical schedules, air conditioning design data and major air conditioning equipment items.

SCHEDULE OF CAPACITIES FOR THE STEAM TO WATER CONVERTOR

CONVERTOR NUMBER	WATER SIDE					STEAM SIDE		
	G.P.M.	WATER ENT. °F.	WATER LVS. °F.	BTU/HR	MAX.PD.FT.	PRESSURE	LBS/HR	TRAP CAP

SCHEDULE OF CAPACITIES FOR THE WATER TO WATER CONVERTOR

CONVERTOR NUMBER	HIGHER TEMPERATURE WATER SIDE					LOWER TEMPERATURE WATER SIDE		
	G.P.M.	WATER ENTERS °F.	WATER LVS. °F.	BTU/HR	MAX.PD.FT.	G.P.M.	WAT.ENT. °F.	WAT.LV. °F.	BTU/HR

FIGURE 16-4

Mechanical schedules, converters.

HEATING-VENTILATING & AIR CONDITIONING UNIT FANS (BUILT UP SYSTEM)

UNIT NO.	LOCATION	FAN CFM	O.A. CFM	FAN ARRANGEMENT	FAN MOTOR		TOTAL S.P. OF SYSTEM	TYPE WHEEL	MIN. WHEEL DIA.
					NOM. H.P.	PHASE-VOLT.			

PREFILTERS

CFM	SYSTEM	MAX. FACE VELOCITY	TYPE	MIN. EFFICIENCY NBS COTTRELL TEST	MAX. CLEAN FILTER SP	CALCULATED SP FOR SYSTEM

FINAL FILTERS

CFM	SYSTEM	CARTRIDGES			MIN. EFFICIENCY NBS. ATMOSPHERIC DUST SPOT TEST	MAX. CLEAN FILTER SP	CALCULATED SP FOR SYSTEM
		NUMBER	SIZE	ARRANGEMENT			

FIGURE 16-5

Mechanical schedules, HVAC fans and filters.

EXPANSION TANK SYSTEM

Symbol	System	Approx. System Vol. Gal.	System Temp. Range	Initial Press in Tank PSIG	PRV.Fill Press at Tank PSIG	Max.Oper.Pressure Relief Valve PSIG	Max.Oper.Pressure At Tank PSIG	Tank Size Gal.	Air Separator Capacity GPM	Air Separator Max P.D. FT H$_2$O	Air Separator Strainer	Pipe Size To Tank	Cold Water Size

AIR FILTERS, EXTENDED AREA (BAG TYPE)

FILTER NO.	TYPE	GRADE	CFM	SYSTEM	CARTRIDGES NUMBER	CARTRIDGES SIZE

FIGURE 16-6

Mechanical schedules, expansion tanks and bag-type filters.

CIRCULATING PUMPS

LOCATION	SYSTEM	CFM	HEAD FT.	FLUID TEMP °F.	% EFF.	DOUBLE OR SINGLE SUCTION	NOM.HP	MOTOR PHASE-VOLT.	RPM

CHILLED WATER COOLING COILS

COIL ROWS	SYSTEM	CFM	MAX. FACE VEL. FPM.	MAX. S.P.	ENT.AIR °F. DB.	ENT.AIR °F. WB.	LVG. AIR °F. DB.	LVG. AIR °F. WB.	CIRC. WATER TEMP. IN °F.	CIRC. WATER TEMP. OUT °F.	CIRC. WATER GPM	MAX. P.D. FT.WATER	BTUH

STEAM HUMIDIFIER

HUMID. NO.	SYSTEM	CFM	WINTER ENT. AIR °F. DB.	ENT. AIR °F. WB.	LVG. AIR °F. DB.	LVG. AIR °F. WB.	STEAM PSIG	CONTROL VALVE LBS/HR	TRAP LBS/HR

FIGURE 16-7

Mechanical schedules, pumps, coils, humidifier.

311

CHILLED WATER COOLING COILS

COIL NO.	SYSTEM	CFM	MAX. FACE VEL. FPM	MAX. S.F.	ENT. AIR °F		LVG. AIR °F		CIRC. WATER				BTU/H
					Db	Wb	Db	Wb	TEMP. IN °F	TEMP. OUT °F	GPM	MAX. P.D. FT. WATER	

CIRCULATING PUMPS

PUMP NO.	LOCATION	SYSTEM	GPM	HEAD FT.	FLUID TEMP. °F	% EFF.	DOUBLE OR SINGLE SUCTION	MOTOR		
								NOM. (H.P.)	PHASE-VOLT	RPM

PACKAGE AIR COOLED RECIPROCATING CHILLER UNIT

EQUIPMENT	LOCATION	PERFORMANCE MINIMUM CAPACITY	MOTOR		REMARKS
			NOM. K.W.	PHASE VOLT	
COMPRESSOR		TONS			COMPRESSOR SHALL BE CAPABLE OF CAPACITY REDUCTION TO %
CHILLER		GPM 54° F WATER IN 44° F WATER OUT			FT. MAX. WATER P.D.
AIR COOLED CONDENSER		AIR COOLED COMPRESSOR SHALL PROVIDE ABOVE PERFORMANCE WITH 35° F AMBIENT TEMP.			SHALL PERFORM SATISFACTORY AT MIN. COMPRESSOR CAPACITY WITH 35° F AMBIENT TEMP.

PACKAGE AIR COOLED RECIPROCATING CHILLER UNIT SHALL BE PROVIDED WITH MANUFACTURERS RECOMENDED HOT GAS BYPASS TO PROVIDE SATISFACTORY OPERATION DOWN TO 0% CAPACITY WITH A 35° F AMBIENT TEMPERATURES. STARTERS REQUIRED SHALL BE FUNISHED BY MANUFACTURER OF EQUIPMENT.

FIGURE 16-8

Mechanical schedules, coils, pumps, packaged units.

Air Conditioning Equipment: Figure 16-8 covers chilled water coils, circulating pumps, and a packaged chiller unit that are common to the design of an air conditioning system. Since this is a simple job with a compact specification, we have added a few notes that should either be in your specification or on the plan. If your system requires only one pump and one chiller, we hardly see the need for a pump or chiller schedule.

School Systems: Figures 16-9 and 16-10 cover items commonly found in most school design projects. The vast majority of the time the unit ventilator and fin radiation systems shown on Fig. 16-9 use hot water as the heating medium. On occasion some of the radiation may be fed by low pressure steam. To cover all bases we show two fin radiator schedules, one steam and one hot water. Figure 16-10 depicts a fan coil unit schedule and a fresh air intake schedule. The fresh air intake schedule probably should have been located

SCHEDULE OF CAPACITIES OF THE CLASSROOM TYPE UNIT VENTILATOR, HOT WATER HEATING

TOTAL CFM	RET. AIR CFM	MAX. FRESH AIR C.F.M.	HEATING							MOTOR				FAN			PIPE RUNOUT SZ.			REMARKS
			AIR		HOT WATER					BTU/HR	HP	Ø	MIN. RPM	MAX. O.V.	MAX. RPM		S.	R.	DR.	
			ENT.	LVG.	ENT.	LVG.	GPM	P.D.												

FIN TUBE RADIATION SCHEDULE

SYMBOL	BTU/HR	WATER ENTERS GPM	WATER LEAVES	HEATING ELEMENT				ENCLOSURE			RUNOUT SIZES	VALVE SIZES	REMARKS
				TUBE	FINS	FINS LENGTH	ROWS	D	H				

FIN TUBE RADIATION SCHEDULE

SYMBOL	BTU/HR	STEAM PSI	HEATING ELEMENT				ENCLOSURE			RUNOUT SIZES		VALVE SIZE	TRAP SIZE	REMARKS
			TUBE	FINS	FINS LENGTH	ROWS	D	H	STEAM	RET			

FIGURE 16-9

Mechanical schedules, unit ventilator, finned radiation.

FAN - COIL UNIT SCHEDULE

| LOCATION | ENTERING WATER TEMPERATURE | | | | | | | | TEMPERATURE DROP | | | |
	QUANTITY	SIZE	TYPE	HEATING CAPACITY	S & R	CFM	O.V.	S.P.	RPM	HP	NOTES

FRESH AIR INTAKE SCHEDULE

UNIT NUMBER	THROAT SIZE	THROAT AREA	MAX. CFM	MAX. THROAT VEL.

FIGURE 16-10

Mechanical schedules, fan-coils, fresh air intakes.

315

elsewhere or have a note that it relates to the unit ventilator schedule. It does relate but that fact is not clear because we had to separate our schedules to fit the pages of this book.

Office Buildings: Three details cover some of the special items commonly used in office building systems. Figure 16-11 covers the wall-hung induction unit, Fig. 16-12 covers the dual-duct air handling unit, and Fig. 16-13 covers the dual coil, usually a four-pipe fan coil unit. Keep in mind these are still, like all the other schedules, performance schedules, and they contain the minimum amount of information you

SCHEDULE OF CAPACITIES OF THE WALL HUNG INDUCTION UNIT, COMBINATION COIL WITH WATER CONTROL

SYMBOL	CFM	NOZZ. PRESS	COOLING CAPACITY		HEATING CAPACITY		WATER FLOW			
			SENS. CAP.	PRI. AIR CAP	TOTAL CAP.	AIR-ON CAP.	GRAVITY CAP	GPM	MAX. PD	C.W. ENT.	C.W. LVS.

WATER FLOW		CONNECTION SIZES		
H.W. ENTERS	H.W. LEAVES	PRIM. AIR	WATER SUP.	WATER RET.	DRAIN

FIGURE 16-11

Mechanical schedules, induction units.

SCHEDULE OF CAPACITIES OF THE DUAL DUCT AIR HANDLING UNIT												
UNIT NO.	TOTAL CFM	RETURN CFM	OUTDOOR CFM	MAX. CFM COLD DECK	MAX. CFM HOT DECK	COOLING COIL						
						AIR ENT	AIR LVS	BTU/HR	GPM	WAT. ENT.	WAT. LVS.	MAX. WATER PD

MATCH LINE

MATCH LINE

HEATING COIL							TOT. STATIC PRESSURE INCH. W.G.	FAN MOTOR			
AIR ENT.	AIR LVS.	BTU/HR	GPM	WAT. ENT.	WAT. LVS.	MAX. WAT. P.D.		HP	VOLTS	PHASE	CYCLE

FIGURE 16-12
Mechanical schedules, dual duct air handling unit.

DUAL COIL FAN COIL UNIT SCHEDULE										
				COOLING COIL CAPACITY						
SYM.	CFM	MOTOR HP	MAX. WATTS	GPM	MAX.PD	SENSIBLE CAP.	LATENT CAP.	TOTAL CAP.	WATER ENT.	WATER LVS.

MATCH LINE

MATCH LINE

HEATING COOLING CAPACITY					RUNOUTS & VALVE SIZE	
GPM	MAX. P.D.	HEATING CAP.	WATER ENT.	WATER LVS.	COOLING	HEATING

FIGURE 16-13
Mechanical schedules, dual duct fan coil unit.

should portray. You can add to, or elaborate on, any of the items when you feel other information would be helpful.

Motors and Filters: Figure 16-14 shows a motor and motor control schedule and a schedule of louver sizes. We thought that, if nothing else, we would put the two controversial schedules on one page. A large number of consultants, designers, and engineers feel that the motors are already covered in other schedules, that the louver sizes are already shown on the plans, and that these two schedules are useless. We buy some of that argument to some degree, at least the part about the louvers. However, this motor schedule belongs on the electrical detail plan if not on the mechanical detail

MOTOR AND MOTOR CONTROL SCHEDULE																
MOTOR	LOCATION	HP	VOLTS	PHASE	AMPS		STARTER				IND. LTS.	CONTROL VOLTAGE	MANUAL CONTROL		AUTOMATIC CONT.	
					FL.	LR.	TYPE	NEMA SIZE	CIR. BKR.	MTG.			DEVICE	LOCATION		

MATCH LINE

MATCH LINE

INTERLOCKING			ITEM	REMARKS
SAFETY DEVICE	NO. CONTACTS			
	N.O.	N.C.		

SCHEDULE OF LOUVER SIZES					
LOUVER NO.	WIDTH	HEIGHT	MATERIAL	SCREEN	REMARKS

FIGURE 16-14
Mechanical schedules, motor, control, and louvers.

GRILLE AND DIFFUSER SCHEDULE

DESIGNATION	MODEL	SIZE	CFM	TYPE	FINISH	DUCT.SIZE

SCHEDULE OF MIXING BOXES

SYMBOL	CFM	MIN. Ps	MAX. Ps	SOUND RATING NC	REMARKS

FIGURE 16-15
Mechanical schedules, grille, diffuser, and mixing box.

plan. We feel strongly that the motor, starter, and control should very definitely be spelled out somewhere.

Grilles and Diffusers: Figure 16-15 contains a grille and diffuser schedule and a mixing box schedule to use if you have a dual-duct high velocity air system. These are the schedules we especially referred to in the opening discussion of this chapter; they can cover a large number and variety of situations and a large quantity of items. This is where the demand for a count commonly occurs, and this is where we especially say: do not provide any count. Last minute changes still constantly change the count and create limitless possibilities for error.

Plumbing: Our last detail, Fig. 16-16 depicts a plumbing fixture schedule and a plumbing equipment schedule. We operated for many years as professional consultants without any such schedule on our plans. At the last minute there are always changes in plumbing fixtures and equipment, and decisions on some of these fixtures and equipment do not become final until literally the last minute. This involves endless proof-reading and last minute specification alterations and corrections. We developed this schedule to suit our own needs and to solve this problem. Changes are quick and easy, and including the size of the coil, vent, and hot and cold water connections greatly simplifies the same notations which are usually squeezed into the little space available on the plan. We feel this is a very workable set of schedules.

PLUMBING FIXTURE SCHEDULE

FIXTURE	SIZE – TYPE	MANUFACTURER AND MODEL	SOIL	VENT	CW	HW	REMARKS, ACCESSORIES, NOTES

PLUMBING EQUIPMENT SCHEDULE

ITEM	SIZE – TYPE	MANUFACTURER AND MODEL	SOIL	VENT	CW	HW	RATING	INPUT	OUTPUT	NOTES

FIGURE 16-16

Mechanical schedules, plumbing fixtures and equipment.

————————————	Waste or drain piping above grade
——————WBG——————	Waste or drain piping below grade
— — — — — — — —	Vent piping
— · — · — · —	Cold water
— ·· — ·· — ·· —	Hot water
— ··· — ··· — ··· —	Recirculating hot water
——————G——————	Gas
——————ACID——————	Acid waste
——————DW——————	Drinking water flow
——————DWR——————	Drinking water return
——————VAC——————	Vacuum (air)
——————A——————	Compressed air
——————NAME——————	Chemical supply pipes
●RL	Rainwater leader
○	Waste or vent riser
▣FD	Floor drain
▣RD	Roof drain
○	Cold water riser

x	Hot water riser
⊔	Running trap
⊔	Trap
⊓	Drum trap
⊤	Sillcock
‖	Cleanout
⟱	Gate valve
⊶	Globe valve
⊿	Check valve
▯	Thermometer
⊙	Pressure gauge
⊟	Relief valve
╫	Union
▢	Square head cock
△	Air eliminator
⟍	Strainer
Ⓜ	Water meter
⟶⟩	Shower head

FIGURE 16-17
Plumbing symbol list.

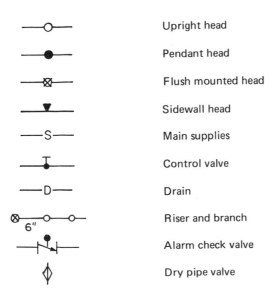

—○—	Upright head
—●—	Pendant head
—⊗—	Flush mounted head
—▼—	Sidewall head
—S—	Main supplies
—⊤—	Control valve
—D—	Drain
⊗——○——○— 6"	Riser and branch
—⊩—	Alarm check valve
◇	Dry pipe valve

FIGURE 16-18
Sprinkler piping symbols.

Symbols and Legends

Generally symbol lists are limited to one generalized category of work, and our lists are arranged somewhat in this fashion. Figures 16-17 and 16-18 cover plumbing and sprinkler piping. To fit the lists in this book they are grouped together. In your work the plumbing list should go with the plumbing drawings and the sprinkler list should go with the sprinkler or fire protection drawings. Note that valves are shown differently for each category.

Figure 16-19 lists the symbols for various types of heating lines, as well as specialty lines such as oil, compressed air, and the like. Our opening remarks on symbols indicated that every type of line should be covered by an appropriate symbol. This is not to imply that designations such as dots, dashes, or crosshatches with one to four strokes are not acceptable. They certainly are, provided that is your standard system. Generally the letter designation we have shown is easier to follow and less prone to error. Figure 16-20 depicts air conditioning and refrigeration piping symbols. Probably the biggest difference from what you have seen or may use is in the suction and discharge symbols. In place of RD and RS the most commonly used symbols are L and S.

Figures 16-21 and 16-22 show valve symbols and what we call special symbols. Actually these two figures could have been grouped under a single head entitled "HVAC Symbols" since they really apply and are most often used in HVAC work.

HEATING PIPING SYMBOLS

——— HPS ———	High pressure steam
——— MPS ———	Medium pressure steam
——— LPS ———	Low pressure steam
——— HPR ———	High pressure return
——— MPR ———	Medium pressure return
——— LPR ———	Low pressure return
——— BBD ———	Boiler blowdown
——— CP ———	Condensate pump discharge
——— VPD ———	Vacuum pump discharge
——— MU ———	Makeup water
——— V ———	Air relief line (vent)
——— FOF ———	Fuel oil flow
——— FOR ———	Fuel oil return
——— FOV ———	Fuel oil tank vent
——— HWS ———	Low temperature hot water supply
——— MTWS ———	Medium temperature hot water supply
——— HTWS ———	High temperature hot water supply
——— HWR ———	Low temperature hot water return
——— MTWR ———	Medium temperature hot water return
——— HTWR ———	High temperature hot water return
——— A ———	Compressed air
——— VAC ———	Vacuum (air)
——— (Name)E ———	Existing piping
—✕—✕— (Name) —✕—✕—	Pipe to be removed

FIGURE 16-19

AIR CONDITIONING AND REFRIGERATION PIPING SYMBOLS

——— RD ———	Refrigerant discharge
——— RS ———	Refrigerant suction
——— B ———	Brine supply
——— BR ———	Brine return
——— C ———	Condenser water supply
——— CR ———	Condenser water return
——— CHWS ———	Chilled water supply
——— CHWR ———	Chilled water return
——— FILL ———	Fill line
——— H ———	Humidification line
——— D ———	Drain

FIGURE 16-20

VALVE SYMBOLS

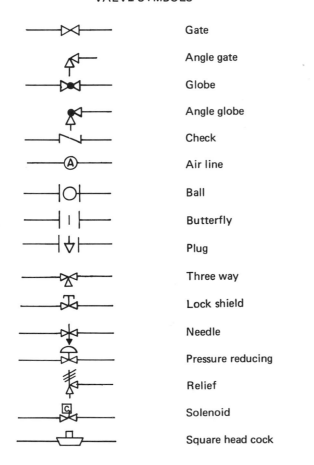

	Gate
	Angle gate
	Globe
	Angle globe
	Check
	Air line
	Ball
	Butterfly
	Plug
	Three way
	Lock shield
	Needle
	Pressure reducing
	Relief
	Solenoid
	Square head cock

FIGURE 16-21

SPECIAL SYMBOLS

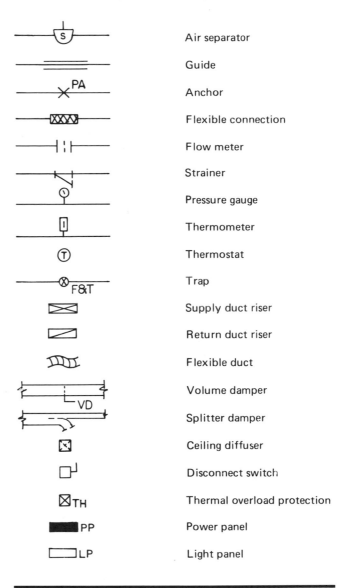

	Air separator
	Guide
	Anchor
	Flexible connection
	Flow meter
	Strainer
	Pressure gauge
	Thermometer
	Thermostat
	Trap
	Supply duct riser
	Return duct riser
	Flexible duct
	Volume damper
	Splitter damper
	Ceiling diffuser
	Disconnect switch
	Thermal overload protection
	Power panel
	Light panel

FIGURE 16-22

Index

ABOUT THE AUTHOR

Jerome F. Mueller has been a practicing consulting engineer in the mechanical and electrical disciplines for over thirty years. A Professional Engineer in many states, he has designed mechanical and electrical systems for government departments, universities and colleges, cities and towns, the military, architectural firms, and manufacturing companies. He also has extensive experience in industrial process engineering and process design.

Mr. Mueller is a leader in the design application of computer-controlled installations and in automation control application. He was an early leader in fluorescent applications lighting and a pioneer in Connecticut's usage of electrical heating systems. He also has special expertise in project evaluation, cost estimation, and contract administration. He is the author of *Standard Application of Electrical Details* and *Standard Mechanical and Electrical Details* (both McGraw-Hill).